170
Advances in Polymer Science

Advances in Polymer Science

Recently Published and Forthcoming Volumes

Advanced Computer Simulation Approaches for Soft Matter Sciences I
Volume Editors: Holm, C., Kremer, K.
Vol. 173, 2004

Microlithography · Molecular Imprinting
Vol. 172, 2004

Polymer Synthesis
Vol. 171, 2004

NMR · Coordination Polymerization · Photopolymerization
Vol. 170, 2004

Long-Term Properties of Polyolefins
Volume Editor: Albertsson, A.-C.
Vol. 169, 2004

Polymers and Light
Volume Editor: Lippert, T. K.
Vol. 168, 2004

New Synthetic Methods
Vol. 167, 2004

Polyelectrolytes with Defined Molecular Architecture II
Volume Editor: Schmidt, M.
Vol. 166, 2004

Polyelectrolytes with Defined Molecular Architecture I
Volume Editor: Schmidt, M.
Vol. 165, 2004

Filler-Reinforced Elastomers · Scanning Force Microscopy
Vol. 164, 2003

Liquid Chromatography · FTIR Microspectroscopy · Microwave Assisted Synthesis
Vol. 163, 2003

Radiation Effects on Polymers for Biological Use
Volume Editor: Kausch, H.
Vol. 162, 2003

Polymers for Photonics Applications II
Nonlinear Optical, Photorefractive and Two-Photon Absorption Polymers
Volume Editor: Lee, K.-S.
Vol. 161, 2003

Filled Elastomers · Drug Delivery Systems
Vol. 160, 2002

Statistical, Gradient, Block and Graft Copolymers by Controlled/ Living Radical Polymerizations
Authors: Davis, K.A., Matyjaszewski, K.
Vol. 159, 2002

Polymers for Photonics Applications I
Nonlinear Optical and Electroluminescence Polymers
Volume Editor: Lee, K.-S.
Vol. 158, 2002

Degradable Aliphatic Polyesters
Volume Editor: Albertsson, A.-C.
Vol. 157, 2001

Molecular Simulation · Fracture · Gel Theory
Vol. 156, 2001

New Polymerization Techniques and Synthetic Methodologies
Vol. 155, 2001

Polymer Physics and Engineering
Vol. 154, 2001

NMR · 3D Analysis · Photopolymerization

With contributions by
N. Fatkullin · T. Ikehara · H. Jinnai · S. Kawata · R. Kimmich
T. Nishi · Y. Nishikawa · H.-B. Sun

Springer

The series presents critical reviews of the present and future trends in polymer and biopolymer science including chemistry, physical chemistry, physics and material science. It is addressed to all scientists at universities and in industry who wish to keep abreast of advances in the topics covered.

As a rule, contributions are specially commissioned. The editors and publishers will, however, always be pleased to receive suggestions and supplementary information. Papers are accepted for "Advances in Polymer Science" in English.

In references Advances in Polymer Science is abbreviated Adv Polym Sci and is cited as a journal.

The electronic content of APS may be found at http://www.springerLink.com

ISSN 0065-3195
ISBN 978-3-662-14555-5 ISBN 978-3-540-40000-4 (eBook)
DOI 10.1007/978-3-540-40000-4

Library of Congress Control Number 2004105251

springeronline.com

Originally published by Springer-Verlag Berlin Heidelberg New York in 2004
Softcover reprint of the hardcover 1st edition 2004

Typesetting: Stürtz AG, Würzburg
Cover: Künkellopka GmbH, Heidelberg; design&production GmbH, Heidelberg

Printed on acid-free paper 02/3020/kk – 5 4 3 2 1 0

Editorial Board

Advances in Polymer Science Also Available Electronically

For all customers who have a standing order to Advances in Polymer Science, we offer the electronic version via SpringerLink free of charge. Please contact your librarian who can receive a password for free access to the full articles by registering at:

http://www.springerlink.com

If you do not have a subscription, you can still view the tables of contents of the volumes and the abstract of each article by going to the SpringerLink Homepage, clicking on "Browse by Online Libraries", then "Chemical Sciences", and finally choose Advances in Polymer Science.

You will find information about the

- Editorial Board
- Aims and Scope
- Instructions for Authors
- Sample Contribution

at http://www.springeronline.com using the search function.

Contents

APS (2004) Single Contribution:1–113
DOI 10.1007/b12766

Polymer Chain Dynamics and NMR

Rainer Kimmich[1] (✉) · Nail Fatkullin[2]

[1] Sektion Kernresonanzspektroskopie, Universität Ulm, 89069 Ulm, Germany
rainer.kimmich@physik.uni-ulm.de
[2] Department of Molecular Physics, Kazan State University, 420008 Kazan, Tatarstan, Russia

Abstract The universal features of polymer dynamics are specifically represented by laws for (anomalous) segment diffusion and chain relaxation modes. Nuclear magnetic resonance (NMR)-based techniques provide direct access to these phenomena. This in particular refers to NMR relaxation and diffusion studies. Methods suitable for this purpose are described in detail. Three basic classes of polymer dynamics models, namely the Rouse model, the tube/reptation model, and the renormalized Rouse models are outlined and discussed with respect to predictions for NMR measurands. A wealth of experimental NMR data are reviewed and compared with predictions of the model theories. It is shown that characteristic features of all three types of models can be verified in great detail provided that the model premisses are suitably mimicked in the experiments. Rouse dynamics is shown to be relevant for polymer melts with molecular weights below the critical value and for solutions of diminished entanglement effect. Features specific for the renormalized Rouse model reveal themselves in the form of high- and low-mode-number limits of the spin–lattice relaxation dispersion. These results are considered to mirror the analytical structure of the Generalized Langevin Equation. Finally, anomalous-diffusion and relaxation laws characteristic for the tube/reptation model can be perfectly reproduced in experiment if the polymer chains are confined in a nanoporous, solid matrix whereas bulk melts are not in accord with these predictions. The dynamics of chains confined in artificial tubes can be treated analytically assuming a harmonic radial potential for the polymer/wall interaction. These results derived for a real tube closely render the characteristic features of the original Doi/Edwards model predicted for a fictitious tube.

Keywords NMR · Relaxation · Diffusion · Rouse model · Reptation · Renormalized Rouse model

Abbreviations

BWR	Bloch/Wangsness/Redfield theory
CLF	contour length fluctuation
CPMG	Carr/Purcell/Meiboom/Gill
FID	free-induction decay
GLE	generalized Langevin equation
NMR	nuclear magnetic resonance
ODF	order director fluctuations
PB	poly(butadiene)
PDES	poly(diethylsiloxane)
PDMS	poly(dimethylsiloxane)
PE	poly(ethylene)
PEO	poly(ethylene oxide)
PGSE	pulsed gradient spin echo method
PHEMA	poly(hydroxyethylmethacrylate)
PIB	poly(isobutylene)
PIP	poly(isoprene)
PMMC	polymer mode-mode coupling model
PS	poly(styrene)
PTHF	poly(tetrahydrofurane)
PU	phenylurazole
RF	radio frequency
RRM	renormalized Rouse model
SGSE	steady gradient spin echo method
ThRRM	three times renormalized Rouse model
TRRM	twice renormalized Rouse model
WLF	Williams/Landel/Ferry

1 Introduction

The objective of this review is to outline the basic theories and models of linear polymer chain dynamics [1], and to demonstrate the potential of NMR (nuclear magnetic resonance) techniques for probing the characteristics of molecular motions in such systems. From the methodological point of view we will mainly focus on field-cycling NMR relaxometry and field-gradient NMR diffusometry, but will also make some reference to transverse relaxation and the residual dipolar interaction occurring in entangled polymer systems. This combination of methods turned out to be most powerful for the examination of chain modes and the translational diffusion properties of polymer segments and the coil center-of-mass. The techniques referred to are best suited to shed light on almost every aspect of polymer dynamics.

Particular emphasis will be laid on the experimental verification of features of standard theories of polymer dynamics such as the Rouse model, the tube/reptation model [1] or the renormalized Rouse models. Based on the special conditions under which predictions of these theories match experimental findings well, the application limits and the deficiencies of these models for more general scenarios become obvious and help to ameliorate the underlying model ansatz for chain dynamics.

In order to elucidate the *universal* character of polymer chain dynamics, we will attempt to consider as many different polymer species as possible for which NMR relaxometry and diffusometry data are available. On the other hand this means that we must restrict ourselves predominantly to melts of linear homopolymers with a narrow molecular weight distribution to keep the range of literature referred to at manageable levels. Only where enlightening information is expected will we refer to solutions, networks, and mesomorphic systems in addition to isotropic melts. Empirically the conclusions suggested by the huge amount of experimental NMR data available in the literature provide a complete picture rendering dynamical features in great detail.

This is in contrast to the stage of development of model theories where the "many-entangled-excluded-volume-chain" problem still has to be overcome, although promising approaches exist. In particular it will turn out that the theoretical concepts available for this sort of system are unable to describe many experimental details quantitatively. On the other hand, predictions of all model theories referred to in this context will be shown to be compatible with certain experimental findings provided that the sample and measuring conditions are adapted to the scenarios that were anticipated in the theoretical derivations.

Apart from the introductory section, the article is subdivided into four major sections: *NMR Methods; Modeling of Chain Dynamics and Predictions for NMR Measurands; Experimental Studies of Bulk Melts, Networks, and Concentrated Solutions;* and *Chain Dynamics in Pores*. First, the NMR techniques of interest in this context will be described. Second, the three fundamental polymer dynamics theories, namely the *Rouse model*, the *tube/reptation model*, and the *renormalized Rouse theories* are considered. The immense experimental NMR data available in the literature will be classified and described in the next section, where reference will be made to the model theories wherever possible. Finally, recent experiments, analytical treatments, and Monte Carlo simulations of polymer chains confined in pores mimicking the basic premiss of the tube/reptation model are discussed.

2 NMR Methods

NMR provides a rich variety of methods for polymer studies. This in particular refers to the chemical structure of such compounds. Here we deliberately restrict ourselves to investigations on length scales, where chemical bonds and interactions merely matter indirectly via parameters such as segment lengths or friction coefficients. That is, we are dealing with dynamic properties of Kuhn segment chains for times or frequencies beyond those relevant for fluctuations within Kuhn segments. Any chemical shift-based spectroscopy is beyond the scope of this review. Figure 1 shows a schematic representation of the fluctuation time or angular frequency ranges covered by the methods of interest in this context. We will be referring mainly to field-cycling NMR relaxometry combined with conventional high-field methods, to field-gradient NMR diffusometry, and, to a minor degree, to dipolar broadening-based techniques such as the dipolar correlation effect. For an introduction to these methods the reader is referred to monographs like Ref. [2].

Diffusion in polymers can manifest itself in "normal" or "anomalous" form depending on the time or displacement length scale probed in the experiment. The criterion for this distinction is the time dependence of the mean squared displacement

$$\langle R^2 \rangle \propto t^{\kappa} \tag{1}$$

valid in the form of a power law in random unrestricted media. An exponent $\kappa=1$ indicates normal diffusion as is expected for center-of-mass diffusion, i.e., root mean square displacements much larger than the Flory radius. On the other hand, $\kappa<1$ is the signature of anomalous (subdiffusive) displacements. This is of particular interest for segment motions within the length scale of the random polymer coil, where all time dependences are governed by chain modes. That is, characteristic laws are expected for this so-called segment diffusion behavior that may be decisive for the validity of chain dynamics theories. It is therefore of paramount importance to have experimen-

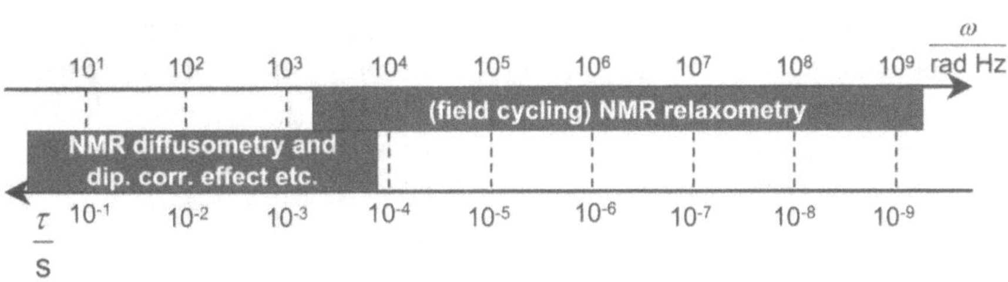

Fig. 1. Schematic representation of the ranges of fluctuation time constants and angular frequencies accessible by the two main NMR techniques under consideration in this article

tal tools in hand probing displacements and chain dynamics on the right time and length scales. The NMR techniques to be described turned out to be particularly powerful in this respect.

2.1
NMR Diffusometry

The standard *pulsed gradient spin echo* (PGSE) and *steady gradient spin echo* (SGSE) methods are based on spatially constant field gradients $\vec{G} = \vec{\nabla} B_0 = const$, where B_0 is the magnitude of the main magnetic field. In the following, all main field gradients are assumed along the z direction so that the total magnetic flux density at a position z is given by $B(z)=B_0+Gz$. The spin echoes to be considered here are the result of ordinary coherence refocusing of the Hahn type [2].

The attenuation factor of the echo amplitude due to translational diffusion is given by

$$A_{diff}(T_E) = \langle e^{i\varphi(T_E)} \rangle . \qquad (2)$$

T_E is the echo time and φ is the (precession) phase a spin possesses at time $t=T_E$. The brackets represent an ensemble average over all spins in the sample. In the following, this expression will be analyzed for different pulse sequences.

Any echo attenuation by relaxation occurs independently of diffusive attenuation and is of marginal interest in this context. It can readily be separated by either keeping all pulse intervals constant (and varying the gradient strength) or by compensating for relaxation losses in experimental protocols especially designed for this purpose.

2.1.1
Hahn Spin Echo in the Short Gradient Pulse Limit ($\delta \ll \Delta$)

We consider the coherence evolution of uncoupled spins $I=1/2$ during the pulsed gradient Hahn spin echo sequence schematically shown in Fig. 2a. A suitable basis for the treatment is the spherical product operator formalism. Explanations, definitions, and rules of the spin operator formalism needed in this context can be found in Ref. [2]. Times just before and immediately after RF (radio frequency) and field gradient pulses will be indicated by minus and plus signs, respectively.

The initial reduced density operator is given by the equilibrium density operator,

$$\sigma(0-) = \sigma_0 = I^{(0)} . \qquad (3)$$

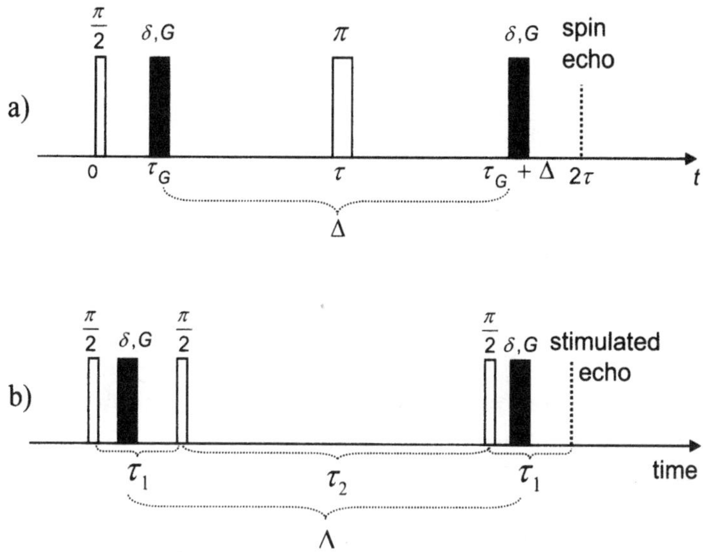

Fig. 2a, b. Radio frequency pulse sequences for the Hahn spin echo (**a**) and the stimulated spin echo (**b**) combined with magnetic field gradient pulses (*black*). Echoes other than the stimulated echo are not indicated in sequence (b). The stimulated echo variant of the pulsed gradient spin echo (PGSE) method is of interest in viscous systems with small diffusion coefficients and, as a consequence, transverse relaxation times much shorter than the spin–lattice relaxation time. Note that in order to obtain distinct echoes at the indicated positions a small but finite background gradient is necessary

The sequence begins with a 90° pulse of width τ_{90} and a rotating-frame phase direction assumed along the x axis. The rotating-frame Hamiltonian during this RF pulse is defined by $H_{90}=-\hbar\gamma B_1 I_x$, where B_1 is the amplitude of the rotating flux density component of the RF field. The reduced density operator immediately after this pulse reads

$$\sigma(0+)=e^{-\frac{i}{\hbar}H_{90}\tau_{90}}\sigma(0-)e^{+\frac{i}{\hbar}H_{90}\tau_{90}}=\frac{i}{\sqrt{2}}\left(I^{(+1)}+I^{(-1)}\right)=\sigma(\tau_G-). \tag{4}$$

During the field gradient pulses, the coherences evolve under the action of the rotating-frame Hamiltonian $H_G=-\hbar\gamma G z_j I_z$, where z_j is the position coordinate of the spin-bearing particle on the gradient axis during the j-th gradient pulse. This produces

$$\sigma(\tau_G+)=e^{-\frac{i}{\hbar}H_G\delta}\sigma(\tau_G-)e^{+\frac{i}{\hbar}H_G\delta}=\frac{i}{\sqrt{2}}\left(I^{(+1)}e^{i\varphi_1}+I^{(-1)}e^{-i\varphi_1}\right)=\sigma(\tau-), \tag{5}$$

where $\varphi_1=\gamma G z_1\delta$. The subsequent 180° RF pulse of length τ_{180} and a rotating-frame phase direction assumed along the y direction produces

$$\sigma(\tau+) = e^{-\frac{i}{\hbar}H_{180}\tau_{180}}\sigma(\tau-)e^{+\frac{i}{\hbar}H_{180}\tau_{180}} = \frac{i}{\sqrt{2}}\left(I^{(-1)}e^{i\varphi_1} + I^{(+1)}e^{-i\varphi_1}\right)$$
$$= \sigma(\tau_G + \Delta-) \tag{6}$$

with $H_{180}=-\hbar\gamma B_1 I_y$. The second gradient pulse causes further phase shifts according to

$$\sigma(\tau_G + \Delta+) = e^{-\frac{i}{\hbar}H_G\delta}\sigma(\tau_G + \Delta-)e^{+\frac{i}{\hbar}H_G\delta}$$
$$= \frac{i}{\sqrt{2}}\left(I^{(-1)}e^{i(\varphi_1-\varphi_2)} + I^{(+1)}e^{-i(\varphi_1-\varphi_2)}\right) = \sigma(2\tau), \tag{7}$$

where $\varphi_2=\gamma G z_2 \delta$.

Coherence evolution in the presence of a field gradient winds the transverse magnetization up along the gradient direction in the form of a helix of pitch $2\pi/(\gamma G\delta)$. The periodicity implied in this helix may be characterized by a "wave number"

$$k = \gamma G \delta. \tag{8}$$

The spin echo amplitude of a spin ensemble being at z_1 during the first gradient pulse and at z_2 during the second is thus represented by the reduced density operator

$$\sigma(2\tau) = \frac{i}{\sqrt{2}}\left(I^{(-1)}e^{ik(z_1-z_2)} + I^{(+1)}e^{-ik(z_1-z_2)}\right). \tag{9}$$

The phase factors depend on the particle displacement $Z=z_1-z_2$ which is distributed after the gradient pulse interval Δ according to the propagator $P(Z,\Delta)$, i.e., the probability density that a particle has been displaced a distance Z along the gradient direction in a time Δ. The average over all particles is given by

$$\langle e^{\pm ikZ}\rangle = \int_{-\infty}^{+\infty} P(Z,\Delta)\exp\{\pm ikZ\}dZ = \int_{-\infty}^{+\infty} P(Z,\Delta)\cos\{kZ\}dZ, \tag{10}$$

where we have anticipated that $P(Z,\Delta)$ is an even function of Z and that k is constant within the sample.

2.1.1.1 Gaussian Propagators

The propagator for normal displacements, Z, along the gradient direction with the diffusion coefficient D is given by the Gaussian function

$$P(Z,\Delta) = \frac{1}{(4\pi D\Delta)^{1/2}}\exp\left\{-\frac{Z^2}{4D\Delta}\right\}. \tag{11}$$

Displacements along other space directions are not directly probed in this experiment and are therefore irrelevant. Evaluating Eq. 10 based on this propagator results in $\langle e^{\pm ikz}\rangle=\exp\{-k^2D\Delta\}$ so that Eq. 9 becomes

$$\sigma(2\tau)=I_y e^{-k^2 D\Delta}. \tag{12}$$

Equation 2 takes the form

$$A_{diff}(T_E)=\left\langle e^{i\varphi(T_E)}\right\rangle \equiv A_{diff}(k,\Delta)=\exp\left\{-k^2 D\Delta\right\}. \tag{13}$$

Inserting $\langle Z^2\rangle=2D\Delta$ in Eq. 13 provides a direct relation between echo attenuation and the mean squared displacement in the diffusion time Δ:

$$A_{diff}(T_E)=\exp\left\{-\frac{1}{2}k^2\langle Z^2\rangle\right\}\overset{\text{isotropic}}{\underset{\text{system}}{=}}\exp\left\{-\frac{1}{6}k^2\langle R^2\rangle\right\} \tag{14}$$

which refers to both normal and anomalous displacements provided the propagator is Gaussian.

2.1.1.2
Non-Gaussian Propagators

In contrast to center-of-mass diffusion, segment diffusion is characterized by non-Gaussian propagators. The attenuation of spin echoes is then no longer governed by unspecific equations like Eqs. 13 or 14. However, in special cases such as the reptation model, a formalism for the evaluation of the echo attenuation can be set up [3].

If such an analytical formalism is not available, the mean squared displacement, i.e., the second moment of the propagator, can be evaluated approximately from the experimental echo attenuation function by restricting oneself to the low wave number limit. The phase factor given in Eq. 2 can be expanded according to

$$A_{diff}=\left\langle e^{\pm ikZ}\right\rangle=\left\langle 1\pm\frac{ikZ}{1!}-\frac{k^2Z^2}{2!}\mp\frac{ik^3Z^3}{3!}+\cdots\right\rangle. \tag{15}$$

Keeping in mind that in the absence of flow the propagator is an even function of Z, one finds in the lowest nontrivial approximation for small wave numbers

$$A_{diff}=\left\langle e^{\pm ikZ}\right\rangle\overbrace{\approx}^{k^2\langle Z^2\rangle\ll 1}1-\frac{1}{2}k^2\langle Z^2\rangle\overset{\text{isotropy}}{=}1-\frac{1}{6}k^2\langle R^2\rangle\approx\exp\left\{-\frac{1}{6}k^2\langle R^2\rangle\right\}, \tag{16}$$

where $\langle R^2\rangle$ is the three-dimensional mean squared displacement in the gradient pulse interval Δ. Plotting the echo amplitude as a function of k^2 and evaluating the initial slope of the decay curve thus provides the diffusion time dependence of the mean squared displacement, $\langle R^2\rangle=\langle R^2(\Delta)\rangle=6D(\Delta)\Delta$, where $D(\Delta)$ is the time-dependent diffusion coefficient.

2.1.2
Hahn Spin Echo for Long Gradient Pulses

We are again referring to the pulse scheme shown in Fig. 2a. The gradient pulse width, δ, is now assumed to be comparable to the interval Δ. Displacements during the gradient pulses are therefore no longer negligible. That is, the phase shift accumulated by a spin depends on the trajectory of the spin-bearing particle during the gradient pulse. For constant gradients along the z axis, this can be expressed by

$$\varphi_j(\delta) = \gamma G \int_0^\delta z_j(t')dt', \tag{17}$$

where the subscript j=1,2 indicates the first or the second gradient pulse. The attenuation factor, Eq. 2, thus becomes

$$A_{diff}(T_E) = \langle \exp\{i\varphi(T_E)\}\rangle = \left\langle \exp\left\{ i\gamma G\left[\int_0^\delta z_1(t')dt' - \int_0^\delta z_2(t')dt'\right]\right\}\right\rangle \tag{18}$$

in analogy to the treatment in the previous section. The analytical form of this attenuation function is best obtained by solving the Bloch/Torrey equation [4], that is, Bloch's equations supplemented by a diffusion term according to

$$\frac{dM_+(z,t)}{dt} = -i\gamma G z M_+(z,t) - \frac{M_+(z,t)}{T_2} + D\frac{d^2}{dz^2}M_+(z,t), \tag{19}$$

where $M_+=M_x+iM_y$ is the local complex transverse magnetization. The resulting attenuation factor by ordinary diffusion (Gaussian propagator) is [5]

$$A_{diff}(k,\Delta) = \exp\{-k^2 D(\Delta - \delta/3)\}, \tag{20}$$

which takes the form of the short pulse limit given in Eq. 13 for $\delta<<\Delta$.

2.1.3
Stimulated Echo

The longest diffusion time (which is essentially given by Δ) that can be probed with the Hahn echo methods is limited by the transverse relaxation time T_2. As a consequence, the echo attenuation achievable in this time limit may not be sufficient for very slow diffusion even for the strongest field gradients technically feasible. Also, in the case of anomalous diffusion it may be desirable to probe a diffusion time range as wide as possible in order to acquire the time dependence of the mean square displacement. Since the spin–lattice relaxation time T_1 tends to be much longer than T_2 in viscous media, it is of interest to employ echo signals the relaxation decay of which is partly governed by T_1. The stimulated echo is a favorable example of this sort, although a factor of two in the signal strength is sacrificed relative to the Hahn echo [6].

Figure 2b shows a typical RF pulse sequence combined with field gradient pulses for probing diffusion with the aid of the stimulated echo. In the first RF pulse interval the transverse magnetization is attenuated by transverse relaxation, and the spin coherences evolve under the action of the magnetic field gradient. The second RF pulse "stores" half of the magnetization in the z direction in spatially modulated form (often called "magnetization grid" or "grating") in complete analogy with the optical "forced Rayleigh scattering" technique [7]. The modulation degree of the grating is leveled off by diffusion and spin–lattice relaxation. The third RF pulse converts the z magnetization back to spin coherences which then evolve into the stimulated echo. A detailed description of the spin coherence evolution can be found in Ref. [2].

The total attenuation of the stimulated echo in liquids after two free-evolution intervals τ_1 and the grating storage interval τ_2 is given by

$$A_{stim.echo} = A_{relax} A_{diff}, \tag{21}$$

where

$$A_{relax} = \exp\{-2\tau_1/T_2\} \exp\{-\tau_2/T_1\}. \tag{22}$$

A_{diff} represents any of the formulas given in Eqs. 13–16 and Eq. 20 for the respective limits. If $\tau_1 << T_2 << T_1 > \tau_2$, the maximum diffusion time can be adjusted much longer than in the Hahn echo case.

2.1.4
Spin Echoes and Steady B_0 Gradients

The efficiency of spatial encoding by field gradients depends on the wave number given in Eq. 8, that is the "area" $G\delta$ of the gradient pulses in the free-evolution intervals of the pulse sequences shown in Fig. 2. To generate short and intense gradient pulses with extremely short rise and decay times is technically difficult and, even worse, is susceptible to motional artifacts due to pulsed magnetic forces on the gradient coil system. For very short transverse relaxation times (which are usually accompanying slow translational diffusion) it may therefore be advisable to use steady gradients instead of pulsed ones. Any gradient switching is then avoided and the gradient encoding efficiency in the coherence free-evolution intervals is optimal.

Figure 3a shows the steady-gradient RF pulse scheme for the Hahn echo. The pulsed-gradient parameters δ and Δ defined in Fig. 2a now take identical values and are equal to the RF pulse spacing,

$$\delta = \Delta = \tau. \tag{23}$$

Equation 20 thus becomes

$$A_{diff}(\tau) = \exp\left\{-\frac{2}{3}k^2 D\tau\right\} = \exp\left\{-\frac{2}{3}\gamma^2 G^2 D\tau^3\right\}, \tag{24}$$

where we have anticipated Gaussian propagators as above. If the propagator does not have a Gaussian form and if no special analytical formalism is available for the evaluation of experimental attenuation data, the low-wave-number limit given in Eq. 16 can again be employed as an approach.

Figure 3b shows the analogous variant for the stimulated echo. This is the preferential method for extremely short transverse relaxation times. In this case the pulsed-gradient parameters turn into $\delta=\tau_1$ and $\Delta=\tau_1+\tau_2$, so that Eq. 20 for Gaussian propagators becomes

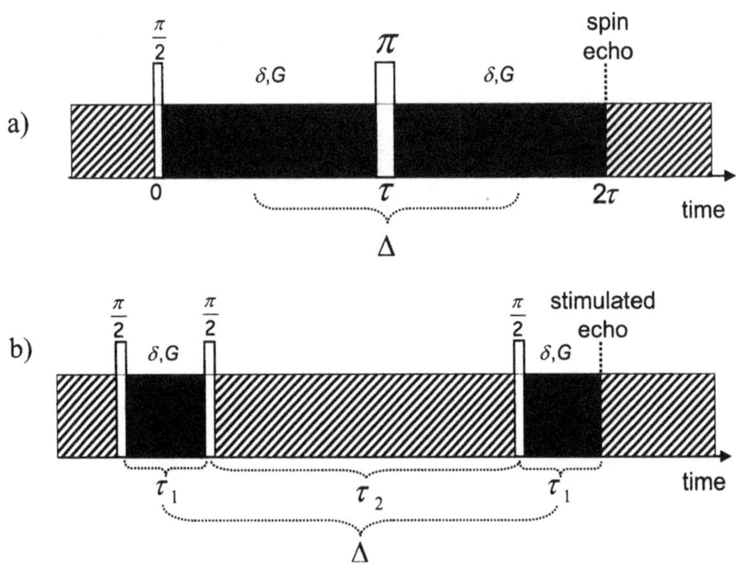

Fig. 3a, b. Radio frequency (RF) pulse sequences for the Hahn spin echo (**a**) and the stimulated spin echo (**b**) combined with steady magnetic field gradients. The *hatched sections* of the gradients are irrelevant for diffusion measurements. Echo attenuation by diffusive displacements is solely caused by the *black sections* of the gradient cut out by the RF pulses and the echo time. The fringe field of superconducting magnets turned out to provide particularly stable and strong gradients

$$A_{diff}(\tau_1,\tau_2)=\exp\left\{-\gamma^2G^2D\tau_1^3\left(\frac{2}{3}+\frac{\tau_2}{\tau_1}\right)\right\}. \tag{25}$$

In the limit $\tau_1 << \tau_2$, this expression takes a form analogous to Eq. 14,

$$A_{diff}(\tau_1 << \tau_2)\approx\exp\left\{-k^2D\tau_2\right\}=\exp\left\{-\frac{1}{6}k^2\langle R^2(\tau_2)\rangle\right\}, \tag{26}$$

where the wave number is now defined by $k=\gamma G\tau_1$. For non-Gaussian propagators, one may again employ the low-wave-number approximation

$$A_{diff}=\left\langle e^{\pm ikZ}\right\rangle \overset{k^2\langle Z^2\rangle\ll 1}{\approx} 1-\frac{1}{2}k^2\langle Z^2\rangle \overset{\text{isotropy}}{=} 1-\frac{1}{6}k^2\langle R^2\rangle\approx\exp\left\{-\frac{1}{6}k^2\langle R^2\rangle\right\}$$
$$=\exp\left\{-k^2D(\tau_2)\tau_2\right\} \tag{27}$$

with the time-dependent diffusion coefficient $D=D(t)$.

Steady gradients of considerable strength (e.g., up to 60 T/m with a conventional 9.4 T wide-bore magnet) are readily available in the fringe field of superconducting magnets [8]. Applications to diffusion studies in polymers have been reported [9–12]. Even stronger steady gradients can be produced with a special Maxwell coil design for superconducting magnets [13].

Since the fringe field gradient has a fixed strength at the position of the probe head, the only experimental parameters of the pulse schemes shown in Fig. 3 that can be varied are the time intervals between the RF pulses. That is, the echo attenuation by relaxation must be considered and compensated for. This can conveniently be done with the aid of self-compensating pulse schemes [14].

Some care must be taken with fringe field methods if motional averaging of dipolar interactions is incomplete. Echoes then tend to be modulated by the so-called dipolar correlation effect [15, 16]. One can account for this phenomenon by dividing the (normalized) echo amplitudes recorded with and without gradient at the same Larmor frequency and at the same pulse intervals [11, 12]. Any influence by relaxation and the dipolar correlation effect can be eliminated in this way.

2.2 NMR Spin–Lattice Relaxometry

The term “relaxometry” is normally used in context with techniques for the measurement of spin–lattice relaxation times in general. Transverse relaxation and effects due to residual dipolar couplings will be considered in the next section.

As indicated in the scheme shown in Fig. 1, diffusometry and relaxometry are complementary to each other with respect to the time or frequency scales they are probing. This is mainly due to the potential of the field-cycling NMR relaxometry technique that permits us to measure relaxation times down to the kHz regime, corresponding to a time scale on which field-gradient diffusometry becomes applicable. Above 10 MHz, the experimental frequency scale can be supplemented by spin–lattice relaxation measurements with conventional NMR spectrometers operating at several hundred MHz by present-day standards. Apart from spin–lattice relaxation in the laboratory frame, we will also present some experimental rotating-frame spin-lattice relaxation data later for comparison.

2.2.1 General Theoretical Background

The studies to be reviewed below mainly refer to proton or deuteron magnetic resonance in melts or concentrated solutions of polymers of molecular masses close to or above the critical value M_c. Under such circumstances the predominant spin–lattice relaxation mechanism of protons (spin 1/2) is based on fluctuating dipole–dipole couplings of like spins, that is, we are dealing with the homonuclear case. Deuterons (spin 1), in contrast, possess a finite electric quadrupole moment which is subject to quadrupole coupling to local molecular electric field gradients which is much more efficient than dipolar interactions.

The relaxation formalisms of dipolar-coupled two-spin 1/2 systems (protons) on the one hand, and of quadrupolar-coupled spin 1 nuclei (deuterons) on the other, have very much in common and lead to largely equivalent analytical expressions [2, 17]. This very much facilitates comparisons of experimental results obtained with either technique.

The reason for this analogy is the fact that the spatial part of the dipolar as well as of the quadrupolar coupling Hamiltonian can be commonly expressed by second-order spherical harmonics $Y_{2,m}(\vartheta,\varphi)$ with $m=0, \pm 1, \pm 2$. For a definition of these functions see Ref. [2], p. 403. The only terms of the Hamiltonian relevant for spin–lattice relaxation in the laboratory frame are those for $m=\pm 1$ and $m=\pm 2$ in connection with spin operator terms inducing single-quantum and double-quantum transitions, respectively, in the two-spin 1/2 (dipolar coupling) or single-spin 1 (quadrupolar coupling) systems [2]. These transitions do not conserve Zeeman spin energy. They are rather connected with energy exchange between Zeeman spin energy and lattice energy as required for spin–lattice relaxation. The term "lattice" comprises all motional degrees of freedom of the spin-bearing molecules.

The spherical harmonics $Y_{2,m}(\vartheta,\varphi)$ are expressed in polar coordinates, that is the polar angle ϑ and the azimuthal angle φ. These coordinates define the orientation of the internuclear vector relative to the external magnetic flux density $\vec{B}_0$ in the case of dipolar coupling, or the orientation of the principal electric field gradient (i.e., of a molecular axis) again relative to $\vec{B}_0$ in the case of quadrupolar interactions, where one usually anticipates rotationally symmetric electric field gradients.

Molecular motions in the sense of reorientations of molecules or chemical groups lead to fluctuating polar coordinates, $\vartheta=\vartheta(t)$, $\varphi=\varphi(t)$. As a consequence, the dipolar or quadrupolar Hamiltonians become time dependent and, hence, tend to induce spin transitions as predicted by time-dependent perturbation theory. With dipolar coupling, there is a third variable fluctuating as a result of molecular motion, namely the internuclear distance $r=r(t)$ of the two-spin 1/2 system. This, however, matters only with intermolecular or intersegment interactions while all intrasegment couplings are associated with constant r values. Below we will see that the comparison of proton relaxation with deuteron relaxation of the same polymer species permits one to distinguish intra- from intermolecular interactions and to elucidate the dynamical implications connected with these interactions on this basis.

In the frame of the Bloch/Wangsness/Redfield (BWR) relaxation theory [2, 17], the fluctuations of the spin Hamiltonians are described with the aid of (preferably normalized) autocorrelation functions of the type

$$G_m(t) = \left\langle \frac{Y_{2,m}(\vartheta_0,\varphi_0)\,Y_{2,-m}(\vartheta_t,\varphi_t)}{r_0^3 r_t^3} \right\rangle \Bigg/ \left\langle \frac{|Y_{2,m}(\vartheta_0,\varphi_0)|^2}{r_0^6} \right\rangle \text{(dipolar coupling)},$$

$$G_m(t) = \left\langle Y_{2,m}(\vartheta_0,\varphi_0)\,Y_{2,-m}(\vartheta_t,\varphi_t) \right\rangle \text{(quadrupolar coupling)}. \tag{28}$$

The subscripts 0 and t of the spatial variables indicate the time at which they are to be taken. Actually the expressions given in Eq. 28 are statistically stationary functions, so that only the time interval matters rather than the absolute time. The brackets stand for an ensemble average over all spin systems in the sample.

According to time-dependent perturbation theory, the transition probability per time unit is proportional to the spectral density (or intensity function) of the fluctuating coupling inducing the transition. The spectral density is given as the Fourier transform of the autocorrelation functions,

$$I(\omega) = \int_{-\infty}^{\infty} G_m(t)\,e^{-i\omega t}\,dt. \tag{29}$$

Note that the spectral density defined in this way is independent of the subscript m for isotropic systems since it is based on the normalized autocorrelation functions given in Eq. 28.

The spin–lattice relaxation rate directly reflects the spin transition probabilities per time unit for single- and double-quantum transitions, and hence is proportional to a linear combination of spectral densities in the form

$$\frac{1}{T_1} = C_{coupl}[I(\omega) + 4I(2\omega)], \tag{30}$$

where $\omega = \gamma B_0$ is the resonance angular frequency depending on the gyromagnetic ratio γ of the nuclei and the external magnetic flux density B_0. The analytical form of Eq. 30 is valid for systems of two like, dipolar-coupled spins 1/2 as well as for spins 1 quadrupolar coupled to local electric field gradients. The prefactor C_{coupl} is a constant specific for the type of the dominating spin coupling. The first spectral density term in the brackets on the right-hand side of Eq. 30 refers to single-quantum, the second to double-quantum transitions. The latter consequently is a function of twice the single-quantum resonance frequency.

Later in the review we will distinguish intrasegment from intersegment dipolar couplings giving rise to two-proton spin–lattice relaxation contributions according to

$$\frac{1}{T_1} = \frac{1}{T_1^{\text{intra}}} + \frac{1}{T_1^{\text{inter}}}. \tag{31}$$

The analytical form of Eq. 31 anticipates stochastic independence of the two types of fluctuating couplings. This appears to be a reasonable assumption since intersegment couplings are modulated by translational segment diffusion, whereas the variations of intrasegment interactions originate from chain modes in the time/frequency range of interest. As a consequence, intrasegment couplings fluctuate much faster than intersegment interactions. The latter therefore matter only at low frequencies. Note that the relaxation of quadrupole nuclei such as deuterons is dominated at any frequency solely by intrasegment interactions to local electric field gradients. Deuteron relaxation is therefore a favorable means for the discrimination of intra- from intersegment relaxation mechanisms.

The overall correlation time of the fluctuating spin couplings is defined by

$$\tau_c = \int_0^\infty G_m(t)dt. \tag{32}$$

The BWR theory resulting in Eq. 30 is valid only in the limit

$$T_1 >> \tau_c. \tag{33}$$

Below we will discuss numerous NMR relaxometry applications to polymers. The information on the type of segment–internal fluctuations, chain modes or center-of-mass displacements is contained in the autocorrelation functions in the time domain or, equivalently, in the spectral density in the frequency domain according to Eq. 29. In order to probe characteristic features of polymer dynamics it is therefore of interest to measure the frequency dependence of the spin–lattice relaxation time or rate, which then directly reflects the spectral density related to the type of segment or chain motion. In the next section we describe methods suitable for the recording of such spin–lattice relaxation "dispersion" curves.

2.2.2
The Field-Cycling NMR Relaxometry Technique

Spin–lattice relaxation dispersions according to Eq. 30 can be recorded over several decades of the frequency with the aid of field-cycling NMR relaxometry [18–21]. Combined with ordinary high-field NMR relaxometry, the accessible ranges for protons and deuterons are

$$10^3\,\mathrm{Hz} < \frac{\omega}{2\pi} < 10^9\,\mathrm{Hz} \;\text{ and }\; 10^2\,\mathrm{Hz} < \frac{\omega}{2\pi} < 10^8\,\mathrm{Hz}, \tag{34}$$

respectively. The high-frequency limit is given by the available high-field magnets. At low frequencies several factors may restrict the applicability of the field-cycling technique: (1) the "local field" representing the secular part of the spin couplings may exceed the external flux density in the relaxation interval of the field cycle; (2) the compensation of the earth field (or other magnetic stray fields in the lab) at the sample position may be imperfect; (3) the low-field spin–lattice relaxation times may become too short to permit field switching fast enough for reliable measurements; and (4) the low-field spin–lattice relaxation time may become shorter than the correlation time of the longest correlation function component so that the validity condition of the BWR formula given in Eq. 33 is violated.

Figure 4 schematically shows a typical field cycle. The external magnetic flux density is cycled in a sequence of three different values: the polarization field, the relaxation field, and the detection field. The sample is polarized in the polarization field, B_p, which is chosen as high as compatible with the cooling device of the resistive magnet coil [21]. This takes a couple of spin-lattice relaxation times taken at this particular field value. After that the magnetization is equal to the Curie equilibrium magnetization $M_0(B_p)$.

The magnetic flux density is then electronically switched down to the preselected relaxation field, B_r, at which spin–lattice relaxation is to be examined. On the one hand, the field-switching rate must be large enough to avoid excessive relaxation losses of the magnetization during the switching process. On the other hand, it should be slow enough to permit adiabatic field changes in case the relaxation field is perceptibly superimposed by local fields (of arbitrary directions other than that of $\vec{B}_r$).

In the relaxation field the magnetization is aligned along the external magnetic field direction and is initially equal to the Curie equilibrium magnetization in the polarization field, $M(0)=M_z(0)=M_0(B_p)$, where we have ignored potential relaxation losses during the switching interval. It then relax-

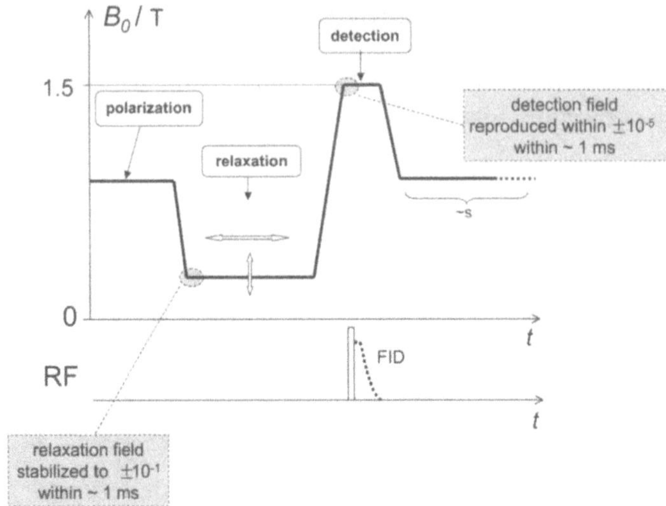

Fig. 4. Schematic representation of a typical cycle of the main magnetic field B_0 employed with field-cycling NMR relaxometry. The free induction decay (FID) is recorded after a 90° radio frequency (RF) pulse in the detection field. The repetition time amounts several times the spin–lattice relaxation time in the polarization field. The most critical sections of the cycle are brought out by *gray boxes*

es toward the new Curie equilibrium magnetization of the relaxation field, $M_0(B_r)$, so that the magnetization evolves according to the following solution of Bloch's equation:

$$M_z(\tau) = M_0(B_r) + \left[M_0\left(B_p\right) - M_0(B_r)\right] \exp\{-\tau/T_1(B_r)\}. \tag{35}$$

The magnetization remaining after the relaxation interval τ is finally detected with the aid of a 90° RF pulse or a suitable spin echo sequence in the form of a NMR signal after switching the magnetic flux density to a fixed detection field. Since the acquisition of a free-induction signal takes only a few milliseconds at most, the detection field period can be kept extremely short and the detection flux density, hence, correspondingly strong without overloading the magnet coil. After having recorded the signal, the flux density is switched back to the polarization field value, and after a period of a couple of spin–lattice relaxation times the whole cycle begins again.

Incrementing the relaxation interval τ thus permits one to record the relaxation curve for a given relaxation flux density B_r. The spin–lattice relaxation dispersion is then scanned point by point by stepping B_r through a series of discrete values spread over the desired flux density (or frequency) range.

It can be shown that relaxation losses in the finite switching intervals diminish the dynamic range of the variation of the relaxation decay and hence the experimental accuracy, but that they do not cause any systematic experimental error provided that the passages between the different field levels are reproducible when incrementing the relaxation interval τ for a given relaxation flux density B_r [2]. However, some care must be taken to avoid experimental artifacts that may arise at the critical field crossover points marked in Fig. 4.

At low fields, proton and especially deuteron spin–lattice relaxation times of viscous polymer systems may easily fall short of a millisecond. That is, coming from the large polarization field of typically 0.5 T, relaxation fields that can be as low as 10^{-5} T must be reached *and* settled and stabilized with a desired accuracy of better than 10% within a total passage time in the order of one millisecond. The short settling time is a stringent condition which is not easy to fulfill practically. Likewise the passage from the relaxation field to the detection field should occur in a transition time of the same order. In particular, the detection flux density needed for magnetic resonance must be hit and reproduced with an accuracy of about 10^{-5} in subsequent transients.

It is therefore of paramount importance to check and calibrate the field cycle with the aid of a test device providing the accuracy, the time resolution and, in particular, the dynamic range required. A fast field probe placed at the sample position connected to a 12 to 16 bit transient recorder with a MHz bandwidth turned out to be a safe way for avoiding any experimental artifacts by imperfections of the field cycle.

Field-cycling NMR relaxometry experiments can be favorably supplemented by ordinary high-field relaxation measurements employing the inversion–recovery or saturation–recovery variants. Comparative spin–lattice relaxation experiments "in the rotating frame" ($T_{1\rho}$) are also of interest particularly in the presence of molecular order [22]. A detailed description of the diverse NMR methods referred to can be found in Ref. [2].

The experimental frequency/temperature "window" opened by NMR relaxometry is schematically shown in Fig. 5. Considering temperatures above the glass transition and below thermal degradation suggests a typical range between 300 and 400 K. The proton/deuteron frequency range accessible by the field-cycling technique in combination with high-field relaxometry is within 10^2 and 10^9 Hz. This window almost completely matches the dynamic range of chain modes of polymers with molecular masses up to 10^5. Only at low temperatures and high frequencies does some influence of local segment–internal fluctuations tend to show up. On the other hand, at high temperatures and low frequencies, center-of-mass motions of polymers with molecular masses below 10^5 may manifest themselves in the form of low-frequency cutoffs of the dispersion curves. However, for the most part, spin–lattice relaxation in the experimental window is dominated by chain modes irrespective of the molecular mass.

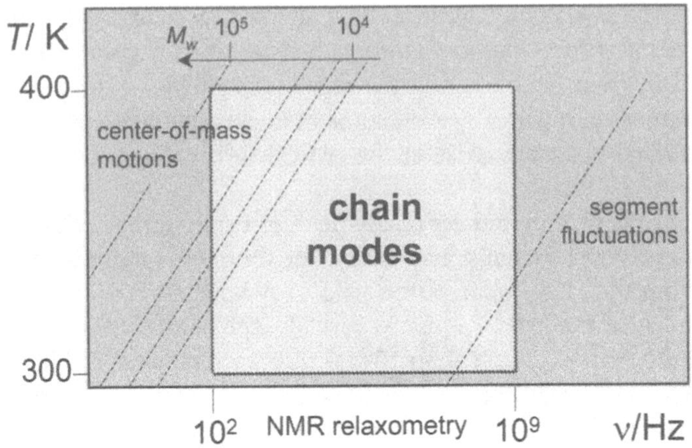

Fig. 5. Schematic representation of the experimental temperature/frequency "window" conveniently accessible by field-cycling NMR relaxometry in combination with conventional high-field techniques. It specifically addresses the chain-mode regime of typical polymers, whereas local segment fluctuations and center-of-mass motions can only be probed at low temperatures/high frequencies and high temperatures/low frequencies/low molecular weights combinations, respectively

2.3
Residual Dipolar Broadening and Transverse Relaxation

The standard BWR theory for transverse relaxation anticipates complete motional averaging of all local fields due to dipolar interaction (or by any other spin coupling) on a time scale corresponding to the linewidth in the absence of molecular motions. This time scale is defined by the reciprocal rigid-lattice linewidth, $(\Delta\omega)_{rl}^{-1}$. The so-called *motional narrowing* condition reads [2]

$$(\Delta\omega)_{rl}\tau_c \ll 1, \tag{36}$$

where τ_c is the correlation time defined in Eq. 32 for m=0.

In the motional narrowing limit, transverse relaxation is governed by spin transitions induced by fluctuating (both secular and nonsecular) spin interactions rather than by spin evolution in secular local fields of the same origin, but without being connected with spin transitions. The distinction between transverse relaxation by spin transitions on the one hand, and by evolution in secular local fields on the other, is best done by considering the spin operator terms in the Hamiltonians of the spin interactions (see the tables given in Chap. 46 of Ref. [2]). Spin operators inducing non-spin-energy conserving transitions are nonsecular by definition. Operators that do not induce spin transitions and operator terms that induce only spin-energy-conserving transitions are secular. Secular spin interactions cause a distribution of the eigenenergies and, hence, a distribution of the local Larmor frequencies. The latter scenario becomes relevant if the motional narrowing condition given by Eq. 36 is violated. With all macromolecules, this will most likely be the case.

The most prominent spin interaction in polymers is dipolar coupling. If motional narrowing according to Eq. 36 is absent or incomplete, one speaks of *dipolar broadening* or *residual dipolar broadening*, respectively. Residual dipolar broadening contains some useful information on chain dynamics. Numerous techniques in principle suitable for retrieving this information have been suggested.

The free-induction decay of transverse magnetization has been analyzed in terms of polymer dynamics [23–26]. A solid echo technique was employed for the same purpose [27, 28]. The so-called dipolar correlation effect on the stimulated echo turned out to be a particularly simple and robust tool in this context too [15, 29, 30]. Finally, double-quantum NMR spectroscopy was suggested [31, 32] as a means of probing features of chain dynamics.

At temperatures far above the glass transition residual dipolar couplings in flexible polymers are getting negligible. Also, heteronuclear species such as ^{13}C are subject to secular dipolar coupling only to a minor degree. Under such circumstances, the (average) transverse relaxation can be studied using standard techniques. Corresponding data reveal characteristic molecular weights and temperatures of entangled polymer systems in a quite spectacular way [33–35]. The crossover from unentangled to entangled chain dynamics was visualized clearly this way. Finally it should be mentioned that transverse relaxation curves are suitable to distinguish the special dynamics of chain-end blocks compared to that of segments in the middle of the polymer [36] by analyzing relaxation curve components representing different segment mobilities.

2.3.1 The Anderson/Weiss Approach

Transverse relaxation under incomplete motional narrowing conditions in multispin systems can be treated with the aid of the Anderson/Weiss approximation [2]. Precession phase shifts caused by spin interactions with many coupling partners is assumed to be subject to the central limit theorem. That is, the distribution of the local fields B_{loc}, i.e., the distribution of the corresponding angular frequency offsets $\Omega=\gamma B_{loc}$, and hence the distribution of the phase shifts accumulated in a given time interval t, is described by a Gaussian function. With this approximation one finds for the normalized transverse relaxation decay function

$$S(t) = \exp\left\{-\langle\Omega^2\rangle_{rl}\int_0^t (t-\tau)g(\tau)d\tau\right\}, \tag{37}$$

where $\langle\Omega^2\rangle_{rl}$ is the mean squared angular frequency offset (or the *second moment* or the *variance* of the distribution of angular frequency offsets) in the absence of motions (i.e., for the "rigid lattice" limit). The normalized autocorrelation function of the frequency offsets is given by

$$g(\tau) = \frac{\langle\Omega(0)\Omega(\tau)\rangle}{\langle\Omega^2\rangle_{rl}}. \tag{38}$$

The brackets indicate ensemble averages. An analogous formalism for the dipolar correlation effect can be found in Refs. [2, 15, 16].

Expressing the autocorrelation function given in Eq. 38 by its Fourier transform, the spectral density $I(\omega)$,

$$g(\tau) = \frac{1}{2\pi}\int_{-\infty}^{\infty} I(\Omega)e^{i\Omega\tau}d\Omega, \tag{39}$$

leads to [2]

$$S(t) = \exp\left\{-\frac{1}{\pi}\int_0^{\infty}\frac{\langle\Omega^2\rangle_{rl}}{\Omega^2}I(\Omega)[1-\cos(\Omega t)]d\Omega\right\}. \tag{40}$$

With incomplete motional narrowing, the transverse relaxation decay obviously reflects fluctuations with rates $\Omega lt;\sqrt{\langle\Omega^2\rangle_{rl}}$, that is a range complementary and partly overlapping with that of field-cycling NMR relaxometry. The representation given in Eq. 40 is particularly useful for the distinction of different dynamic components (see also Fig. 18).

3
Modeling of Chain Dynamics and Predictions for NMR Measurands

Polymers are molecules subject to complex intra- and intermolecular interactions combined with many intramolecular degrees of motional freedom. The degree of polymerization ϵ typically ranges from 10^2 to 10^6. Assuming only three rotamers per monomer, the number of possible conformations already takes the astronomical number value of the order 3^ϵ. The time constants characterizing the molecular motions are spread over up to six orders of magnitude. In view of these facts, it is a rather demanding task to model polymers and their molecular dynamics.

At room temperatures $T \approx 300$ K, conformational transitions typically take $\tau_s = 10^{-11} - 10^{-9}$ s. All elementary conformational changes displace atoms of the polymer on a distance length scale $l \lesssim 1$ nm. These characteristic quantities permit the distinction of "local" and "global" properties.

A detailed description of statically or dynamically local properties must be based on the stereochemical composition of the monomers and their potential energy landscape [1, 37–44]. On the other hand, the global, i.e., large-scale, properties of macromolecules tend to be insensitive to the stereochemical details of the monomers. Therefore, a coarse-grained description becomes adequate by considering model chains. All chemical specificity can then be represented by a few characteristic parameters to be specified in the following.

A number of more or less equivalent chain models have been suggested [1, 42]. We will discuss only the simplest and most popular one, that is the "random walk" or "freely jointed" chain. Macromolecules are represented (i.e., approximated) by a linear sequence of Kuhn segments of length b (see Fig. 6). The number of monomers per Kuhn segment is taken just large enough to permit the neglect of any stereochemical restriction of the orientation of the Kuhn segments relative to each other. There is no mutual orientation correlation. Merely the linear order of segments, that is the "linear memory", is retained. The conformational statistics of such a freely jointed chain of N Kuhn segments $N >> 1$ is mathematically equivalent to random walks in three-dimensional space.

In the Rouse model to be discussed below, each Kuhn segment is subdivided into a "bead" (where the segment mass is concentrated) and a massless "spring". The elasticity of these springs is based on the conformational entropy of the segments which depends on how far they are stretched. This bead/spring model is characterized by the following basic parameters partially also illustrated in Fig. 6.

The *Kuhn segment length* is defined by the root mean squared value

$$b \equiv \sqrt{\langle x^2 \rangle} \tag{41}$$

of the lengths x of the segments in a chain at a given instant. Likewise the *Flory radius* represents the root mean squared value

$$R_F \equiv \sqrt{\langle \vec{R}_{ee}^2 \rangle} = \sqrt{N b^2} \tag{42}$$

of the chain end-to-end vector $\vec{R}_{ee}$ in an ensemble of independent polymer chains with N segments each. Note that $\langle \vec{R}_{ee} \rangle = 0$ in an isotropic distribution of Kuhn segments, i.e., in the absence of an external force field or any mesomorphic order. Equations 41 and 42 anticipate that the length distribution is Gaussian.

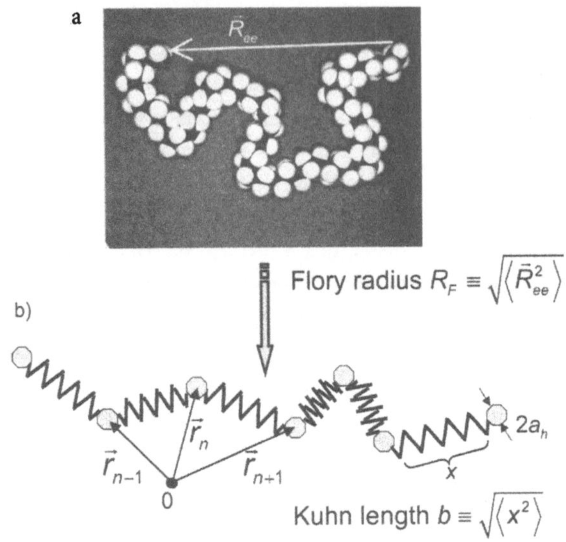

Fig. 6. Freely jointed chain of Kuhn segments and bead/spring chain model. For coarse-grain treatments for long time and length scales, the polymer may be modeled by a chain of beads and massless entropic springs. The beads are assumed to have a hydrodynamic radius a_h. The Kuhn segment length, b, is defined as the root mean squared length of the springs. The root mean squared chain end-to-end distance is called the Flory radius R_F

The freely jointed segment chain model provides a satisfactory description of large-scale conformational properties of real chains in polymer melts and concentrated solutions, where effects of excluded-volume distortions of the (Gaussian) conformational statistics can be neglected [1, 37–44]. Typical orders of magnitude for flexible polymers such as polyethylene (PE), polyethylene oxide (PEO), polystyrene (PS) etc. are $b \simeq 0.5 \div 2$ nm, $N \simeq 10 \div 10^4$, $R_F \simeq 2 \div 10^2$ nm. A Kuhn segment roughly represents 2–5 monomers. Numbers like these can be estimated based on the conformational statistics expected under the restrictions of the conformational degrees of freedom, that is, for the angles between the valence bonds in the real chain. They can also be evaluated from experimental data as empirical characteristic parameters of the macromolecule [37–39].

3.1 The Rouse Model

The dynamics of an isolated Kuhn segment chain in its bead-and-spring form is considered in a viscous medium without hydrodynamic backflow or excluded-volume effects. The treatment is based on the Langevin equation generalized for Brownian particles with internal degrees of freedom. A first, crude formalism of this sort was reported by Kargin and Slonimskii [45]. Independently of this work, Rouse published later a more detailed version [46] which is nowadays referred to as the "Rouse model".

The effective intramolecular interactions between the segments are approximated by entropic harmonic interactions, reflecting the Gaussian character of the large-scale chain conformation. Intermolecular interactions (with the surrounding viscous medium) are taken into account by friction and stochastic forces acting on the segments.

The *entropic spring constant* and the *friction coefficient* of a Kuhn segment are given by

$$K = \frac{3k_B T}{b^2} \tag{43}$$

and

$$\zeta = 6\pi\eta a_h, \tag{44}$$

respectively, where k_B is Boltzmann's constant, T is the absolute temperature, η is the viscosity of the medium surrounding the segment, and a_h is the hydrodynamic radius of the segments.

The equation of motion of the nth segment (see Fig. 6b) reads

$$\frac{\partial}{\partial t}\vec{p}_n(t) = K(2\vec{r}_n - \vec{r}_{n+1} - \vec{r}_{n-1}) - \zeta\frac{\partial \vec{r}_n}{\partial t} + \vec{f}_n^L(t), \tag{45}$$

where $\vec{p}_n(t)$ is the momentum, and $\vec{f}_n^L(t)$ is the Langevin stochastic force acting on this segment. In the continuum limit, the segment number n can be treated as a continuous variable ranging from 0 to N, so that Eq. 45 can be rewritten in approximate form as

$$\frac{\partial}{\partial t}\vec{p}_n(t) \approx K\frac{\partial^2 \vec{r}_n}{\partial n^2} - \zeta\frac{\partial \vec{r}_n}{\partial t} + \vec{f}_n^L(t). \tag{46}$$

The boundary conditions reflecting the existence of only one neighbor segment at the chain ends are

$$\left.\frac{\partial}{\partial n}\vec{r}_n(t)\right|_{n=0,N} = 0. \tag{47}$$

The first term on the right-hand side of Eq. 46 represents an intramolecular force whereas the second and third terms are forces of an intermolecular nature.

The inertial force term $\frac{\partial}{\partial t}\vec{p}_n(t)$ on the left-hand side of Eq. 46 is important only for times $t << m/\zeta$, where m is the mass of a Kuhn segment. For typical experimental situations with polymer melts one estimates $m/\zeta \div 10^{-13}$–10^{-12} s. That is, the inertial force term can be neglected on the time scale relevant for chain dynamics studies. Equation 46 thus reads

$$\zeta \frac{\partial}{\partial t}\vec{r}_n(t) \approx \frac{3k_BT}{b^2}\frac{\partial^2 \vec{r}_n(t)}{\partial n^2} + \vec{f}_n^L(t). \tag{48}$$

The differential operator $\frac{3k_BT}{b^2}\frac{\partial^2}{\partial n^2}$ represents entropically elastic forces and has a discrete set of eigenfunctions. In this context, these eigenfunctions are called the *Rouse normal coordinates* defined by

$$\vec{X}_p(t) = \frac{1}{N}\int_0^N dn \cos\left(\frac{\pi p}{N}n\right)\vec{r}_n(t), \tag{49}$$

where p=0, 1, 2, ... is the normal-mode number. Strictly speaking, this series of eigenfunctions is infinite. However, normal coordinates with mode numbers p=N, N+1 ... must be considered as a mathematical artifact owing to the replacement of a discrete set of ordinary differential equations (see Eq. 45) by partial differential equations in the continuum limit, Eq. 48. Fortunately for most physical quantities of interest here, it does not matter whether the set of normal coordinates is limited or not provided that $N>>1$.

The equation of motion given by Eq. 48 can be solved by combining it with the normal coordinate expression defined in Eq. 49. Forming the correlation function $\langle \vec{X}_p(t)\cdot\vec{X}_{p'}(0)\rangle$ of the normal coordinates leads then to

$$\langle \vec{X}_p(t)\cdot\vec{X}_{p'}(0)\rangle = \delta_{pp'}\frac{Nb^2}{2\pi^2p^2}\exp\left\{-\frac{t}{\tau_p}\right\}, \tag{50}$$

where we have employed the fact that the Cartesian components α and β (where $\alpha,\beta = x,y,z$) of the Langevin stochastic forces acting on the nth and k^{th} segments at times t_2 and t_1, respectively, are correlated as

$$\left\langle f_{\alpha n}^L(t_2) f_{\beta n}^L(t_1)\right\rangle = 2\zeta k_BT\delta_{\alpha\beta}\delta(t_2-t_1)\delta(n-k). \tag{51}$$

Equation 50 represents the Rouse modes in the proper sense with the relaxation time of the pth mode

$$\tau_p = \tau_s\left(\frac{N}{p}\right)^2. \tag{52}$$

The relaxation time of the Nth Rouse mode,

$$\tau_s = \frac{b^2\zeta}{3\pi^2k_BT}, \tag{53}$$

is called the *segmental relaxation time*, referring to the mode with the shortest relaxation time in the frame of the Rouse model. On the other hand, the longest relaxation time associated with the mode number p=1 is the so-called *terminal chain relaxation time for the Rouse model* or simply *Rouse relaxation time*. It is given by

$$\tau_1 \equiv \tau_R = \tau_s N^2. \tag{54}$$

The Kronecker symbol $\delta_{pp'}$ in Eq. 50 indicates that the normal modes are mutually orthogonal to each other. That is, the Rouse modes are in reality referring to the autocorrelation functions

$$C_p(t) = \langle \vec{X}_p(t) \cdot \vec{X}_p(0) \rangle \tag{55}$$

of the normal coordinates. This is the function we will refer to in the treatments of experimental NMR parameters of interest in this context.

3.1.1 Translational Segment Diffusion of a Rouse Chain

The mean squared displacement of the nth Kuhn segment can be expressed in the following way:

$$\begin{aligned}\langle (\vec{r}_n(t) - \vec{r}_n(0))^2 \rangle &= \left\langle \left(\vec{X}_0(t) - \vec{X}_0(0)\right)^2 \right\rangle + 8\sum_{p=1}^{N-1} \left[C_p(t) - C_p(0)\right] \cos^2\left(\frac{\pi}{N} pn\right) \\ &\equiv g_{cm}(t) + g_{rel}^n(t). \end{aligned} \tag{56}$$

The first term on the right-hand side represents the mean squared displacement of the chain center-of-mass, $g_{cm}(t)$, the second term refers to displacements of segment n relative to the center-of-mass, $g_{rel}^n(t)$. Averaging Eq. 56 over all segment positions n gives

$$\begin{aligned}\langle \vec{R}^2(t) \rangle \equiv \langle (\vec{r}(t) - \vec{r}(0))^2 \rangle &= \left\langle \left(\vec{X}_0(t) - \vec{X}_0(0)\right)^2 \right\rangle + 4\sum_{p=1}^{N-1} \left[C_p(t) - C_p(0)\right] \\ &\equiv g_{cm}(t) + g_{rel}(t), \end{aligned} \tag{57}$$

where $g_{rel}(t) = \frac{1}{N} \sum_{n=1}^{N-1} g_{rel}^n(t)$. Combining Eqs. 57 and 50 leads to the following limits for the mean squared segment displacement in the frame of the Rouse model:

$$\langle \vec{R}^2(t) \rangle = \begin{cases} \frac{2}{\sqrt{\pi^3}} b^2 \sqrt{\frac{t}{\tau_s}} & \text{for} \quad \tau_s \ll t \ll \tau_R \\ 6 D_R t & \text{for} \quad t \gg \tau_R \end{cases}, \tag{58}$$

where

$$D_R = \frac{1}{3\pi^2} \frac{b^2}{\tau_s N} \tag{59}$$

is the Rouse chain diffusion coefficient. On a time scale shorter than the Rouse relaxation time, segment diffusion is subdiffusive, and the mean squared displacement follows a power law $\langle R^2\rangle \propto N^0 t^{1/2}$. This is in contrast to the behavior at times longer than τ_R where the relation $\langle R^2\rangle \propto N^{-1} t^1$ holds.

3.1.2 Spin–Lattice Relaxation of a Rouse Chain

For the treatment of spin–lattice relaxation, we need an expression for the autocorrelation function of the Kuhn segment tangent vector, which in the continuum limit is given by

$$\vec{b}_n(t) = \frac{\partial}{\partial n}\vec{r}_n(t) = -\frac{2\pi}{N}\sum_{p=1}^{N-1}\vec{X}_p p \sin\left(\frac{\pi}{N}pn\right). \tag{60}$$

The autocorrelation function of this vector is readily found as an expression based on the autocorrelation function of the normal coordinates, Eq. 55,

$$\left\langle \vec{b}(t)\cdot\vec{b}(0)\right\rangle = \frac{2\pi^2}{N^2}\sum_{p=1}^{N-1}p^2 C_p(t), \tag{61}$$

where we have averaged over all segments n, and where the orthogonality of the normal coordinates was taken into account again. In analogy to Eq. 58 the following limits can be derived for this tangent vector autocorrelation function:

$$\left\langle \vec{b}(t)\cdot\vec{b}(0)\right\rangle = \begin{cases} \frac{\sqrt{\pi}}{2}b^2\left(\frac{\tau_s}{t}\right)^{1/2} & \text{for } \tau_s \ll t \ll \tau_R \\ \frac{b^2}{N}\exp\left\{-\frac{t}{\tau_R}\right\} & \text{for } t \gg \tau_R \end{cases}. \tag{62}$$

According to Refs. [47–49], the spin-lattice relaxation rate is related to the tangent vector autocorrelation function by

$$\frac{1}{T_1} \propto \int_{-\infty}^{\infty}\left\langle \vec{b}(t)\cdot\vec{b}(0)\right\rangle^2 e^{-i\omega t}dt. \tag{63}$$

This expression is based on a variant of Eq. 28 especially suited for the description of chain modes [47]. Inserting Eq. 62 we find the limits

$$\frac{1}{T_1} \propto \begin{cases} -\tau_s \ln(\omega\tau_s) & \text{for } \tau_R^{-1} \ll \omega \ll \tau_s^{-1} \\ \omega^0 \tau_s \ln N & \text{for } \omega \ll \tau_R^{-1} \end{cases}. \tag{64}$$

The logarithmic frequency dependence [50, 51] in the limit $\tau_R^{-1} << \omega << \tau_s^{-1}$ will be of particular interest for experimental tests of the model.

Equations 63 and 64 are valid for intrasegment spin interactions. Since the Rouse model is experimentally relevant mainly for modes of relatively short chains or short chain sections, any contribution from intersegment dipolar interactions are negligible on this time/frequency scale, and need not be treated explicitly.

3.2 The Tube/Reptation Model

The Rouse model describes the dynamical properties of melts of macromolecules of a relatively small number of Kuhn segments, $N \leq N_c$, reasonably well. The critical number N_c is the number of Kuhn segments for the critical molecular mass M_c. Flexible polymers have critical Kuhn segment numbers typically in the range $N_c = 40 \div 60$ [1, 42–44, 52]. On the other hand, chain dynamics in concentrated systems of polymers with $N >> N_c$ is much slower than expected on the basis of the Rouse model. Alluding to chain entanglements that are considered to become relevant in this case, one speaks of "entangled" dynamics. For example, experimental terminal relaxation times and center-of-mass self-diffusion coefficients scale as $\tau_t \propto N^{3 \div 3.4}$ and $D \propto N^{-2 \div 2.5}$, whereas the Rouse model predicts much weaker dependences, $\tau_R \propto N^2$ and $D \propto N^{-1}$, respectively.

The term "entanglement" is understood as the global representation of interchain interactions, which have not been considered in the Rouse equation of motion, Eq. 46. The Rouse equation of motion is the simplest way to describe effectively single-chain dynamics. In the frame of this model, intermolecular interactions are merely accounted for by way of a molecular mass independent local friction coefficient ζ. The fact that the Rouse theory predicts a chain length-dependent relaxation time, $\tau_R = \tau_s N^2$, already indicates some internal inconsistency, because this molecular weight dependence in combination with the uncrossability property of polymer chains (the "connectivity restriction") suggests a molecular weight dependent friction. That is, the applicability of the Rouse model is rather limited (see also Ref. [58] and the literature cited therein). In the following, two of the most prominent attempts to take entanglement interactions into account will be discussed in more detail. The first model of interest is the tube/reptation model.

The tube concept and the reptation model were originally introduced by Edwards [53] and de Gennes [54], respectively, and then investigated in more detail by Doi and Edwards [1, 55] and many other authors [56–66]. Entanglement interactions between neighboring chains are modeled by a fictitious tube around the chain referred to. This chain is also termed the "tagged chain" in an ensemble of otherwise identical chains. All other macromolecules form the so-called matrix.

Let us assume for the moment that the matrix is frozen while the tagged chain is still mobile. The tagged chain cannot cross contours of matrix chains due to intermolecular *excluded volume interactions* and the *chain connectivity*. It appears intuitively self-evident that the motions of the tagged chain will be restricted to a tube-like region of space around the initial chain conformation. After some time, the only degree of translational freedom, namely displacements of the tagged chain in this tube back and forth, will thread the tagged chain from the "initial tube" into a "new tube". The diameter of this sort of tube would be equal to typical nearest neighbor distances between two monomers from different chains, d=0.3–0.5 nm.

In real polymer melts the matrix is as mobile as the tagged chain. In the tube/reptation model this is attempted to be taken into account by assuming a fictitious tube of a much larger effective diameter (typically 5–7 nm). It is postulated that segment motions become highly anisotropic for times $t >> \tau_e = \tau_s N_e^2$, where $N_e \approx N_c/2$. Intermolecular entanglement forces effectively localize the macromolecule in a space region in the form of a curved tube of diameter $d = bN_e^{1/2}$ around the *contour line* or *primitive path* of the chain (see Fig. 7). The tube conformation corresponds to that of a freely jointed chain, that is, the tube conformation has a Gaussian end-to-end length distribution. The characteristic time

$$\tau_e \equiv \tau_s N_e^2 \approx \frac{1}{4} \tau_s N_c^2 \tag{65}$$

is called the *entanglement time*, and N_e is the number of Kuhn segments effectively corresponding to an "entanglement".

Segments moving inside the tube are considered to move in a viscous medium as it is anticipated in the Rouse model. In the interior of the tube, entanglement interactions are assumed to be absent. However, as soon as a segment approaches the tube "walls", entanglement forces become infinitely large so that the polymer is effectively confined by the topological constraints imposed by the tube. The chain-end segments are excepted from this restraint: they are not subject to entanglement forces and can therefore move isotropically. The consequence is that the chain can migrate in a reptile manner (*reptation*) into a new conformation completely uncorrelated to that of the initial chain coil.

3.2.1 The Doi/Edwards Limits

There are four dynamic processes characterized by the four characteristic time constants of the tube/reptation model (see Fig. 7): the segment fluctuation time τ_s (given in Eq. 53), the entanglement time τ_e (given in Eq. 65), the (longest) Rouse relaxation time τ_R (given in Eq. 54), and the (tube) disengagement time τ_d (given below in Eq. 71). With these four time constants, four dynamic limits can be defined, which will be termed the Doi/Edwards limits $(I)_{DE}$ to $(IV)_{DE}$. Table 1 shows the limiting laws for the mean squared

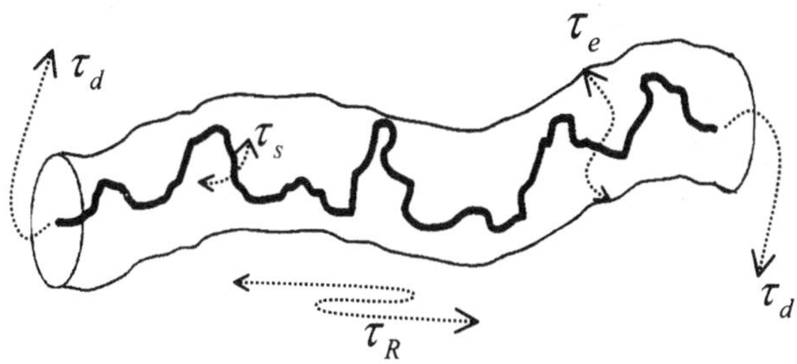

Fig. 7. Illustration of a polymer chain (the tagged chain) confined in the fictitious tube of diameter d formed by the matrix. The contour line of the tube is called the primitive path having a random-walk conformation with a step length a=d. The four characteristic types of dynamic processes (*dotted arrow lines*) and their time constants τ_s, τ_e, τ_R, and τ_d defined in the frame of the Doi/Edwards tube/reptation model are indicated

segment displacement and the spin–lattice relaxation time commonly expected on this basis. Figure 7 illustrates the chain processes dominating in the four limits.

Limit (I)$_{\mathrm{DE}}$ ($\tau_s << t << \tau_e$) represents Rouse dynamics. Entanglement interactions are not yet effective. Segment displacements are isotropic. On this time scale the segments of the tagged chain can be considered to be surrounded by an isotropic viscous medium just as assumed in the Rouse model. The length scale of the Rouse modes is however restricted by the diameter of the fictitious tube. If a regime like this exists, then it should reveal itself by the mean squared segment displacement and the (intrasegment) spin–lattice relaxation time laws

$$\langle R^2 \rangle \propto M^0 t^{1/2} \tag{66}$$

and

$$T_1 \propto -M^0 / \ln(\omega\tau_s), \tag{67}$$

respectively (Table 1).

On a longer time scale, in **limit (II)$_{\mathrm{DE}}$** ($\tau_e << t << \tau_R$), the tube constraints become effective, and chain modes are consequently connected with elongations along the tube contour (the "primitive path"). The distribution of relaxation times extends up to the value known as the (longest) Rouse relaxation time given in 54. That is, segment motions are anisotropic on the time scale of this limit. Since segments distant enough from each other are fluctuating still independently, the mean squared segment displacement remains molecular mass independent. Two particularly characteristic power laws for limit (II)$_{\mathrm{DE}}$ refer to the mean squared segment displacement,

$$\langle R^2 \rangle \propto M^0 t^{1/4}, \tag{68}$$

Table 1. Theoretical dependences on time (t), angular frequency (ω), and molecular mass (M) predicted by the tube/reptation model for the mean squared segment displacement and the intrasegment spin-lattice relaxation time in the four Doi/Edwards limits. The factors C_I, C_{II}, C_{III}, and C_N are frequency and molecular mass independent constants

	Limit	Mean squared segment displacement $\langle R^2 \rangle =$	Refs	Intrasegment spin-lattice relaxation time $T_1^{\text{intra}} =$	Refs
$(I)_{DE}$	$\tau_s << (t, 1/\omega) << \tau_e$	$\frac{2}{\pi^{3/2}} b^2 \left(\frac{t}{\tau_s}\right)^{1/2}$ $\propto M^0 t^{1/2}$	[1]	$-C_I \frac{M^0}{\tau_s \ln(\omega \tau_s)}$	[47, 50]
$(II)_{DE}$	$\tau_e << (t, 1/\omega) << \tau_R$	$b^2 N_e^{1/2} \left(\frac{t}{\tau_s}\right)^{1/4}$ $\propto M^0 t^{1/4}$	[1, 54]	$C_{II} M^0 \omega^{3/4}$	[54, 70]
$(III)_{DE}$	$\tau_R << (t, 1/\omega) << \tau_d$	$\frac{2}{\pi} b^2 \left(\frac{N_e}{3N} \frac{t}{\tau_s}\right)^{1/2}$ $\propto M^{-1/2} t^{1/2}$	[1, 54]	$C_{III} M^{-1/2} \omega^{1/2}$	[54]
$(IV)_{DE}$	$\tau_d << (t, 1/\omega)$	$2 \frac{k_B T}{\zeta} \frac{N_e}{N^2} t$ $\propto M^{-2} t^1$	[1,54]	$C_{IV} M^{-(1.5...2.0)} \omega^0$	[123]

and to the (intrasegment) spin-lattice relaxation time,

$$T_1 \propto M^0 \omega^{3/4} \tag{69}$$

(Table 1). Equations 68 and 69 are totally equivalent as can be seen from a simple probability consideration. In the frame of this and the following time limit, the correlation function given in Eq. 28 can essentially be considered to represent the probability that the spin-bearing segment is still or again in the tube section where it was initially [54, 70, 71]. That is,

$$G_m(t) \propto \frac{a^2}{\langle R^2(t) \rangle}, \tag{70}$$

where $a = b\sqrt{N_e}$ is the step length of the primitive path. Inserting Eq. 68 and carrying out the Fourier transform required by Eqs. 29 and 30 directly leads to the proportionality given in Eq. 69.

Reptation in the proper sense begins to show up only beyond the Rouse time scale, namely in **limit (III)$_{DE}$** defined by $\tau_R << t << \tau_d$. The disengagement time is defined by

$$\tau_d = 3\tau_s \frac{N^3}{N_e}, \tag{71}$$

that is, $\tau_d >> \tau_R$ for $N >> N_e$. Segment motions are still anisotropic in the sense of correlated fluctuations along the tube contour. The chain moves coherently as a whole and effectively like a single random walker back and forth. The mean squared segment displacement in curvilinear coordinates along the tube contour is a linear function of time. However, if considered in Euclidean space it is proportional to the square root of time as a consequence of the Gaussian tube conformation. Since the chain moves as a whole and, hence, is subject to friction proportional to the chain length, the mean squared curvilinear displacement is now inversely proportional to the chain length, whereas mean squared displacements measured in the Euclidean space are inversely proportional to the square root of the chain length. The latter is again a consequence of the random coil character anticipated for the tube. The power laws characteristic for this limit are

$$\langle R^2 \rangle \propto M^{-1/2} t^{1/2} \tag{72}$$

and

$$T_1 \propto M^{-1/2} \omega^{1/2}, \tag{73}$$

where the latter refers to intrasegment spin interactions (Table 1). The complete equivalence of Eqs. 72 and 73 can be shown the same way as above with the aid of Eq. 70.

Finally, for $t >> \tau_d$ **limit (IV)$_{\mathbf{DE}}$** comes into play. This is the regime of normal diffusion, when all correlations with the initial conformation are lost, and the mean squared displacement obeys the Einstein relation

$$\langle R^2 \rangle = 6 D_c t, \tag{74}$$

where

$$D_c = \frac{k_B T}{3\zeta} \frac{N_e}{N^2} \tag{75}$$

is the isotropic center-of-mass self-diffusion coefficient. The dispersion of the spin–lattice relaxation time becomes independent of the frequency in the form of a low-frequency plateau, as is expected for isotropic rotational tumbling of the spin-bearing object in the so-called extreme narrowing limit reading $\omega \tau_d << 1$ in the present case.

Note that all limits of the time and molecular weight dependences of the mean squared segment displacement discussed so far are reflected by corresponding limits of the frequency and molecular weight dependences of the spin–lattice relaxation time (see Table 1). Experimental tests based on both phenomena therefore appear to be particularly favorable.

3.2.2
Echo Attenuation by Reptation

The evaluation formulas given in Eq. 20 or 25 for attenuation of the stimulated echo by diffusion anticipates Gaussian propagators. As illustrated in Fig. 8, waiting time delays matter in NMR diffusion measurements with polymers subject to reptation [72, 73]. This follows from the fact that segments diffusing along the curvilinear path of the randomly coiled tube contour are intermittently "trapped" in loops that contribute little to displacements along the field gradient direction. Effectively non-Gaussian propagators are the consequence for the Doi/Edwards limits $(II)_{DE}$ and $(III)_{DE}$. One may try an approximate evaluation according to Eq. 16. However, it turned out to be more indicative for the characteristic reptation features to consider a special formalism specifically developed for the reptation problem [3].

In the short gradient "pulse" limit of the fringe-field NMR diffusometry technique, which is of particular interest in context with slowly diffusing polymers (see Fig. 3), the effective diffusion time is given by $t \approx \Delta = \tau_2$. Under this prerequisite the stimulated-echo attenuation factor can be analyzed according to

$$A_{diff}(k,t) = \left\langle e^{i\vec{k}\cdot\vec{R}(t)} \right\rangle_R = \underbrace{\left\langle e^{i\vec{k}\cdot\vec{R}_s(t)} \right\rangle_{R_s}}_{\text{segments; } A_s(k,t)} \underbrace{\left\langle e^{i\vec{k}\cdot\vec{R}_c(t)} \right\rangle_{R_c}}_{\text{center-of-mass; } A_c(k,t)} , \tag{76}$$

where we have anticipated that segment and center-of-mass displacements are independent of each other because of the different time scales on which these processes take place.

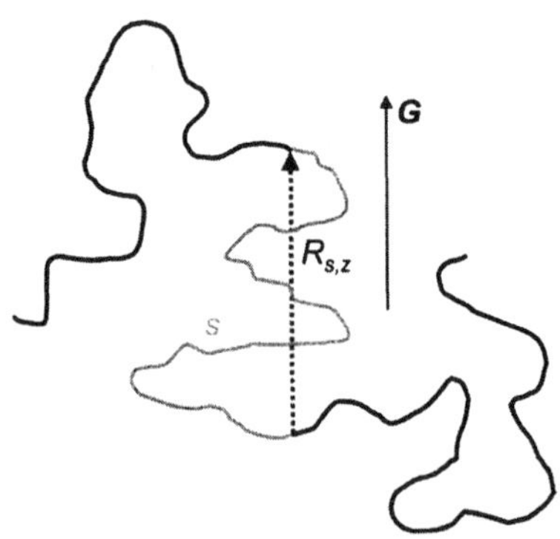

Fig. 8. Polymer segments in a "tube" are displaced along the randomly coiled contour line resulting in the "curvilinear displacement" s in the experimental diffusion time (drawn in *gray*). Field-gradient NMR diffusometry measures displacements $R_{s,z}$ along the gradient direction (G). Contour loops act as "traps" for segments the displacements of which are monitored along G

The ensemble averages indicated by the brackets refer to displacements along the field gradient direction. The subscripts of the ensemble averages indicate displacements R in general, displacements R_s relative to the center-of-mass by segment diffusion (limits $(I)_{DE}$, $(II)_{DE}$, and $(III)_{DE}$), and displacements of the center-of-mass R_c (limit $(IV)_{DE}$), respectively.

Attenuation by center-of-mass diffusion (limit $(IV)_{DE}$) is described by the standard expression given in Eqs. 13 or 26

$$A_c(k,t) = \exp\left\{-k^2 D_c t\right\}, \tag{77}$$

where D_c is the center-of-mass diffusion coefficient of the whole chain (see Eq. 75). In limits $(II)_{DE}$ and $(III)_{DE}$, displacements achieved in the interval of limit $(I)_{DE}$ due to free Rouse modes are relatively small and can safely be neglected. The spin-echo attenuation factor is then dominated [3] by

$$A_s(k,t) = \underbrace{\left\langle \underbrace{\int \left(\frac{4\pi}{3} a|s|\right)^{-3/2} e^{R_s^2/a|s|} e^{i\vec{k}\cdot\vec{R}_s(t)} d^3\vec{R}_s}_{\text{average over all } R_s \text{ for a given } s} \right\rangle}_{\text{average over all } s}$$

$$= \exp\left\{\frac{k^4 a^2 \langle s^2(t)\rangle}{72}\right\} \operatorname{erfc}\left\{\frac{k^2 a \sqrt{\langle s^2(t)\rangle}}{6\sqrt{2}}\right\} \tag{78}$$

where the mean squared *curvilinear* segment displacement,

$$\langle s^2(t)\rangle = \frac{2D_0 t}{N + \frac{12a^2 D_0 t}{N^2 b^4}} + \frac{2b\sqrt{D_0 t}}{\sqrt{3\pi} + 18\frac{\sqrt{D_0 t}}{Nb}} \tag{79}$$

combines the expressions given in [1] for limits $(II)_{DE}$ and $(III)_{DE}$. The step length of the so-called primitive path is denoted by $a = b\sqrt{N_e}$. The quantity $D_0 = k_B T/\zeta$ is the (fictitious) diffusivity of a Kuhn segment in the absence of chain connectivity.

3.2.3
General Remarks on the Tube/Reptation Model

The tube diameter, $d = a = bN_e^{1/2}$, is the main phenomenological parameter of the model. Actually this is a fitting parameter for the description of the viscoelastic plateau on the basis of the tube/reptation formalism [1, 42]. Typical values for melts of flexible polymers are in the mesoscopic range of d = 5–7 nm. This is much larger than the intermolecular distance of typically σ=0.3–0.5 nm. The inequality $d >> \sigma$ partly spoils the uniformity and the logic of the picture inherent to the tube/reptation model: that large tubes contain many chains and not just a single "tagged chain". Nevertheless,

Rouse displacement dynamics is anticipated on a length scale up to the tube diameter. That is, the chain connectivity constraint and excluded-volume interactions are ignored for this limit. In the frame of this model, these restraints are taken into account only outside the tube but not inside.

A further inconsistency refers to the treatment of shear stress relaxation. The Doi/Edwards tube cannot exist without long-living intermolecular correlations pertaining to a time scale of the order $\tau_d = \frac{1}{3}\tau_s N^3/N_e$. Nevertheless, in the calculation of the stress tensor on the basis of the tube/reptation model all intermolecular correlations are considered to be fast decaying so that their neglect appears to be justified. That is, the stress tensor is treated on the basis of intramolecular interactions only by ignoring all intermolecular contributions [66–69, 74–77].

A number of modified versions of the tube/reptation model have been suggested in the literature [56, 57, 59–65] in order to improve the description of experimental data. Terms like "contour length fluctuation", "reptons", "double reptation", "constraint release" etc. have been introduced. The main goal of these modifications was to account for finite-length effects in the sense that segments near the chain ends behave in a different way compared to segments in the middle, and that a tube formed of chains of finite lengths cannot be really static. Nevertheless, all of these modifications adopt the analytical features of the original tube/reptation model in the limit $N\rightarrow\infty$. The changes are expected to become effective for the Doi/Edwards limits $(III)_{DE}$ and $(IV)_{DE}$, but not for $(I)_{DE}$ and $(II)_{DE}$. These corrections unavoidably require the introduction of new phenomenological elements or parameters lacking a clear, mathematically well-defined, microscopic basis. The improved coincidence between theory and experiment is thus paid for by complications and ambiguities in the formalism.

Numerous computer simulations have been carried out in order to examine the transition from Rouse to reptation dynamics [70, 78–88]. Entanglement effects on chain dynamics clearly showed up. However, the discussion as to what extent the characteristic features of the reptation model for concentrated polymer liquids are verified by these simulations still remains controversial. It also should be mentioned that a series of phenomenological "nonreptative" models were published recently [89–94]. They mainly focus on viscoelastic properties of entangled polymer systems.

Below we will come back to the reptation model in context with the dynamics of polymers confined in tube-like pores formed by a solid matrix. For a system of this sort the predictions for limits $(II)_{DE}$ and $(III)_{DE}$ (see Table 1) could be verified with the aid of NMR experiments [11, 95] as well as with an analytical formalism for a harmonic radial potential and a Monte Carlo simulation for hard-pore walls [70]. The latter also revealed the crossover from Rouse to reptation dynamics when the pore diameter is decreased from infinity to values below the Flory radius.

3.3 Memory Function Formalisms

The illustrative and intuitive nature of the tube/reptation model is juxtaposed by its partly phenomenological character that makes it impossible to trace the formalism back to elementary principles on a molecular level. The strategic idea to project the whole multiparticle complexity of an entangled polymer system to the assumption of a fictitious tube, so to speak, is the beginning and the end of our thinking in the frame of this model.

The memory function approaches to be outlined in the following are of a more formal nature but nevertheless are relying on some assumptions that are not yet founded on basic principles. At the present state of the art some phenomenological premiss is unavoidable. However, memory function approaches are analytically more flexible and can be adapted to experimental data by scrutinizing the general laws anticipated for short-time segmental dynamics. On the other hand, they certainly do not inspire our imagination in the same way as the reptation model does.

In this section we will first refer to the formalism originally introduced by Zwanzig and Bixon [96, 97], which was then applied to polymer dynamics by Schweizer [98, 99] and others [100–108]. This sort of theory is based on the Zwanzig/Mori projection operator technique in connection with treatments of the *Generalized Langevin Equation* [109–115]. It should be noted that this equation can be considered as the "microscopic" basis of phenomenological approaches based on the memory function formalism [116–122].

3.3.1 The Generalized Langevin Equation (GLE)

Polymers can be described adequately in terms of classical formalisms. The set of all position vectors and momentums of all segments in the system to be treated is denoted by γ, that is a point in phase space. In principle any physical property of the system can be described by an appropriate function $A(\gamma)$ of phase variables. Consider now two arbitrary physical quantities corresponding to the functions $A(\gamma)$ and $B(\gamma)$. The scalar product of these functions is given by

$$\langle A|B\rangle \equiv \int d\gamma A^*(\gamma)B(\gamma)\rho_{eq}(\gamma) \equiv \langle A^*(\gamma)B(\gamma)\rangle_{eq}, \tag{80}$$

where $A^*(\gamma)$ is the conjugate complex of $A(\gamma)$, and $\rho_{eq}(\gamma)$ is the equilibrium Gibbs canonical distribution function of the whole system. The set of all functions of this sort forms the so-called Liouville space L.

One is now interested in the kinetics of a certain set of physical quantities $A_1, A_2, \ldots A_n$, the so-called quantities of interest. The set of all linear combinations of these quantities forms the linear subspace $L^n\{A_i\} \in L$. The projection operator with respect to $L^n\{A_i\}$ is defined by

$$\hat{P} \equiv \sum_{k,m} |A_k\rangle \| \langle A|A\rangle \|^{-1}_{km} \langle A_m|, \tag{81}$$

where $\| \langle A|A\rangle \|^{-1}_{\mathrm{km}}$ is an element of the matrix inverse to the static correlation matrix $\| \langle A_k|A_m\rangle \|$, the ket $|A_m\rangle$ is the vector in the Liouville space L corresponding to the quantity A_m (in fact $|A_m\rangle = A_m(\gamma)$), and the bra $\langle A_k|$ is the vector in the Liouville space L corresponding to the quantity A_k (in fact $\langle A_k| = A_k^*(\gamma)$).

Any physical quantity obeys the equation of motion:

$$\frac{d}{dt} A_n(t) = \{A_n; H\} \equiv i\hat{L} A_n \,, \tag{82}$$

where $\{f;g\}$ is the Poisson bracket of quantities f and g, $\hat{L} \equiv i\{H; ...\}$ is the Liouville operator, and H is the Hamiltonian of the system. Note that $A_n(t)$ is a short notation for $A_n(\gamma)$ in the Heisenberg representation. It is defined by

$$A_n(t) \equiv \exp\left\{ i\hat{L}t \right\} A_n(\gamma). \tag{83}$$

The Mori transformation can be used to formally obtain an exact system of equations for "quantities of interest":

$$\frac{d}{dt} A_n(t) = \sum_k i\omega_{nk} A_k(t) - \sum_k \int_0^t d\tau K_{nk} A_k(t-\tau) + F_n^Q(t), \tag{84}$$

where $\omega_{nk} = \sum_m \| \langle A|A\rangle \|^{-1}_{km} \langle A_m | \hat{L} | A_n \rangle$ is the frequency matrix. The quantity

$$F_n^Q(t) = \exp\left\{ i\hat{Q}\hat{L}\hat{Q} \right\} i\hat{Q}\hat{L} |A_n\rangle \tag{85}$$

is a generalized stochastic Langevin force associated with the physical quantity A_n. The operator given by $\hat{Q} = 1 - \hat{P}$ projects on the subspace orthogonal to $L^n\{A_r\}$. Finally, the memory matrix is defined by

$$K_{nk}(\tau) \equiv \sum_m \| \langle A|A\rangle \|^{-1}_{km} \left\langle F_m^Q(0) | F_n^Q(\tau) \right\rangle. \tag{86}$$

Equation 84 is often called the generalized Langevin equation (GLE) [109, 115].

Note that the time evolution of experimental observables $A_n(t)$ is governed by "real dynamics" which is determined according to Eq. 83 by the real propagator $\exp\left\{ i\hat{L}t \right\}$. The situation with the time evolution of the stochastic force $F_n{}^Q(t)$ and the memory matrix $K_{nk}(t)$ is much more complicated. According to Eq. 85, their evolution is governed by "projected dynamics" the propagator of which is given by $\exp\left\{ i\hat{Q}\hat{L}\hat{Q}t \right\}$.

3.3.2
The GLE for the Tagged Macromolecule

The position and momentum vectors of the segments of the tagged chain are denoted by $\vec{r}_1, \vec{r}_2, \ldots \vec{r}_N, \vec{p}_1, \vec{p}_2, \ldots \vec{p}_N$, where N is the number of segments. The dynamical state of the tagged chain is given by its phase space point

$$\gamma_N \equiv \{\vec{r}_1, \vec{r}_2, \ldots \vec{r}_N, \vec{p}_1, \vec{p}_2, \ldots \vec{p}_N\}. \tag{87}$$

The "quantities of interest" (analogous to $A_k(\gamma)$) are the single-chain *phase densities* determined by the following relation:

$$f(\Gamma_N; \gamma_N) \equiv \delta(\Gamma_N - \gamma_N) \equiv \prod_{i=1}^{N} \delta(\vec{R}_i - \vec{r}_i)\delta(\vec{P}_i - \vec{p}_i), \tag{88}$$

where $\Gamma_N \equiv \{\vec{R}_1, \vec{R}_2, \ldots, \vec{R}_N; \vec{P}_1, \vec{P}_2 \ldots, \vec{P}_N\}$ is a general phase space point, and δ is Dirac's delta function. The variables forming Γ_N are called the *field variables.* In the following we will use the short notation

$$f(\Gamma_N; \gamma_N) \equiv f(\Gamma_N). \tag{89}$$

The physical meaning and importance of the phase density $f(\Gamma_N)$ is elucidated by its mean equilibrium value

$$\langle f(\Gamma_N) \rangle = \rho_N^*(\Gamma_N) \tag{90}$$

which is the single-chain distribution function over all segment coordinates and momentums.

The effective single-chain Hamiltonian, or effective single-chain free energy of the tagged macromolecule, can be determined by the relation

$$H_N^* \equiv H_N^*(\Gamma_N) = -kT \ln \rho_N^*(\Gamma_N). \tag{91}$$

The effective Hamiltonian δH^*_N can be represented as

$$H_N^* = \sum_{i=1}^{N} \frac{P_i^2}{2m} + W^*(\{\vec{R}_i\}), \tag{92}$$

where $W^*(\{\vec{R}_i\})$ is the so-called effective intramolecular potential energy, or the potential of the mean force. The projection operator on the Liouville space of the tagged macromolecule can be represented by

$$\begin{aligned} \hat{P}A(\gamma) &= \frac{1}{\rho_N^*(\gamma_N)} \int d\gamma' \delta(\gamma_N - \gamma'_N) A(\gamma') \rho_{eq}(\gamma') \\ &= \int d\gamma_M A(\gamma) \exp\{-\beta \delta H(\gamma)\} \equiv \langle A(\gamma) \rangle_{\gamma_N}^*, \end{aligned} \tag{93}$$

where γ_M is the short notation for all position vectors and momentums of the matrix segments. That is $\gamma_N + \gamma_M = \gamma$ and $\delta H(\gamma) = H(\gamma) - H_N^*(\gamma_N)$. $A(\gamma)$ is an arbitrary property of the system. $\rho_{eq}(\gamma)$ is the equilibrium distribution function of the total system.

The quantity $\delta H(\gamma)$ does not depend on the momentums of the tagged-chain segements in accordance with Eq. 92, and can be therefore considered as the Hamiltonian of the matrix, which is moving under the condition that the conformation of the tagged chain is fixed. The quantity $\langle A(\gamma)\rangle_{\gamma N}^*$ can be regarded as a conditionally averaged value of $A(\gamma)$ under the condition that γ_N has a fixed value.

The generalized Langevin equation for phase densities has the following structure:

$$\frac{\partial}{\partial t} f(\Gamma_N;t) = \int d\Gamma'_N i\Omega\left(\Gamma_N \middle| \Gamma'_N\right) f(\Gamma'_N;t) - \\ - \int_0^t d\tau \int d\Gamma'_N K\left(\Gamma_N \middle| \Gamma'_N;\tau\right) f\left(\Gamma'_N;t-\tau\right) + F^Q(\Gamma_N;t), \quad (94)$$

where $\Omega(\Gamma_N|\Gamma'_N)$ is the frequency matrix, $K(\Gamma_N|\Gamma'_N;\tau)$ is the memory matrix, and $F^Q(\Gamma_N;t) \equiv \exp\{i\hat{Q}\hat{L}_\gamma\hat{Q}t\} i\hat{Q}\hat{L}_\gamma f(\Gamma_N)$ is the generalized Langevin force associated with the phase density $f(\Gamma_N) = \delta(\Gamma_N - \gamma_N)$. The subscript γ indicates that the Liouville operator acts on the variables of the phase space point γ. $\hat{Q} = 1 - \hat{P}$ is the projection operator on the subspace orthogonal to L_N.

The frequency matrix can be expressed as

$$\Omega\left(\Gamma_N \middle| \Gamma'_N\right) = i\left\{H_N^*\left(\Gamma'_N\right); \delta\left(\Gamma_N - \Gamma'_N\right)\right\}_{\Gamma'_N}, \quad (95)$$

where the subscript Γ'_N means that the differentiation involved in the Poisson bracket {;} is to be done over variables forming Γ'_N only. The generalized stochastic Langevin force $F^Q(\Gamma_N;t)$ is written in the form

$$F^Q(\Gamma_N;t) \equiv \exp\{i\hat{Q}\hat{L}_\gamma\hat{Q}t\}\{\delta(\Gamma_N - \gamma_N); \delta H(\gamma)\}_\gamma. \quad (96)$$

The memory matrix reads

$$K(\Gamma_N|\Gamma'_N;t) = \frac{1}{\rho_N^*(\Gamma'_N)} \Big\langle \{\delta(\Gamma'_N - \gamma); \delta H(\gamma)\}_\gamma \exp\{i\hat{Q}\hat{L}_\gamma\hat{Q}t\} \cdot \\ \cdot \{\delta(\Gamma_N - \gamma); \delta H(\gamma)\}_\gamma \Big\rangle_{eq}. \quad (97)$$

Note that the time evolution of the generalized stochastic Langevin force $F^Q(\Gamma_N;t)$ and the memory matrix are governed by so-called projected dynamics determined by the projected propagator $\exp\{i\hat{Q}\hat{L}\hat{Q}t\}$, which must be distinguished from the real dynamics determined by the real propagator.

Let the function $a(\gamma_N)$ be an arbitrary characteristic quantity of the tagged chain for the moment. The generalized Langevin equation for $a(\gamma_N)$ is then obtained from Eq. 94. To do this one needs to multiply both sides of Eq. 94 by $a(\Gamma_N)$ and integrate over Γ_N. After some calculations the following result is found:

$$\frac{d}{dt}a(t) = \{a(\gamma(t)); H_N^*(\gamma_N(t))\} + \exp\{i\hat{Q}\hat{L}_\gamma\hat{Q}t\}\{a;\delta H\} + \\ + \int_0^t \frac{d\tau}{\rho_N^*(\gamma_N(t-\tau))} \sum_k \frac{\partial}{\partial \vec{p}_k(t-\tau)} \cdot \\ \cdot \left\langle \delta\vec{F}_k^{\text{inter}} \delta(\gamma_N(t-\tau) - \gamma_N) \exp\{i\hat{Q}\hat{L}_\gamma\hat{Q}\tau\}\{a;\delta H\}_\gamma \right\rangle_{eq}, \tag{98}$$

where $\delta\vec{F}_k^{\text{inter}} = \hat{Q}\vec{F}_k = -\frac{\partial}{\partial \vec{r}_n} H(\gamma) + \frac{\partial}{\partial \vec{r}_n} W^*\{\vec{r}_i\}$ is the fluctuating part of the intermolecular force acting on the nth segment of the tagged macromolecule.

Equation 98 is now considered for $a(\gamma_N) = \vec{r}_n$ and $a(\gamma_N) = \vec{p}_n$ leading to

$$\frac{d}{dt}\vec{r}_n(t) \equiv \vec{v}_n(t) = \frac{1}{m}\vec{p}_n(t) \tag{99}$$

$$\frac{d}{dt}\vec{p}_n = -\frac{\partial}{\partial \vec{r}_n} W^*(\{\vec{r}_i\}) - \sum_k \int_0^t d\tau \Gamma_{nk}^{\alpha\beta}(\tau; t-\tau) v_k^\beta(t-\tau)\vec{e}_\alpha + \vec{F}_n^Q(t), \tag{100}$$

where $\vec{e}_\alpha$ is the unit vector aligned along the axis labeled α. The memory matrix is given by

$$\Gamma_{nk}^{\alpha\beta}(\tau; t-\tau) = \frac{1}{kT\rho_N^*(t-\tau)} \left\langle F_k^{Q\beta}(0)\delta(\gamma_N(t-\tau) - \gamma_N) F_n^{Q\alpha}(\tau) \right\rangle_{eq}. \tag{101}$$

The quantity $V^\beta{}_k(t-\tau)$ is the β component of the kth segment's velocity at a time moment $t-\tau$. The function $F_n^{Q\alpha}(\tau) \equiv \exp\{i\hat{Q}\hat{L}\hat{Q}\tau\}\hat{Q}F_n^\alpha(\gamma)$ is the α component of the generalized stochastic Langevin force acting on the segment number n at time τ. The Dirac delta function in Eq. 101 allows us to rewrite this expression in a form of conditional averaging over the matrix variables under the condition that phase variables of the tagged chain have constant values equal to their real values $\gamma_N(t-\tau)$:

$$\Gamma_{nk}^{\alpha\beta}(\tau; t-\tau) = \frac{1}{kT} \left\langle F_k^{Q\beta}(0) F_n^{Q\alpha}(\tau) \right\rangle_{\gamma_N(t-\tau)}^*. \tag{102}$$

The system of Eqs. 99 and 100 is an exact representation of segment dynamics in a system of many entangled Kuhn chains. However, this system

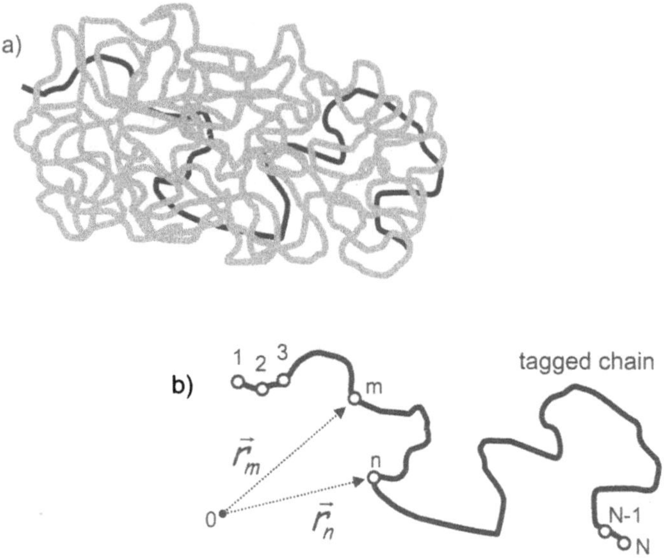

Fig. 9a, b. The global effect of the "matrix chains" (*gray*) on the dynamics of the "tagged chain" (*black*) is represented by the memory function (**a**). The Kuhn segments of the tagged chain are numbered consecutively (**b**)

cannot be solved rigorously, even not in principle, because it is not closed. That is, it contains two new, unknown objects, namely the potential of the mean force $W^*\{\vec{r}_i\}$ defined by Eq. 92 and the memory matrix $\Gamma_{nk}^{\alpha\beta}(\tau;t-\tau)$ given in Eq. 102. All one can do is to replace them by an intuitive ansatz or some physically plausible approximation. These two functions represent the global "entanglement" effect of the matrix on the tagged chain (for an illustration see Fig. 9).

3.3.2.1
The Potential of the Mean Force and the Memory Matrix

The potential of the mean force $W^*\{\vec{r}_i\}$ in principle is determined by equilibrium statistical mechanics. The coarse grain configuration of polymer chains in melts obeys Gaussian statistics with respect to the end-to-end distance, for instance. This suggests the following approximation of the potential of the mean force:

$$W^*\{\vec{r}_i\} = \frac{3k_BT}{2b^2}\sum_{n=1}^{N}(\vec{r}_n - \vec{r}_{n-1})^2. \tag{103}$$

Substituting this expression in Eq. 100 reproduces the effectively intramolecular entropic force acting on the nth segment, which is exactly equal to the first term on the right-hand side of the Rouse equation of motion given by Eq. 45.

The analytical form of the memory matrix $\Gamma_{nk}^{\alpha\beta}(\tau;t-\tau)$ given in Eq. 101 is much less clear. The memory matrix $\Gamma_{nk}^{\alpha\beta}(\tau;t-\tau)$ contains all the (unknown) information about intermolecular interactions, i.e., all consequences of entanglement effects. The fluctuating part of the intermolecular force $\vec{F}_n(\tau)$ acting on the nth segment of the tagged chain can be expressed by matrix density fluctuations around this segment:

$$\vec{F}_n^{Q}(\tau) = \int d^3\vec{r}\vec{f}(\vec{r})\delta\rho_n^{Q}(\vec{r},\tau,\gamma_n(t-\tau)), \tag{104}$$

where $\vec{f}(\vec{r}) = -\frac{\partial}{\partial\vec{r}}U(\vec{r})$ is the intermolecular force which a matrix segment exerts on the segment of the tagged chain at a distance $\vec{r}$. The function $\delta\rho_n^{Q}(\vec{r},\tau,\gamma_n(t-\tau))$ is given by

$$\delta\rho_n^{Q}(\vec{r},\tau,\gamma_n(t-\tau)) = \exp\left\{i\hat{Q}\hat{L}\hat{Q}\tau\right\}\left(\rho_n(\vec{r},\gamma_n(t-\tau)) - \rho_n^{*}(\vec{r},\gamma_n(t-\tau))\right), \tag{105}$$

where $\rho_n(\vec{r},\gamma_n(t-\tau))$ is the density of matrix segments at a distance $\vec{r}$ from the nth segment of the tagged chain with a conformation given by $\gamma_n(t-\tau)$. The function $\rho_n^{*}(\vec{r},\gamma_n(t-\tau))$ is a conditional equilibrium density of the matrix segments at a distance $\vec{r}$ from the nth segment of the tagged chain under the condition that the conformation of the tagged chain is fixed and given by $\gamma_n(t-\tau)$.

On this basis the memory matrix can be expressed as an integral over the time-dependent autocorrelation function of matrix density fluctuations for projected dynamics:

$$\Gamma_{nk}^{\alpha\beta}(\tau,t-\tau) = \frac{1}{k_B T}\int d^3\vec{r}\int d^3\vec{r}\,' f_n^{\alpha}(\vec{r}) f_k^{\beta}(\vec{r}\,')\left\langle\delta\rho_k^{Q}(0;\vec{r}\,')\delta\rho_n^{Q}(\tau;\vec{r})\right\rangle_{\gamma_n(t-\tau)}^{*}, \tag{106}$$

where the dependence of the density fluctuations $\delta\rho_k^{Q}(0;\vec{r}\,')$ and $\delta\rho_n^{Q}(\tau;\vec{r})$ around segments k and n on the tagged chain variables at time $t-\tau$ via $\gamma_n(t-\tau)$ was omitted for simplicity.

3.3.2.2
The Rouse Equation of Motion as a Special Case of the GLE

The Rouse equation of motion can be treated as a special case of Eq. 100. To show this, the memory function given in Eq. 106 is subjected to the following approximations:

(i) The time-dependent correlation function $\left\langle \delta\rho_k^Q(0;\vec{r}\,')\delta\rho_n^Q(\tau;\vec{r})\right\rangle^*_{\gamma_n(t-\tau)}$ for projected dynamics is a function of the conformation of the tagged chain and its Kuhn segment momentums at time $t-\tau$. We make use of the approximation

$$\left\langle \delta\rho_k^Q(0;\vec{r}\,')\delta\rho_n^Q(\tau;\vec{r})\right\rangle^*_{\gamma_n(t-\tau)} \approx \left\langle \delta\rho_k^Q(0;\vec{r}\,')\delta\rho_n^Q(\tau;\vec{r})\right\rangle_{eq}, \tag{107}$$

that is, we replace the conditional average $\langle ... \rangle^*_{\gamma_n(t-\tau)}$by an equilibrium average $\langle ... \rangle_{eq}$. The correlation function then does not depend on the chain conformation and its segment momentums at time $t-\tau$. Since flexible polymer melts are isotropic, the memory function becomes a scalar with respect to space rotations, i.e., $\Gamma_{nk}^{\alpha\beta}(\tau,\, t-\tau) \sim \delta_{\alpha\beta}$, where $\delta_{\alpha\beta}$ is the Kronecker symbol.

(ii) The memory matrix given in Eq. 106 has a nonlocal character with respect to the Kuhn segment number. There can be nonnegligible correlations in the density fluctuations between distant tagged-chain segments n and k with $|n-k|>>1$. The simplest approximation is to neglect this nonlocality, which is mathematically equivalent to setting $\Gamma_{nk}^{\alpha\beta} \propto \delta_{nk}$.

(iii) The most crucial approximation is the assumption that the memory function decays fast enough to become negligibly small on a molecular mass independent time scale. In accordance with Eq. 106 this means that the local matrix density distribution around the nth segment of the tagged chain equilibrates on a microscopic time scale.

Taking all approximations together, the memory matrix can be expressed as

$$\Gamma_{nk}^{\alpha\beta}(\tau; t-\tau) = 2\zeta\delta_{\alpha\beta}\delta_{nk}\delta(\tau). \tag{108}$$

Accounting for this and Eq. 103, Eq. 100) can be shown to coincide with the Rouse equation of motion given by Eq. 45.

Fast equilibration of the local matrix density fluctuations around the segments of the tagged chain contradicts chain connectivity in a sense. Only if chains are allowed to cross each other, can one expect that a local equilibrium distribution of the matrix segments around a certain tagged chain configuration can be reached on a molecular mass independent microscopic time scale. Anyway, very fast local motions do indeed exist, leading to an approximate local equilibrium distribution of the matrix segments around the segments of the tagged chain with a molecular mass independent accuracy.

From an empirical point of view it is well known that the Rouse model describes satisfactorily the dynamical properties of polymer melts with $N<N_C$. It is instructive to extract the fast decaying Rouse contribution from the total memory matrix given by Eq. 102 and rewrite Eq. 100 as

$$\frac{d}{dt}\vec{p}_n = -\frac{3k_BT}{b^2}\frac{\partial^2}{\partial n^2}\vec{r}_n(t) - \zeta\vec{v}_n(t) - \sum_k \int_{\tau_0}^{t} d\tau \Gamma_{nk}^{\alpha\beta}(\tau; t-\tau) v_k^{\beta}(t-\tau)\vec{e}_\alpha + \vec{F}_n^Q(t), \tag{109}$$

where ζ is the "bare" local friction coefficient connected with the fast local motions and $\tau_0 << t$ is the characteristic time associated with these fast local motions. Here the approximation given in Eq. 103 was taken into account.

From Eqs. 100, 102, and 109, we obtain the self-diffusion coefficient

$$D = \frac{k_BT}{\zeta_{eff}^*}, \tag{110}$$

where

$$\zeta_{eff}^* = \frac{1}{3k_BT}\sum_{n,k}\int_0^\infty \langle \vec{F}_k^Q(0)\cdot\vec{F}_n^Q(\tau)\rangle d\tau = N\zeta + \frac{1}{3k_BT}\sum_{n,k}\int_{\tau_0}^\infty \langle \vec{F}_k^Q(0)\cdot\vec{F}_n^Q(\tau)\rangle d\tau \tag{111}$$

is the effective friction coefficient of the macromolecule.

3.4 Renormalized Rouse Models

"Renormalization" in this context means an attempt to find a physically plausible ansatz for the unknown memory matrix given by Eq. 106. In principle one could postulate a power law with an exponent being a fitting parameter to experimental data after having derived expressions of observables on this basis. However, in the frame of the renormalized Rouse models (RRM) a somewhat less formal and less phenomenological approach is possible.

The preaveraging approximation given in Eq. 107 is again used. The memory matrix then becomes a scalar relative to space rotations, and does not depend on the momentary configuration of the tagged chain:

$$\Gamma_{nk}^{\alpha\beta}(\tau) = \Gamma_{nk}^{\alpha\beta} = \frac{1}{3k_BT}\int d^3\vec{r}\int d^3\vec{r}\,'\vec{f}(\vec{r})\vec{f}(\vec{r}\,')\langle\delta\rho_k^Q(0,\vec{r}\,')\delta\rho_n^Q(\tau,\vec{r})\rangle_{eq}. \tag{112}$$

The general equation of motion (Eq. 109) becomes a linear function of the tagged chain variables, and the system is considered to be isotropic.

Intermolecular interactions determining the force $\vec{f}(\vec{r})$ with which a segment from a chain acts on a segment from another one are approximated by a hard sphere interaction with a certain diameter d comparable with the Kuhn segment length b. The "entanglement effects" or the "topological interactions" are the consequence of chain connectivity and excluded-volume intermolecular interactions.

The matrix density correlation function $\left\langle \delta\rho_k^Q(0,\vec{r}\,')\delta\rho_n^Q(\tau,\vec{r})\right\rangle_{eq}$ for projected dynamics is approximated as a k-space integral (for a detailed argument see Ref. [98]),

$$\Gamma_{nm}(t) = \frac{8}{27}\frac{\rho_m d^6 g^2(d)}{k_B T}\int\limits_0^{b^{-1}} dk k^4 \omega_{nk}^Q(k,t)\hat{S}^Q(k,t), \tag{113}$$

where a mistake in the numerical factor published in Ref. [98] is corrected. The function $g(r)$ is the radial intermolecular distribution function; $\omega_{nk}^Q(k,t)$ and $\hat{S}^Q(k,t)$ are the Fourier transforms of $\omega_{nk}^Q(\vec{R};t)$ and $\hat{S}^Q(\vec{r};t)$, respectively.

The intrachain dynamical structure factor for projected dynamics is defined by

$$\hat{\omega}_{nm}^Q(k,t) = \hat{\omega}_{nm}(k)\exp\left\{-\frac{k^2}{6}\langle\vec{R}^2(t)\rangle_Q\right\}, \tag{114}$$

where $\hat{\omega}_{nm}(k)$ is the static intrachain structure factor. Likewise, the collective dynamical structure factor for projected dynamics of the matrix surrounding the tagged chain is given by

$$\hat{S}^Q(k,t) = \hat{S}(k)\exp\left\{-\frac{k^2}{6}\langle\vec{R}^2(t)\rangle_Q\right\}, \tag{115}$$

where $\hat{S}(k)$ is the static collective structure factor.

Single-chain $\hat{\omega}_{nm}^Q(k,t)$ and collective $\hat{S}^Q(k,t)$ dynamical structure factors for projected dynamics in Eq. 113 reflect the fact that local density fluctuations around the tagged chain are relaxed by the projected motions both of the tagged chain and the matrix chains. The projected nature of the dynamics is hidden in Eqs. 114 and 115 in the projected mean squared displacement, which qualitatively describes typical distances over which elementary density fluctuations become dispersed around the tagged chain during the time t.

The essence of the first and second renormalization ansatzes to be described in the following refers to heuristic replacements of the mean squared segment displacement for projected dynamics, $\left\langle\vec{R}^2(t)\right\rangle_Q$. In the (once) renormalized Rouse model this unknown function is replaced by the result of the ordinary Rouse model, $\left\langle\vec{R}^2(t)\right\rangle_R$, given in Eq. 58 [98]. In the twice renormalized Rouse model, it is substituted by the result of the first renormalization.

3.4.1
The First Renormalization

Equation 109 becomes mathematically closed by equating

$$\langle \vec{R}^2(t) \rangle_Q = \langle \vec{R}^2(t) \rangle_R. \tag{116}$$

From Eq. 109 we find the integrodifferential equation for the autocorrelation function of the normal modes (for a definition see Eq. 49)

$$\frac{\partial}{\partial t} C_p(t) + \int_0^t \Gamma_p(t-\tau) \frac{\partial}{\partial \tau} C_p(\tau) d\tau = -\frac{p^2}{\tau_R} C_p(t), \tag{117}$$

where the contribution of the inertia term was neglected. The memory function with respect to the correlation function of the pth normal mode is defined by

$$\Gamma_p(t) = \frac{1}{\zeta} \int_0^N \Gamma_m(t) \cos\left(\frac{\pi}{N} pm\right) dm, \tag{118}$$

where chain end effects are neglected, so that

$$\Gamma_{nk}^{\alpha\beta}(t) = \delta_{\alpha\beta} \Gamma_{|n-m|}(t). \tag{119}$$

On this basis the following integral representation is obtained for the memory function $\Gamma_p(t)$ [49]

$$\Gamma_p(t) = \frac{16}{3\sqrt{\pi}\pi^2} \frac{\psi b^3}{\langle \vec{R}^2(t) \rangle_Q^{3/2}} \frac{1}{\tau_s} \int_0^{\sqrt{\frac{\langle \vec{R}^2(t) \rangle_Q}{3b^2}}} \frac{q^6 \exp\{-q^2\}}{q^4 + \left(\frac{2\pi p \langle \vec{R}^2(t) \rangle_Q}{N} \right)^2} dq, \tag{120}$$

where $\psi \equiv \rho_m d^3 \left(\frac{d}{b}\right)^3 g^2(d) \hat{S}(0)$ is a dimensionless parameter characterizing the influence of the entanglement effects on the chain dynamics. The quantity $\hat{S}(0)$ is given by $\hat{S}(0) = \rho_m k_B T \kappa_T$, where κ_T is the isothermal compressibility.

The exact solution of Eq. 117 is nonexponential in general. The relaxation time of the pth normal mode can be defined in the following way:

$$\tau_p = \frac{1}{C_p(0)} \int_0^\infty C_p(t) dt = \tau_s \left(\frac{N}{p}\right)^2 \left[1 + \hat{\Gamma}_p(0)\right], \tag{121}$$

where $\hat{\Gamma}_p(0) = \int\limits_0^\infty \Gamma_p(t)dt$. For the self-diffusion coefficient the following relation can be obtained

$$D = \frac{k_B T}{N\zeta\left[1 + \hat{\Gamma}_0(0)\right]}. \tag{122}$$

This treatment of the RRM differs from the original version by Schweizer [98], where the mode-number-dependent term $\left(2\pi p\langle \vec{R}^2(t)\rangle_Q \big/ N\right)^2$ in expressions corresponding to Eq. 120 was neglected. However, this term is quite essential [49, 58, 105] for the normal-mode relaxation of entangled polymers.

In the short time limit, $t \ll \tau'_e \propto \tau_s/\psi$, the integral term in Eq. 109 is negligible, so that chain dynamics tends to approach "unentangled" Rouse behavior. At longer times the integral term starts to dominate over the local friction term, $\zeta\vec{v}_n(t)$. That is, chain dynamics becomes "entangled". The details of the transition unentangled–entangled motion depend on the parameter ψ.

From Eqs. 121 and 120 we find for $\langle \vec{R}^2(t)\rangle_Q = \langle \vec{R}^2(t)\rangle_R$ in the *high-mode-number limit*, $p > N/6\pi$,

$$\tau_p^{RR} = \tau_s \left(\frac{N}{p}\right)^2 \left[1 + 0.05\psi\left(\frac{N}{p}\right)^2\right]. \tag{123}$$

If the entanglement parameter is big enough, $\psi \gg \frac{1}{\pi^2}$, and for normal modes

$$\frac{N}{6\pi} < p < 0.22\sqrt{\psi}N \tag{124}$$

the entanglement effect becomes essential, and Eq. 123 approaches

$$\tau_p^{RR} \simeq 0.05\psi\tau_s \left(\frac{N}{p}\right)^4. \tag{125}$$

For $t \gg \tau'_e \propto \tau_s/\psi$, chains are subject to entangled behavior. The mean squared displacement can be approximated as

$$\langle \vec{R}_n^2(t)\rangle \simeq 0.5b^2 \left(\frac{t}{\psi\tau_s}\right)^{1/4} \tag{126}$$

and the autocorrelation function of the segment tangent vector becomes

$$\left\langle \vec{b}_n(t)\cdot\vec{b}_n(0)\right\rangle \simeq 0.4b^2 \left(\frac{\psi\tau_s}{t}\right)^{1/4}. \tag{127}$$

At longer times, $t >> 0.05(6\pi)^4\psi\tau_s$, the influence of normal modes with numbers $p << N/6\pi$ becomes essential. This limit is called the *low-mode-number limit.* In this case both contributions in the denominator of the integrand are important and the following result for the normal mode relaxation time can be obtained

$$\tau_p^{RR} = \tau_s\left(\frac{N}{p}\right)^2\left[1+2.14\psi\sqrt{\frac{N}{p}}\right] \simeq 2.14\psi\tau_s\left(\frac{N}{p}\right)^{2.5}. \tag{128}$$

The terminal relaxation time of the RRM scales as

$$\tau_1^{RR} \simeq 2.14\psi\tau_s N^{2.5}. \tag{129}$$

In the limit $N\rightarrow\infty$ and $p >> N$, the time dependence of the normal mode relaxation functions can be approximated by an exponential decay as

$$C_p(t) = C_p(0)\exp\left\{-\frac{t}{\tau_p}\right\}. \tag{130}$$

Otherwise Eq. 117 can be solved only numerically. It was found, that the solution of the equation can be approximated by the stretched exponential (Fig. 10)

$$C_p(t) \approx C_p(0)\exp\left\{-\left(\frac{t}{\tilde{\tau}_p}\right)^{\beta_p}\right\}, \tag{131}$$

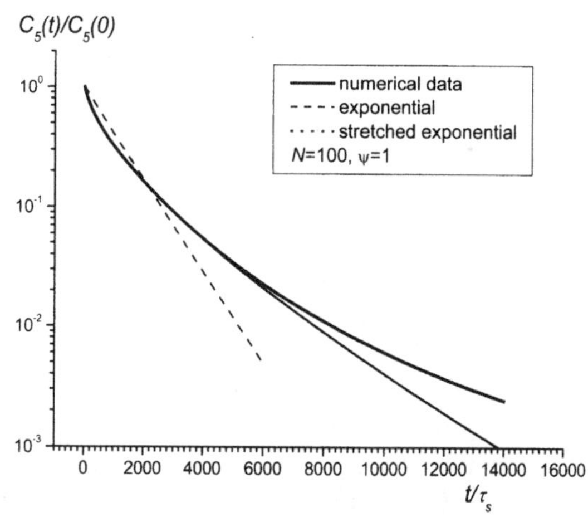

Fig. 10. Numerically evaluated, normalized autocorrelation function for the RRM for p=5 together with a fitted curve for a stretched exponential, Eq. 131, and an exponential curve. The stretched-exponential parameter is β_5=0.66

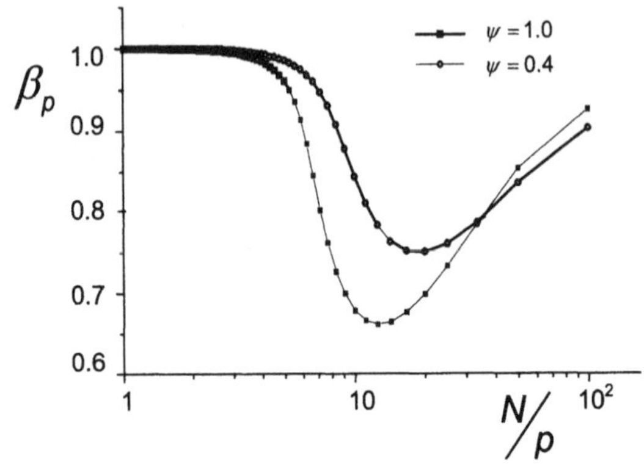

Fig. 11. Mode number dependence of the stretching-parameter β_p for the RRM (see Eq. 131) for different entanglement parameters ψ. The number of Kuhn segments was assumed to be N=100

where $\tilde{\tau}_p = \frac{\beta_p}{\Gamma(\beta_p)}\tau_p$ and β_p is a mode-number-dependent stretching parameter (see Fig. 11). $\Gamma(\beta_p)$ is the gamma function. In the same limit, the mean squared segment displacement obeys

$$\left\langle \vec{R}_n^2(t) \right\rangle = \begin{cases} 0.23 b^2 \left(\frac{t}{\psi \tau_s}\right)^{2/5} & \text{for } t \ll \tau_1^{RR} \\ 6 D_{RR} t & \text{for } t \gg \tau_1^{RR} \end{cases}, \tag{132}$$

where

$$D_{RR} = \frac{0.21}{3\pi^2} \frac{b^2}{\psi \tau_s N^{3/2}} \tag{133}$$

is the center-of-mass diffusion coefficient in the RRM. The correlation function of the segment tangent vector calculated in the Markovian approximation is in the low-mode-number limit given by

$$\left\langle \vec{b}_n(t) \cdot \vec{b}_n(0) \right\rangle \simeq \begin{cases} b^2 \left(\frac{\psi \tau_s}{t}\right)^{2/5} & \text{for } t \ll \tau_1^{RR} \\ \frac{b^2}{N} \exp\left\{-\frac{t}{\tau_1^{RR}}\right\} & \text{for } t \gg \tau_1^{RR} \end{cases}. \tag{134}$$

These results should be asymptotically correct in the limit $N \to \infty$ and $t >> 0.05(6\pi)^4 \psi \tau_s$. For finite N, the exponents predicted in Eq. 132 may change slightly due to the stretched exponential decay of $C_p(t)$ given in Eq. 131. The limiting laws for the observables of interest in the frame of this review, the mean squared displacement and the (intrasegment) spin-lattice relaxation rate, are summarized in Table 2 (compare [49]).

Table 2. Theoretical dependences on time (t), angular frequency (ω), molecular mass (M), or number of Kuhn segments (N) predicted by the (once) renormalized Rouse model for the mean squared segment displacement and the intrasegment spin–lattice relaxation time. The factors C_I, C_I^{RR}, C_{II}^{RR}, and C_{III}^{RR} are frequency and molecular mass independent constants

	Limit	Mean squared segment displacement $\langle R^2 \rangle =$	intrasegment spin–lattice relaxation time $T_1^{intra} =$	Refs
Rouse	$\tau_s << (t, 1/\omega) << \tau_s/\psi$	$\frac{2}{\pi^{3/2}} b^2 \left(\frac{t}{\tau_s}\right)^{1/2}$ $\propto M^0 t^{1/2}$	$-C_I \frac{M^0}{\tau_s \ln(\omega\tau_s)}$	[47, 48]
$(I)_{RR}$	"high – mode – number limit" $\frac{N}{6\pi} < p < 0.22N\sqrt{\psi}$ $\frac{\tau_s}{1.9\psi} \ll (t, 1/\omega) \ll (6\pi)^4 1.9\psi\tau_s$	$0.5b^2 \left(\frac{t}{\tau_s\psi}\right)^{1/4}$ $\propto M^0 t^{1/4}$	$C_I^{RR} M^0 \omega^{1/2}$	[49]
$(II)_{RR}$	"low – mode – number, short – time limit" $p < \frac{N}{6\pi}$ $\frac{(6\pi)^4}{20} \psi\tau_s \ll (t, 1/\omega) \ll \tau_1^{RR}$ $\approx 2.14\psi\tau_s N^{2.5}$	$0.23b^2 \left(\frac{t}{\psi\tau_s}\right)^{2/5}$ $\propto M^0 t^{2/5}$	$C_{II}^{RR} M^0 \omega^{1/5}$	[47, 49]
$(III)_{RR}$	"low – mode – number, long – time limit" $p < \frac{N}{6\pi}$ $\tau_1^{RR} \approx 2.14\psi\tau_s N^{2.5} \ll (t, 1/\omega)$	$6D_{RR}t$ $\approx \frac{0.42}{\pi^2} \frac{b^2}{\psi\tau_s} N^{-3/2} t$	$\left[\frac{C_{III}^{RR} b^4 N^{-2} \tau_1^{RR}}{1 + \left(\omega\tau_1^{RR}/2\right)^2}\right]^{-1}$	present work

3.4.2 The Second Renormalization

The RRM predictions for the chain length dependences of the self-diffusion coefficient, $D^{RR} \propto N^{-1.5}$, and of the terminal relaxation time, $\tau_1^{RR} \propto N^{2.5}$, are weaker than suggested by experiments: $D^{exp} \propto N^{-(2 \div 2.5)}$, $\tau^{exp} \propto N^{3.5}$. This means that, at least at long times, the memory matrix $\Gamma_{nm}(t)$ decays more slowly than predicted by the renormalized Rouse model. That is, the entanglement effect is underestimated. On the other hand, RRM predictions are more realistic for entangled polymers than those of the ordinary Rouse model. This suggests one should attempt another ansatz for projected dynamics on the basis of the RRM results. Substituting $\langle \vec{R}_n^2(t) \rangle_Q = \langle \vec{R}_n^2(t) \rangle_{RR}$ in the formulas given in Eqs. 114 and 115 leads to the twice renormalized Rouse model (TRRM).

The *high-mode-number limit*, $p > N/6\pi$, for the normal mode relaxation time suggests:

$$\tau_p^{TRR} \simeq \tau_s \left(\frac{N}{p}\right)^2 \left[1 + 0.148\psi^2 \left(\frac{N}{p}\right)^2\right] \tag{135}$$

similar to Eq. 123 found for the RRM. In the limit $t \gg \tau'_e = \tau_s/\psi$ and $\psi^2 >> 1/\pi^2$, the TRRM predicts

$$\langle \vec{R}_n^2(t) \rangle \simeq 0.4b^2 \left(\frac{t}{\psi \tau_s} \right)^{1/4} \tag{136}$$

and

$$\left\langle \vec{b}_n(t) \cdot \vec{b}_n(0) \right\rangle \simeq 0.5b^2 \left(\frac{\psi \tau_s}{t} \right)^{1/4} \tag{137}$$

analogous to the relations given in Eqs. 126 and 127.

The difference between the RRM and the TRRM becomes more pronounced in the *low-mode-number limit*, $p<N/6\pi$. The normal-mode relaxation time of order p reads

$$\tau_p^{TRR} = \tau_s \left(\frac{N}{p} \right)^2 \left[1 + 5.52\psi^2 \frac{N}{p} \right] \simeq 5.52\psi^2 \tau_s \left(\frac{N}{p} \right)^3 \tag{138}$$

compared to the expression given in Eq. 128. The terminal relaxation time scales with molecular mass the same way as in the case of the reptation/tube model, namely

$$\tau_1^{TRR} \simeq 5.52\psi^2 \tau_s N^3. \tag{139}$$

However, the dependence on the mode number, $\tau_p \propto (N/p)^3$, differs from that suggested by the reptation model, $\tau_p \propto N^3/p^2$ [1]. Note that the TRRM result in this respect was confirmed by computer simulations [84, 87] in a wide range.

As in the RRM case, the memory function $\Gamma_p(t)$ given by Eq. 120 becomes fast decaying:

$$\Gamma_p(t) \propto \frac{1}{\langle \vec{R}_n^2(t) \rangle^{7/2}} \propto \frac{1}{t^{7/5}}. \tag{140}$$

The Markovian approximation for Eq. 117 therefore becomes correct in the limit $N \to \infty$ and $p<<N$. This means that the solution of Eq. 117 approaches an exponential decay according to

$$C_p(t) = C_p(0) \exp\left\{ -\frac{t}{\tau_p^{TRR}} \right\}. \tag{141}$$

More generally the solutions of Eq. 117 are nonexponential for the TRRM and, as it was found in numerical investigations, can be approximated over two orders of magnitude of $C_p(t)$ by the stretched exponential given in Eq. 131. Only the fitting parameters β_p and $\tilde{\tau}_p$ differ from those found for the RRM. Illustrations are shown in Figs. 12 and 13. The stretched exponen-

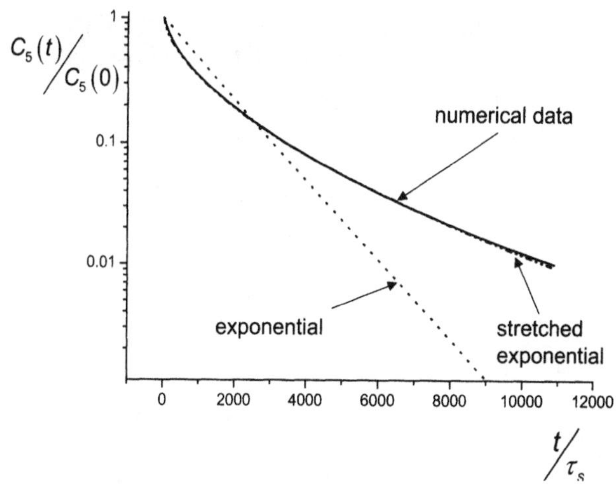

Fig. 12. Numerically evalutated, normalized autocorrelation function for the TRRM for p=5 together with a fitted curve for a stretched exponential, Eq. 131, and an exponential function. The parameters are β_5=0.61, ψ=1

tial decay reproduces the numerical data very well for $C_p(t)/C_p(0)\geq 0.01$ (Fig. 12). The stretching parameter β_p for the TRRM as a function of the normal mode number p shows a minimum and approaches unity for decreasing p (Fig. 13). Qualitatively the same behavior was found in computer

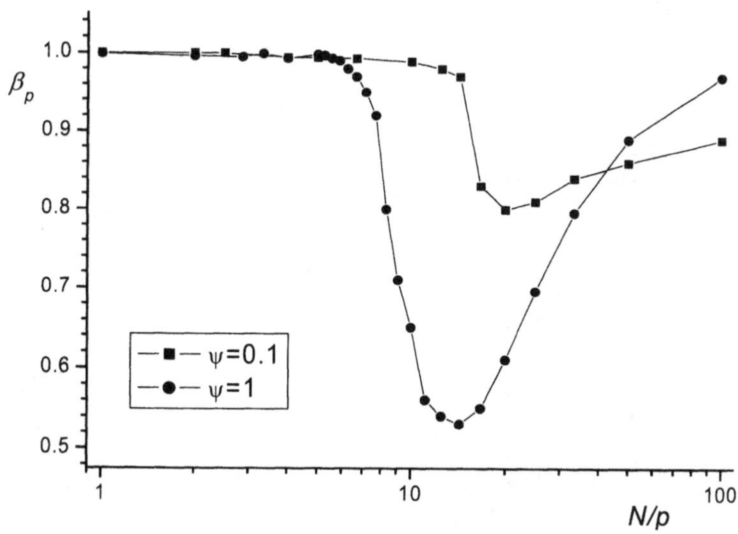

Fig. 13. Mode number dependence of the stretching-parameter β_p (see Eq. 131) for the TRRM for different entanglement parameters ψ=0.1, 1. The number of Kuhn segments was assumed to be N=100

simulations [84, 87]. Comparing Figs. 10 and 11 with Figs. 12 and 13 suggests that the RRM and the TRRM are similar in this respect.

In the low-mode-number limit, the TRRM predicts for the mean squared segment displacement

$$\langle \vec{R}_n^2(t) \rangle_{TRR} = \begin{cases} 0.16 b^2 \left(\frac{t}{\psi^2 \tau_s} \right)^{1/3} & \text{for } \tau_e \ll t \ll \tau_1^{TRR} \\ 6 D_{TRR} t & \text{for } t \gg \tau_1^{TRR} \end{cases}, \tag{142}$$

where

$$D_{TRR} \simeq \frac{0.052}{3\pi^2} \frac{b^2}{\psi^2 N^2 \tau_s}. \tag{143}$$

Remarkably the TRRM gives the same molecular mass dependence for the self-diffusion coefficient as the tube/reptation model.

In the same limit the autocorrelation function of the tangent vector turns out to be

$$\left\langle \vec{b}_n(t) \cdot \vec{b}_n(0) \right\rangle_{TRR} \simeq \begin{cases} 1.58 b^2 \left(\frac{\psi^2 \tau_s}{t} \right)^{1/3} & \text{for } \tau_e \ll t \ll \tau_1^{TRR} \\ \frac{b^2}{N} \exp\left\{ -\frac{t}{\tau_1^{TRR}} \right\} & \text{for } t \gg \tau_1^{TRR} \end{cases}. \tag{144}$$

Table 3. Theoretical dependences on time (t), angular frequency (ω), molecular mass (M), or number of Kuhn segments (N) predicted by the twice renormalized Rouse model for the mean squared segment displacement and the intrasegment spin-lattice relaxation time. The factors C_I, $C_I{}^{TRR}$, $C_{II}{}^{TRR}$, and $C_{III}{}^{TRR}$ are frequency and molecular mass independent constants

	Limit	Mean squared segment displacement $\langle R^2 \rangle =$	Intrasegment spin–lattice relaxation time $T_1{}^{\text{intra}} =$
Rouse	$\tau_s << (t, 1/\omega) << \tau_s/\psi$	$\frac{2}{\pi^{3/2}} b^2 \left(\frac{t}{\tau_s} \right)^{1/2}$ $\propto M^0 t^{1/2}$	$-C_I \frac{M^0}{\tau_s \ln(\omega \tau_s)}$
$(\text{I})_{\text{TRR}}$	"high – mode – number limit" $\frac{N}{6\pi} < p < 0.22 N \sqrt{\psi}$ $\frac{\tau_s}{\psi} \ll (t, 1/\omega) \ll (6\pi)^4 \psi^2 \tau_s$ $\psi \gg \frac{1}{\pi^2}$	$0.4 b^2 \left(\frac{t}{\tau_s \psi} \right)^{1/4}$ $\propto M^0 t^{1/4}$	$C_I{}^{TRR} M^0 \omega^{1/2}$
$(\text{II})_{\text{TRR}}$	"low – mode – number, short – time limit" $p < \frac{N}{6\pi}$ $\frac{(6\pi)^4}{20} \psi \tau_s \ll (t, 1/\omega)$ $\ll \tau_1^{TRR} \approx 5.52 \psi^2 \tau_s N^3$	$0.16 b^2 \left(\frac{t}{\psi^2 \tau_s} \right)^{1/3}$ $\propto M^0 t^{1/3}$	$C_{II}{}^{TRR} M^0 \omega^{1/3}$
$(\text{III})_{\text{TRR}}$	"low – mode – number, long – time limit" $p < \frac{N}{6\pi}$ $\tau_1^{TRR} \approx 5.52 \psi^2 \tau_s N^3 \ll (t, 1/\omega)$	$6 D_{TRR} t$ $\approx \frac{0.104}{\pi^2} \frac{b^2}{\psi^2 \tau_s} N^{-2} t$	$\left[\frac{C_{III}^{TRR} b^4 N^{-2} \tau_1^{TRR}}{1 + \left(\omega \tau_1^{TRR} / 2 \right)^2} \right]^{-1}$

If the entanglement parameter obeys $\psi \leq 1/\pi^2$, only the low-mode-number regime of the entangled motion exists for times $t >> \tau_e = \psi^{-4}\tau_s$ in the TRRM as well as in the RRM. Table 3 summarizes the limiting laws for the observables of interest in the frame of this review.

3.4.3 General Remarks on Renormalized Rouse Models

In principle one could continue the renormalization procedure by constructing an infinite series of n times renormalized Rouse models by equating the mean squared segment displacement for projected dynamics iteratively. However, the renormalization procedure is nothing more than a heuristic way of closure of the equations of motion. It must be distinguished from ordinary iteration procedures serving the solution of *well-defined* equations. The renormalization procedure by way of contrast is an attempt to *define* the equation. This means that an infinite repetition of renormalization steps does not necessarily lead to solutions more appropriate for the problem to be solved. The renormalization steps must rather be considered as trial and error attempts to find good solutions just in a phenomenological sense. However, there is some limitation how far one can go in principle with this iteration strategy.

The three times renormalized Rouse model, ThRRM, means that we set $\langle \vec{R}_n^2(t) \rangle_Q = \langle \vec{R}_n^2(t) \rangle_{TRR}$ for the memory function $\Gamma_p(t)$ in Eq. 120. High-mode-number and low-mode-number limits can be distinguished again for the time scale of entangled dynamics. In the high-mode-number limit, the predictions of the ThRRM coincide with those of the TRRM and the RRM apart from numerical factors (see Eqs. 135–137 and 123, 126, 127, respectively). An essential difference arises only in the low-mode-number limit, where the following scaling predictions are found:

$$\tau_p^{ThRR} \propto \psi^3 \tau_s \left(\frac{N}{p} \right)^{3.5}, \tag{145}$$

$$\langle \vec{R}_n^2(t) \rangle_{ThRR} \propto b^2 \left(\frac{t}{\psi^3 \tau_s} \right)^{2/7} \quad \text{for} \quad \tau_e \ll t \ll \tau_1^{ThRR}, \tag{146}$$

$$\left\langle \vec{b}_n(t) \cdot \vec{b}_n(0) \right\rangle_{ThRR} \propto b^2 \left(\frac{\psi^3 \tau_s}{t} \right)^{2/7} \quad \text{for} \quad \tau_e \ll t \ll \tau_1^{ThRR}, \tag{147}$$

$$D_{ThRR} \propto \frac{b^2}{\psi^3 N^{2.5} \tau_s}, \tag{148}$$

where $\tau_1^{ThRR} \propto \tau_s \psi^3 N^{3.5}$ is the terminal relaxation time for the ThRRM.

If the renormalization procedure is repeated n times, the terminal relaxation time and the self-diffusion coefficient are expected to follow the scaling predictions

$$\tau_1^{nRR} \propto \tau_s \psi^n N^{\frac{n+4}{2}} \tag{149}$$

and

$$D_{nRR} \propto \frac{b^2}{\tau_s \psi^n} N^{-\left(\frac{n+2}{2}\right)}, \tag{150}$$

respectively. The mean squared displacement scales in the low-mode-number limit as

$$\left\langle \vec{R}_n^2(t) \right\rangle_{nRR} \propto b^2 \left(\frac{t}{\psi^n \tau_s} \right)^{-\frac{2}{n+4}}, \tag{151}$$

and the renormalization ansatz is

$$\left\langle \vec{R}_n^2(t) \right\rangle_Q = \left\langle \vec{R}_n^2(t) \right\rangle_{(n-1)RR} \propto b^2 \left(\frac{t}{\psi^{(n-1)} \tau_s} \right)^{-\frac{2}{n+3}}. \tag{152}$$

Then the expression for the memory function $\Gamma_p(t)$ given in Eq. 120 can be approximated as a power law for the limit $N \to \infty$ and $t \to \infty$ while p/N=const,

$$\Gamma_p(t) \propto \frac{1}{\left\langle \vec{R}_n^2(t) \right\rangle} \propto \frac{1}{t^{\left(\frac{7}{n+3}\right)}}. \tag{153}$$

Obviously, if $n \geq 4$ the memory function is slowly decaying, and there is no limit in which a Markovian approximation would be applicable. Therefore the renormalization order n=3 is the boundary between fast and slowly decaying memory functions. That is, renormalization attempts to describe entangled polymer dynamics are realistically restricted to the RRM, TRRM or the ThRRM without any heuristic preference.

It remains to future developments to find a physical, more elementary and less heuristic basis for the renormalization ansatz, or at least some physical arguments and principles making this sort of ansatz more and more plausible. At present the justification of this sort of modeling can only be based on a phenomenological argument, namely the success of describing experimental findings.

Below it will be shown that field-cycling NMR relaxometry studies unambiguously reveal a crossover between high-frequency and low-frequency dispersion regimes that can be identified with the high-mode-number and low-mode-number limits of the renormalized Rouse models. Moreover, the variation of the power law exponents closely corresponds to that predicted by the renormalized Rouse models. These dynamic regimes cannot be explained by the tube/reptation model, for instance. The conclusion is, that the renormalized Rouse models correctly reflect the structure of the generalized Langevin equation with respect to the predicted dynamic limits, while some minor ambiguity is left with respect to the exponent values.

Finally it should be mentioned that Schweizer also suggested the Polymer Mode-Mode Coupling model (PMMC) [99, 100] as a further heuristic ansatz to account for the so-called viscoelastic plateau, a phenomenon not predicted by renormalized Rouse models, at least if based on an effectively intramolecular entropic stress tensor. In a recent review [104] all advantages and deficiencies of the PMMC ansatz are described in detail. The main problem with this sort of model arising in the present context is that a molecular mass dependence is predicted on a time scale shorter than the (longest) Rouse relaxation time. Such a behavior is inconceivable because the polymer segments do not yet "know" the chain length on that time scale. On the other hand, the predictions of the PMMC model for long times, $t \gg \tau_R$, are more realistic than those of the renormalized Rouse models. The future development of memory matrix approaches of any sort will be closely connected with investigations of the nature of "projected dynamics" as a key problem.

4
Experimental Studies of Bulk Melts, Networks and Concentrated Solutions

In this section a series of experimental NMR studies based on the techniques described above will be compared with predictions of the model theories. Of course, any model is based on idealizations approaching reality only under certain conditions. The objective of the first few paragraphs will therefore be to demonstrate the rich variety of phenomena that can influence polymer dynamics. It will be elucidated under what circumstances the essence of the model theories and their predicted limits comes to light. We will first consider three basic features of polymer dynamics, namely the *three dynamic components* governing molecular motions of polymer chains, the *dynamics of chain-end blocks* in contrast to the central segment block, and how *free volume and voids* influence dynamics and the appearance of NMR measurands.

4.1
The Three Components of Polymer Dynamics as Relevant for NMR Relaxometry

The chemical composition of a polymer described by a Kuhn segment chain (see Fig. 6) is exclusively represented by specific parameters such as the segment friction coefficient and the Kuhn segment length. That is, no information referring to a length scale shorter than the Kuhn segment and, hence, to the specific chemistry of the compound is considered in the frame of the polymer theories discussed in this context. In terms of time scales, only molecular motions taking place in the limit $t >> \tau_s$ are regarded, where τ_s is a time constant characteristic for local motions occurring inside a Kuhn segment. In terms of the Rouse model, the segment fluctuation time τ_s is given by Eq. 53. The question to be examined in this section is to what extent NMR relaxation experiments are affected by local motions ($t \leq \tau_s$) on the one hand and by chain modes ($t >> \tau_s$) on the other.

Generally one can distinguish three dynamic components contributing to motions of a chain [35, 48, 123, 124]. *Component A* represents fluctuations occurring within Kuhn segments, that is on a time scale up to $t \approx \tau_s$ and a length scale $R \leq b$. These motions may be supplemented and superimposed by monomer side-group rotational diffusion if such mobile groups exist. *Component B* refers to the hydrodynamic chain-mode regime which is of particular interest in context with chain dynamics models. *Component C*, the cutoff process of component B, finally corresponds to the terminal chain relaxation time τ_t after which all memory of the initial conformation gets lost. In terms of the reptation model, the terminal chain relaxation time is given by the tube disengagement time, $\tau_t = \tau_d$ (see Eq. 71).

4.1.1
Component A

Component A implies anisotropic main-chain-group and, if applicable, side-group reorientations covering a restricted solid angle range of the inter-dipole vector (proton resonance) or the electric field gradient principal axis (deuteron resonance). As concerns the main-chain groups, it is mainly due to rotameric isomerism [126, 127] and may be interpreted even in terms of defect diffusion models [128].

The consequence of the restricted nature of the reorientations by component A is that the dipolar or quadrupolar correlation functions do not decay to zero by these local and molecular weight independent motions. Rather a residual correlation remains at long times that decays further only by chain modes of a hierarchically higher order. The correlation functions given in Eqs. 28 and 38 can therefore be specified for component A as

$$G_A(t) = g_A(t) + G_A(\infty), \tag{154}$$

where $g_A(t >> \tau_s)=0$ and $G_A(\infty)$=const. Experiments suggest that in condensed polymer systems $G_A(\infty)/G_A(0)=10^{-3}...10^{-2}$ only [33, 125,129]. That is, most of the orientation correlation function decays already due to component A. Components B and C can therefore refer only to this small residual correlation.

The consequence of the strong correlation decay due to component A is that the T_1 minimum is predominantly determined by this component and

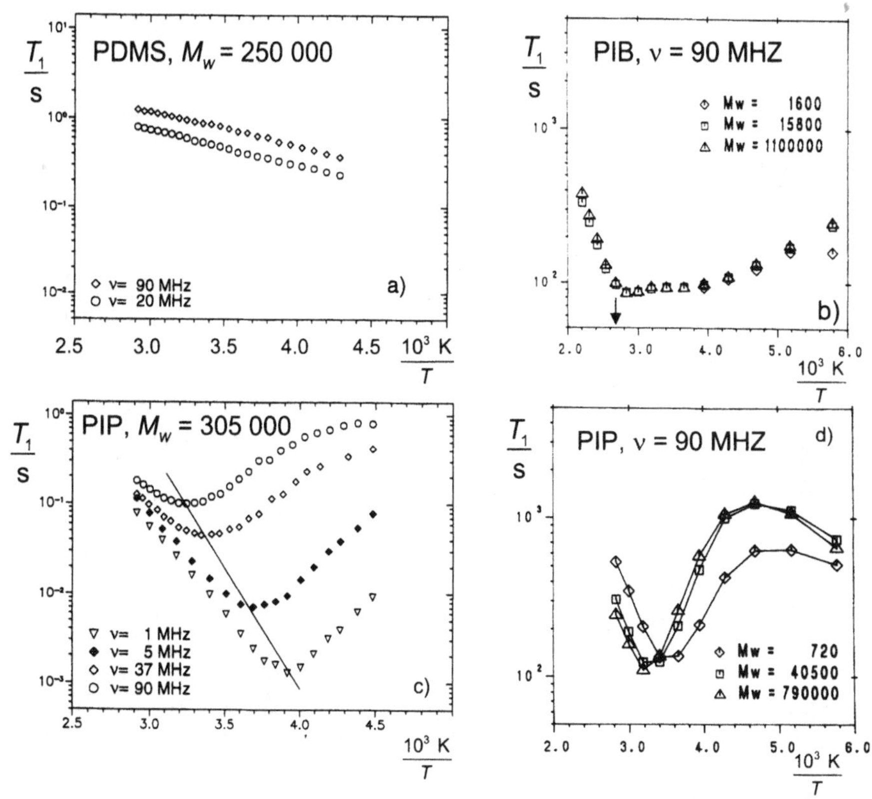

Fig. 14a–d. Proton spin-lattice relaxation time of various polymer melts as a function of the inverse temperature for different molecular masses and frequencies. The conclusion is that field-cycling NMR relaxometry of these polymers at the experimental temperatures is largely dominated by $\omega\tau_s$<<1 as a condition for the observation of components B and potentially C. **a** Polydimethylsiloxane (PDMS) for different frequencies. τ_s is too short in the temperature range indicated to fulfill the minimum condition $\omega\tau_s\approx 1$ even at 90 MHz [125]. **b** Polyisobutylene (PIB) for different molecular masses. The independence of the molecular weight above or below the critical value is obvious. The minimum is broadened by the superposition of rotational diffusion of methyl side groups. The *arrow* indicates the temperature of the field-cycling data shown in Fig. 28 [33]. **c** Polyisoprene (PIP) for different frequencies. The minimum positions (*solid line*) suggest an Arrhenius law τ_s=2×10^{-19} s exp(58 kJ/mol/*RT*) [125]. **d** PIP for different molecular masses

indicates the value of τ_s via the "minimum condition" $\omega\tau_s \approx 1$. A series of experimental data sets are shown in Figs. 14 and 15. The slight shift of the T_1 minimum with the molecular weight observed with polyisoprene (Fig. 14d) can be straightforwardly described in a semiempirical three-component formalism [33].

The values deduced from the T_1 minima corroborate that the frequency range of field-cycling NMR relaxometry largely corresponds to the time limit $t >> \tau_s$ and, hence, addresses the chain-mode regime beyond local seg-

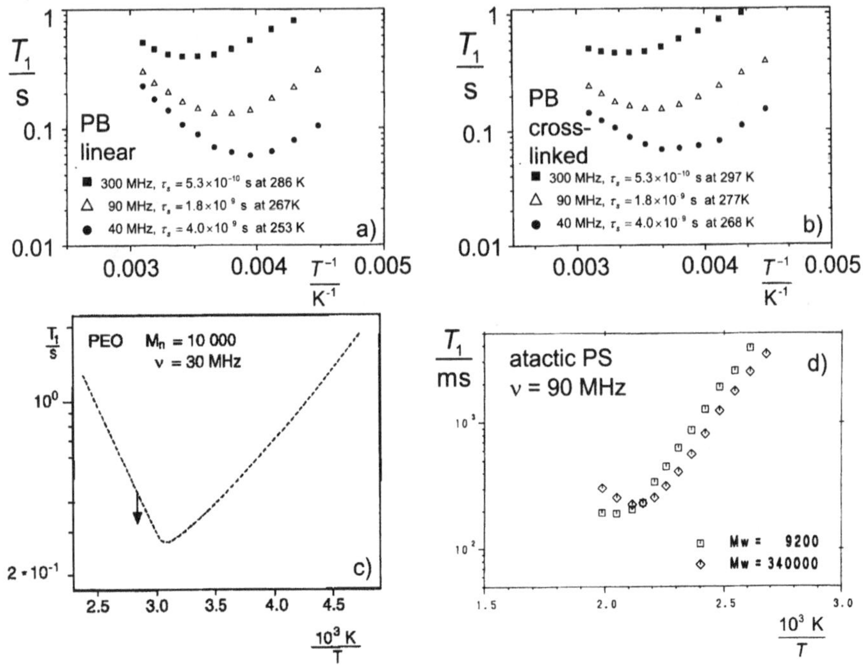

Fig. 15a–d. Temperature dependence of the proton spin-lattice relaxation time of various polymer melts at different frequencies and for different molecular masses. The segment fluctuation times τ_s can be evaluated from the T_1 minimum condition $\omega\tau_s$<<1. The temperatures at which the T_1 minimum occurs are specified for linear and cross-linked PB. It follows again that field-cycling NMR relaxometry at the experimental temperatures is largely dominated by $\omega\tau_s$<<1 as a condition for the observation of components B and potentially C. **a** Linear polybutadiene (PB; M_w=65,000) for different frequencies [131]. **b** Thermoreversibly cross-linked sample prepared from linear PB (M_w=51,000) by adding 37 phenylurazole groups per chain [131]. **c** Polyethyleneoxide (PEO). The *arrow* indicates the temperature of the field-cycling data shown in Fig. 29. M_n is the number average of the molecular mass [48, 130]. **d** Atactic polystyrene (PS) for different molecular masses. In this case, phenyl ring flips are slow enough to dominate the minimum so that segment fluctuations are concealed [33]

ment fluctuations (compare Fig. 5). In terms of the Rouse model, the segment fluctuation time τ_s is given by Eq. 53.

4.1.2
Component B

Component B represents motions in the chain-mode regime between τ_s and the terminal chain relaxation time. On a time scale up to the longest chain-mode relaxation time, which in terms of the Rouse model is given by τ_R (see Eq. 54), it is independent of the molecular mass. A relatively weak molecular weight dependence can only arise beyond τ_R, that is in terms of the reptation model in the time range $\tau_R < t < \tau_d$ (see Table 1).

4.1.3
Component C

Component C finally terminates component B and may become visible in the experimentally accessible frequency window of field-cycling NMR relaxometry in the form of a crossover to an "extreme-narrowing" plateau [2]. Since this crossover is connected with the terminal chain relaxation time τ_t, it will be strongly dependent on the molecular mass. This effect, however, will show up in the experimentally accessible frequency window of NMR relaxometry only for relatively small molecular masses (in contrast to the NMR diffusometry experiments).

4.1.4
The Different Time-scale Approach for the NMR Correlation Function

Representing the correlation function decays by components A, B, and C by the normalized partial correlation functions $G_A(t)$, $G_B(t)$ and $G_C(t)$, and interpreting these functions as probabilities that the respective fluctuations of the spin interactions have not yet taken place at time t, permits one to combine the three partial correlation functions in the total expression

$$G(t) = G_A(t)G_B(t)G_C(t). \tag{155}$$

Combining this function with Eq. 154 and making use of the *different time scale approximation*, that is $g_A(t>\tau_s) \approx 0$, $G_B(t\leq\tau_s) \approx G_B(0)=1$, and $G_C(t\leq\tau_t) \approx G_C(0)=1$ leads to

$$G(t) \approx g_A(t) + G_A(\infty)G_B(t)G_C(t), \tag{156}$$

where τ_t in this case generally represents the terminal chain relaxation time and $G_A(\infty)$ is a constant typically representing less than a few percent of the initial correlation function value.

Figure 16 shows typical proton spin–lattice relaxation dispersion data for polyethylene melts as an illustration of the three-component behavior of polymer melts. For comparison with model theories the chain-mode regime represented by component B is suited best and will be discussed in detail. It will be shown that the NMR relaxometry frequency window of typically 10^3 Hz< ν <10^8 Hz (for proton resonance) almost exclusively probes the influence of chain modes represented by component B (compare Fig. 5). That is, the correlation function experimentally relevant for spin–lattice relaxation dispersion may be identified with component B according to

$$G(t) \approx G_A(\infty)G_C(0)G_B(t) \propto G_B(t) \tag{157}$$

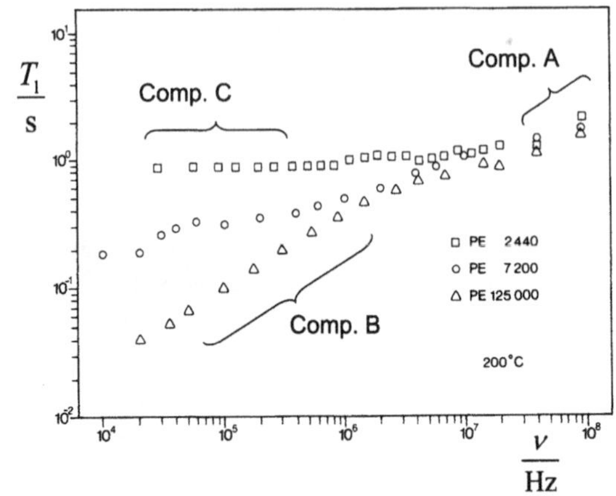

Fig. 16. Visualization of the three component regimes by way of example of proton spin-lattice relaxation dispersion in polyethylene (PE) melts of different molecular masses [123, 132]. The onset of Component A is indicated at the highest frequencies. It represents local fluctuations within a Kuhn segment and is characterized by a time constant τ_s=6.4×10^{-10} s independent of the molecular mass. It causes more than 90% of the decay of the orientation correlation function. At the temperature of the experimental data, 200 °C, the condition $\omega\tau_s \approx 1$ would be fulfilled at ν=2.5×10^8 Hz outside the frequency window shown. Hydrodynamic chain modes reveal themselves in distinct form as component B at low frequencies provided that the molecular mass (i.e., the terminal chain relaxation time) is large enough not to conceal this regime. Component C finally is a manifestation of the terminal relaxation time leading to a crossover to a low-frequency plateau for polymers short enough (compare Fig. 5). Other examples where component C showed up in field-cycling NMR relaxometry are given in Figs. 31b (PDMS, ^{1}H resonance) and 45b (deuterated PEO, ^{2}H resonance)

for polymer melts with $M >> M_c$, T >300 K, and τ_s<10^{-9} s. NMR relaxometry thus offers a unique chance to directly probe predictions by chain-mode model theories.

Before deepening the discussion of the experimental features of component B referring to the chain-mode regime, we will first consider a series of semiempirical studies corroborating the existence of the three dynamic components and other polymer characteristics one has to keep in mind when interpreting experimental polymer data. Such general characteristics can favorably be elucidated with the aid of transverse relaxation.

4.1.5 Average Transverse Relaxation

As a local mechanism, component A is completely independent of the molecular mass. Component B is also expected to be molecular weight independent for $t<\tau_R$, but adopts a weak molecular weight dependence for $\tau_R < t < \tau_t$ under entangled-dynamics conditions. This is in contrast to component C which, as a terminal process, is subject to a strong dependence on the chain length for $t \geq \tau_t$. The crossover between regimes where the more or less molecular weight dependent components dominate can also be visualized by the temperature and molecular weight dependences of the average transverse relaxation rate [33]. Figures 17, 19, and 20 show typical data of this sort plotted as the reciprocal average transverse relaxation rate.

The average transverse relaxation rate is obtained from the initial slope of the (normalized) transverse relaxation curve. Describing the (typically non-exponential) transverse relaxation curve by a linear combination of exponentials,

$$S(t) = \sum_j a_j \exp\left\{-{}^t/_{T_{2,i}}\right\}, \tag{158}$$

with coefficients a_j and time constants $T_{2,j}$, one obtains for the time derivative

$$\frac{dS(t)}{dt} = -\sum_j a_j \frac{1}{T_{2,j}} \exp\left\{-{}^t/_{T_{2,i}}\right\}. \tag{159}$$

In this expression we have neglected any initial Gauss-like decay contribution as it occurs in extreme cases such as networks or linear polymers of high molecular weight close to the glass temperature [133–137] (compare the Anderson/Weiss formalism). With the transverse relaxation data shown in Figs. 17, 19, and 20, this initial Gauss-like decay becomes only perceptible or even prevailing at the lowest temperatures investigated. In that case, a gradual crossover to solid-like behavior occurs leading to average transverse relaxation times as short as 10 μs. Typical transverse relaxation curves of linear polymer melts can be found elsewhere [36, 123, 133, 134, 137]. These data demonstrate that far above the glass temperature, and in particular with polymers having mobile side groups (such as polydimethylsiloxane, PDMS, for example), a data analysis based on the exponential function expression given in Eq. 159 is justified in good approximation. Note also, that part of the data discussed in the following refer to ^{13}C resonance of quaternary carbons (see Fig. 19c), so that a Gauss-like component is excluded in this case anyway.

Under such conditions, the initial slope of the transverse relaxation curve (often evaluated from the decay to $1/e$ of the initial value both in experiment and in theoretical Anderson/Weiss evaluations) approaches the average transverse relaxation rate according to

$$\left.\frac{dS(t)}{dt}\right|_{t\to 0} = -\sum_j a_j \frac{1}{T_{2,j}} \equiv \left\langle \frac{1}{T_{2,j}} \right\rangle. \tag{160}$$

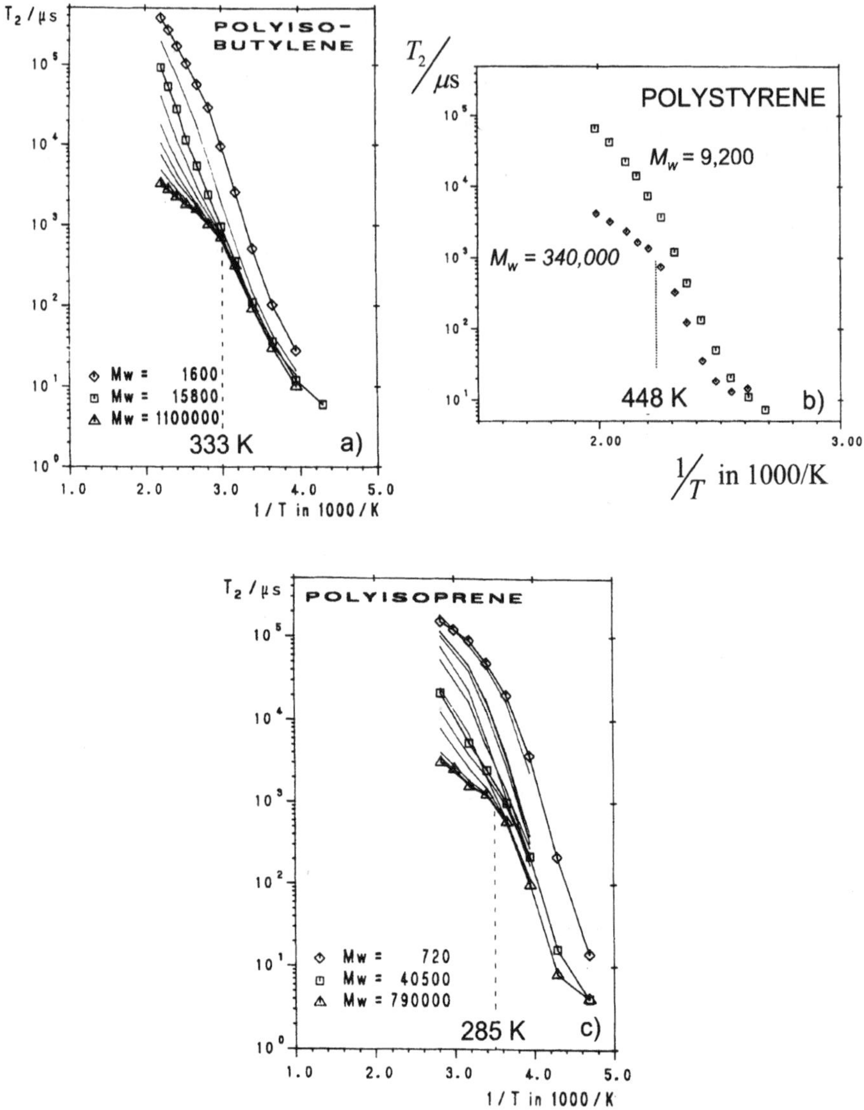

Fig. 17a–c. Effective proton transverse relaxation time $T_2=\langle 1/T_{2,j}\rangle^{-1}$ (see Eq. 160) in different linear polymers of different molecular masses measured at 90 MHz as a function of the reciprocal temperature [33]. A characteristic temperature T_{char} shows up for $M_w \geq M_c$ in the form of a relatively sharp bend in the temperature dependence. This bend disappears when approaching the critical molecular weight from above. For $M_w < M_c$ an additional molecular weight dependence reveals itself as a consequence of the chain length dependent free-volume effects. **a** Polyisobutylene. Symbols for the data points have been plotted for three different molecular weights only, while the data for all other molecular weights are represented by *polygonal lines* for the sake of clearness. From the *top* to the *bottom*, the lines refer to data for M_w=1,600; 5,800; 15,800; 34,000; 55,000; 80,600; 122,000; 182,000; 393,000; 610,000; 830,000; 1,100,000. The critical molecular

The transverse relaxation rate to be discussed in the following will be identified with this average expression, that is, an effective transverse relaxation time is defined as $T_2 = \left\langle T_{2j}^{-1} \right\rangle^{-1}$.

4.1.5.1 Characteristic Temperature

Experimental data of the effective transverse relaxation time are plotted in Fig. 17 as a function of the reciprocal temperature for different polymers. The molecular weight dependence for $M_w<M_c$ is due to free-volume effects. We will come back to this sort of molecular weight effect later in the review.

In contrast to short polymer chains, the molecular weight dependence occurring for $M_w>M_c$ at temperatures above a certain characteristic value T_{char} given in Table 4 is of a completely different nature. The temperature dependence of T_2 for high molecular weights in a sense reflects the hierarchy of segment and chain dynamics in terms of the three components A, B, and C. That is, more and more of the spectral density of dipolar fluctuations is shifted into the motional averaging regime when the temperature is increased (compare the Anderson/Weiss formula in the spectral density variant given in Eq. 40). Figure 18 shows a schematic representation of the low-frequency section of the spectral density that predominantly determines the transverse relaxation rate according to Eq. 40. As a consequence, the spectral densities specific for components A, B, and C come into play only one by one in a more or less combined way as suggested by Eq. 156 upon temperature variation.

At very low temperatures, none of the components is fast enough to be subject to motional averaging. The "rigid-lattice limit" begins to apply (possibly with the exception of methyl group rotation or other fast side group motions if appropriate). Increasing the temperature, the fastest chain motion component, namely component A, is the first one causing motional averaging to some degree. The transverse relaxation time is hence getting longer with increasing temperature. The next slower process is component B, still of no or very little molecular weight dependence. The characteristic

weight is about $M_c \approx 15{,}000$ [35]. The characteristic temperature is $T_{char}=333$ K. **b** Atactic polystyrene. The data refer to $M_w=9{,}200$ ($<M_c \approx 31{,}000$) and $M_w=340{,}000$ ($>>M_c$). The characteristic temperature is $T_{char}=448$ K. **c** Polyisoprene. Symbols for the data points have been plotted for three different molecular weights only, while the data for all other molecular weights are represented by *polygonal lines* for the sake of clearness. From the *top* to the *bottom*, the lines refer to data for $M_w=720$; 1,000; 3,280; 3,450; 5,100; 8,150; 10,800; 19,400; 34,000; 40,500; 58,000; 67,000; 130,000; 205,000; 280,000; 790,000. The critical molecular weight is about $M_c \approx 19{,}000$ [35]. The characteristic temperature is $T_{char}=285$ K

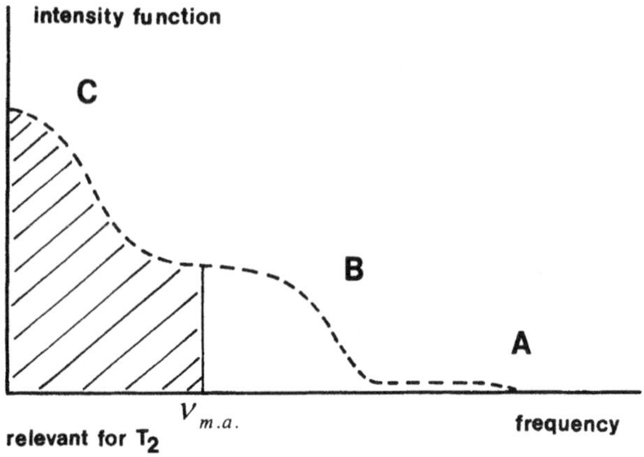

Fig. 18. Schematic representation of the spectral density (or intensity function) for spin couplings in the frame of the three-component analysis of molecular motions in polymers. The dipolar broadening region represented as the *hatched section* at low fluctuation rates is predominantly responsible for the transverse relaxation rate (compare Eq. 40 and Ref. [2]). Variation of the temperature shifts the components across the fluctuation rate defined by the motional-averaging condition, so that the influence of the individual components changes one by one. Variation of the molecular weight or the polymer concentration likewise shifts the molecular weight or concentration dependent components across the motional averaging fluctuation rate $\nu_{m.a}$

temperature T_{char} finally marks the onset of motional narrowing for component C with its strong dependence on the chain length. That is, for $T>T_{char}$ all three components give rise to motional averaging, whereas for $T<T_{char}$ only component A and, with gradually decreasing efficiency, component B contribute to the motion-based averaging of dipolar coupling. A semiempirical three-component formalism using the same evaluation procedure for T_2 and describing this sort of phenomenon in great detail has been reported [33, 35]. Note that the same sort of characteristic temperature was observed with spin-lattice relaxation data in the rotating frame, $T_{1\rho}$, at 1 kHz for polyisoprene. The characteristic temperatures found for a series of different polymer species are listed in Table 4.

Fig. 19a–e. Molecular weight dependence of the effective proton and ^{13}C transverse relaxation times $T_2=\langle 1/T_{2,l}\rangle^{-1}$ (see Eq. 160) in different linear polymers at different temperatures. The data were measured at 2.1 T [35, 138, 139]. The classical critical molecular weight M_c and a value characteristic for the three-component nature of molecular motions in polymers, M_{BC}, reveal themselves as bends in the slopes of the power laws approaching the dependences on the molecular weight. Values of the characteristic molecular weights are listed in Table 4. For $M_w<M_c$ an additional molecular weight dependence

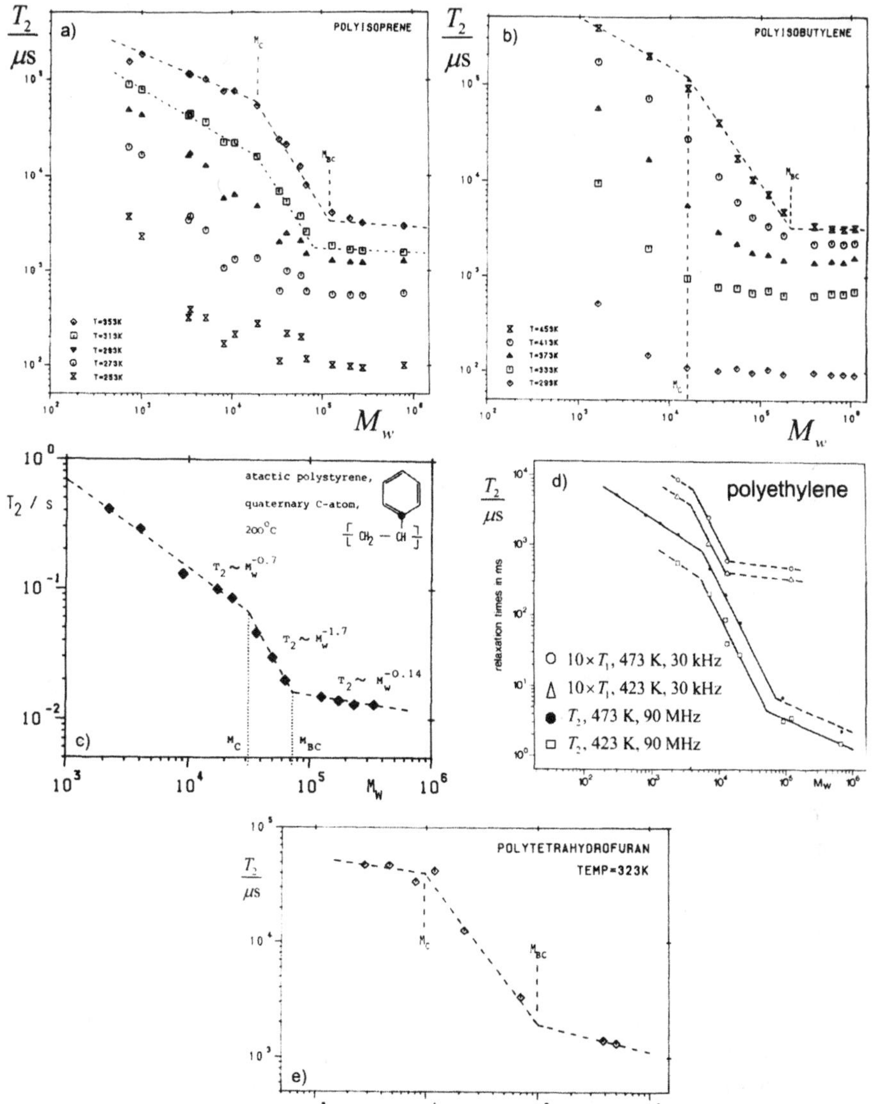

may matter as a consequence of the chain length dependent free-volume effects). **a** Polyisoprene. From the *top* to the *bottom*, these proton data refer to the temperatures 353, 313, 293, 273, 253 K. **b** Polyisobutylene. From the *top* to the *bottom*, these proton data refer to the temperatures 453, 413, 373, 333, 293 K. **c** Atactic polystyrene. These data refer to the ^{13}C line of the quaternary carbon (see *inset*) and were measured at 473 K. Note that the double-bend behavior found with ^{13}C NMR [139] was not observed with proton NMR where only a single bend showed up [138]. That is, M_{BC} is concealed by ring flips in the latter case. **d** Polyethylene. These proton data refer to 423 and 473 K. For comparison, low-frequency T_1 data are also plotted to demonstrate that spin-lattice relaxation at frequencies in a range of those relevant for T_2 is also subject to characteristic molecular weights. **e** Polytetrahydrofuran. These proton data refer to 323 K

Table 4. Characteristic temperatures and molecular weights evaluated from ^{1}H and [13]C (of quaternary carbons) relaxation data of linear polymers [33, 35, 139, 140]. PE, polyethylene; PIP, polyisoprene; PIB, polyisobutylene; PS, atactic polystyrene; PTHF, polytetrahydrofuran; PDMS, polydimethylsiloxane. Literature data for critical molecular weights determined with rheology are also listed for comparison and completeness

	T_{char}/K	M_c	M_{BC}
PE	–	4,000 (T_1, ^{1}H, 30 kHz) 5,600 (T_2, ^{1}H, 90 MHz) 3,800 (rheology [52])	52,000 (423 K, T_2, ^{1}H, 90 MHz) 70,000 (473 K, T_2, ^{1}H, 90 MHz)
PIP	285 K (T_2, ^{1}H, 90 MHz)	19,000 (T_2, ^{1}H, 90 MHz) 20,000 (T_2, ^{13}C, 23 MHz) 10,000 (cis, rheology [52])	119,000 (353 K, T_2, ^{1}H, 90 MHz) 80,000 (313 K, T_2, ^{1}H, 90 MHz) 70,000 (293 K, T_2, ^{1}H, 90 MHz) 140,000 (T_2, 353 K, ^{13}C, 23 MHz)
PIB	333 K (T_2, ^{1}H, 90 MHz)	15,000 (T_2, ^{1}H, 90 MHz) 15,200 (rheology [52])	215,000 (453 K, T_2, ^{1}H, 90 MHz) 133,000 (413 K, T_2, ^{1}H, 90 MHz) 80,000 (373 K, T_2, ^{1}H, 90 MHz)
PS	448 K (T_2, ^{1}H, 90 MHz)	31,000 (T_2, ^{13}C, 23 MHz) 31,200 (rheology [52])	71,000 (T_2, 473 K, ^{13}C, 23 MHz)
PTHF	–	10,000 (T_2, ^{1}H, 90 MHz)	100,000 (323 K, T_2, ^{1}H, 90 MHz)
PDMS	–	20,000 (T_2, ^{1}H, 90 MHz) 24,000 (rheology [52,147])	$\geq$200,000 (353 K, T_2, ^{1}H, 90 MHz)
PEO	–	3,600 (rheology [52])	–

4.1.5.2 Characteristic Molecular Weights

Figure 19 shows typical data for the effective transverse relaxation time $T_2=\langle T^{-1}{}_{2,j}\rangle^{-1}$ as a function of the molecular weight. Analogous to the characteristic temperature discussed above, the molecular weight dependences show two sharp bends defining molecular weights again characteristic for the hierarchical role of the three components of chain dynamics.

At very high molecular weights, it is only the molecular weight independent component A and to some extent component B that contribute to motional averaging. In terms of the schematic representation of the spectral density in Fig. 18, the combined fluctuation intensity of these components is the only one that acts above the motional averaging frequency. Reducing the molecular weight means faster component C fluctuations. This manifests itself at the characteristic molecular weight M_{BC} below which the molecular weight dependent component C gradually increases T_2 with decreasing molecular weight. M_{BC} separates the regimes where component C is subject to motional averaging ($M_w<M_{BC}$) and where not ($M_w>M_{BC}$).

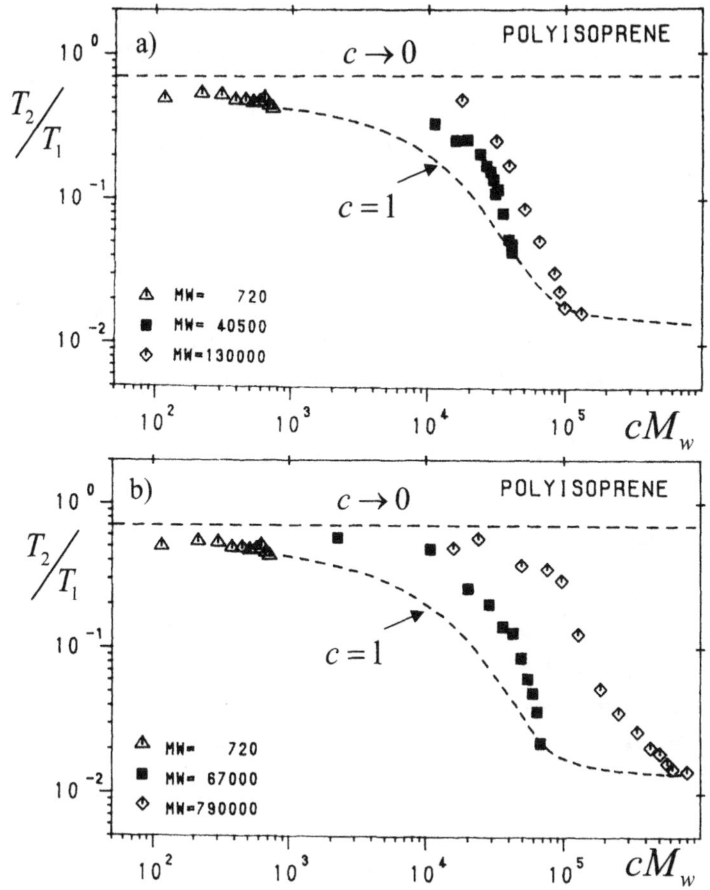

Fig. 20. Combined molecular weight and concentration dependence of T_2/T_1 in solutions of polyisoprene in CCl_4 [34]. The data refer to five different molecular weights (M_w=720; 40,500; 67,000; 130,000; 790,000). All data have been measured at 90 MHz proton resonance and 313 K. The concentration c is given as the volume fraction of the polymer. The *lower broken line* refers to pure melts (c =1; compare Fig. 19), the *upper straight line* to dilute solutions (c<<1)

Before complete motional averaging is reached at very low molecular weights, another phenomenon interferes, namely the crossover from entangled to Rouse-like chain dynamics at the critical molecular weight M_c well-known from rheology [52]. At this second characteristic molecular weight, component B and component C adopt a different character as discussed above in context with the model considerations. For $M_w<M_c$ chain dynamics tends to be governed by Rouse-like dynamics, whereas entangled chain dynamics shows up for $M_w>M_c$.

A semiempirical three-component formalism using the same evaluation procedure for T_2 and describing the molecular weight dependence with both characteristic breaks can be found in Ref. [35]. Note that the same sort of phenomenon also occurs with ^{13}C transverse relaxation [139] and spin–lattice relaxation in the kHz frequency regime [138]. Table 4 summarizes the characteristic molecular weights observed for different polymer species.

Note that the molecular weight dependence of the NMR relaxation times in polymer melts is generally much weaker than that of the viscosity in the same materials [52]. Even below M_{BC} the exponent of power laws for the molecular weight dependence of NMR relaxation times is roughly only half as large. This statement applies to molecular weights both below and above M_c. The reason is that the zero-shear viscosity is dominated by component C of chain dynamics alone, whereas the NMR relaxation times are affected by components B and A as well. In the frame of the three-component concept it was concluded that the molecular weight dependence of the NMR relaxation times for $M_w<M_{BC}$ are governed by an effective correlation time, τ_c^{eff} ,formed as the geometrical average of the time constants of component B, τ_l, and component C,τ_t [35, 138, 139]. That is

$$\tau_c^{eff} = \sqrt{\tau_l \tau_t}. \tag{161}$$

Since τ_l does not or only weakly depend on the chain length, the molecular weight dependence of the effective correlation time for entangled dynamics results in $\tau_c^{eff} \propto M_w^{1.5...2.0}$ [35, 139, 144], whereas the terminal relaxation time of rheology scales as $\tau_1 \propto M_w^{3.4}$ [52]. The effect of the crossover from unentangled to entangled dynamics on spin–lattice relaxation will be discussed later (see Fig. 31b).

4.1.5.3
Crossover from the Melt to Dilute Solutions

Swelling and dissolving polymers in low-molecular solvents changes chain dynamics. This refers to components B and C in particular. Actually, the critical molecular weight that indicates the onset of entangled behavior changes with the polymer concentration according to [52]

$$M_c^{sol} c = const, \tag{162}$$

where c is the polymer concentration and M_c^{sol} is the critical molecular weight of the solution. Decreasing the polymer concentration means increasing the critical molecular weight effective in the solution. Expressed the other way round, the crossover from entangled to unentangled dynamics of a polymer of a given molecular weight $M_w>M_c(melt)$ can be passed through by reducing the polymer concentration. This is visualized in Fig. 20 for polyisoprene of various molecular weights dissolved in CCl_4 [34]. In order to remove any effects by modified local motions (in particular due to the monomeric friction coefficient) and by the free volume that is likely affected by the solvent, the ratio T_2/T_1 is plotted. For c=1, the data refer to unswollen melts. For $c<<1$ a plateau is reached independent of the chain length. This plateau corresponds to the limit of dilute solutions.

The T_2/T_1 data demonstrate the entanglement effect which gradually increases with the molecular weight and the polymer concentration. The data are dominated by local motions (component A) for low concentrations and/or for low molecular weights. On the other hand, the entanglement effect reveals itself for high molecular weights and high polymer concentrations by the correspondingly modified components B and C. This phenomenon can again be described by the semiempirical three-component formalism mentioned before [34]. The dilution effect on spin–lattice relaxation will be discussed later in the review (see Fig. 31d).

4.2
Chain-end Dynamics ("Whip Model")

Segments located in blocks near the chain ends are expected to be subject to molecular motions different from those in the central block provided that the molecular mass of the chain is above the critical value, so that entangled dynamics is established in principle. The dynamics of segments in the chain-end blocks should generally be faster than those in the central block. Actually, in a ^{13}C transverse relaxation study of a polystyrene–polyisoprene–polystyrene three-block copolymer with a monomer ratio of 10:1600:10, this dynamical heterogeneity could be demonstrated qualitatively by selectively measuring ^{13}C signals specific for the polystyrene blocks and for the polyisoprene block [141].

The same sort of three-block dynamics is also expected for homopolymers where different transverse relaxation components on these grounds show up, provided that the molecular weight distribution is practically monodisperse [36]. This suggests experimental investigations in a rather direct way.

Two different scenarios for chain-end dynamics have been suggested. Doi [142] introduced the so-called contour length fluctuation (CLF) as a modification of the tube/reptation model. Due to the stochastic nature of chain modes in the tube, the chain ends are fluctuating back and forth a length proportional to the square root of the chain length and, hence, are subject to tube constraints to a much lower degree than the central part of the chain. The chain-end blocks are therefore expected to have a molecular mass obeying

$$M_{eb} \propto M_w^{1/2} \quad (\text{"CLF model"}). \tag{163}$$

On the other hand, the mobile chain-end blocks may be identified as chain sections not subject to entanglements so that they move in a whip-like fashion. In view of the critical molecular mass M_c indicating the minimum

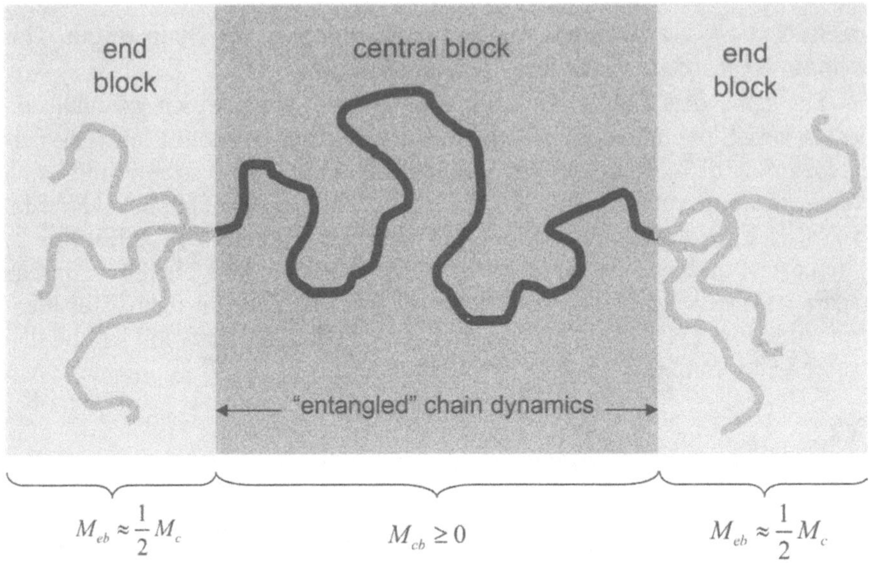

Fig. 21. Schematic representation of the three blocks that can be distinguished in a homopolymer melt with respect to segment dynamics ("whip model") [36]. The two chain-end blocks of a molecular mass M_{eb}≈0.5 M_c each are not expected to be subject to entangled dynamics, whereas the central block of molecular mass $M_{cb} \geq 0$ is. Below the critical molecular mass, $M_w \leq M_c$, no central block exists according to this definition, so that entangled behavior does not arise at all

chain length for entangled dynamics, the chain-end blocks are expected to have a molecular mass independent of the total chain length:

$$M_{eb} \approx \frac{1}{2} M_c = const \quad (\text{"whip model"}). \tag{164}$$

Figure 21 shows an illustration of this sort of three-block distinction.

The transverse relaxation functions can be analyzed for two components related to the central and chain-end blocks according to

$$S(t) \propto g(t) + q \exp\left(-t/T_2^l\right), \tag{165}$$

where T^l_2 is the transverse relaxation time of the longest component which is attributed to the two chain-end blocks. The quantity $q=2M_{eb}/M_w$ is the "amplitude" of the chain-end block component relative to the signal of the whole chain. The first term on the right-hand side, $g(t)$, comprises all faster components including potentially Gauss-like contributions with very long chains and at low temperatures. This fast-decaying part of the transverse relaxation curve is supposed to represent the central chain block. Experimentally an analysis of this sort will be the more reliable the faster $g(t)$ decays relative to the exponential function in Eq. 165. That is, the experiments described in the following refer mainly to $M_w >> M_c$ (compare the molecular weight dependences given in Fig. 19).

For the CLF model, one expects

$$q \propto \frac{\sqrt{M_w}}{M_w} = M_w^{-1/2}. \tag{166}$$

This molecular weight dependence is contrasted by the constant chain-end blocks assumed in the whip model:

$$q = \frac{2M_{eb}}{M_w} = \frac{M_c}{M_w} \propto M_w^{-1}. \tag{167}$$

The objective is now to determine q from experimental data as a function of the molecular mass of the whole chain in order to differentiate the two model predictions.

Figure 22 shows typical transverse relaxation curves. For $M_w < M_c$ the decay is monoexponential (apart from a slight influence of translational diffusion relative to the inhomogeneities of the magnet). The chain-end blocks reveal themselves only above the critical molecular weight M_c as visualized in Fig. 22 by the transverse relaxation curve for $M_w >> M_c$.

Fitting the second term on the right-hand side of Eq. 165 to the long-time tail of experimental transverse relaxation curves permits one to determine

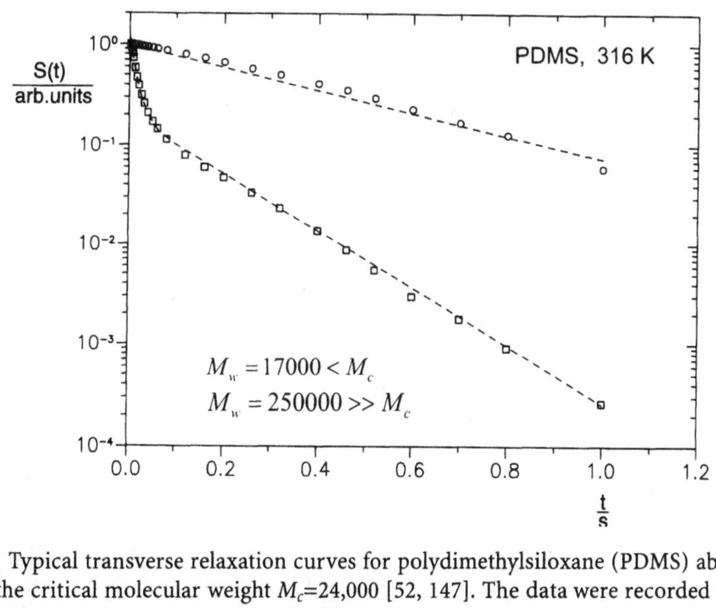

Fig. 22. Typical transverse relaxation curves for polydimethylsiloxane (PDMS) above and below the critical molecular weight M_c=24,000 [52, 147]. The data were recorded with the Hahn echo at 316 K [36]. The slowly decaying exponential tail indicative for the fast chain-end block dynamics reveals itself only above M_c. The *broken lines* have been calculated with the aid of Eq. 165 and a formalism described in Ref. [125]

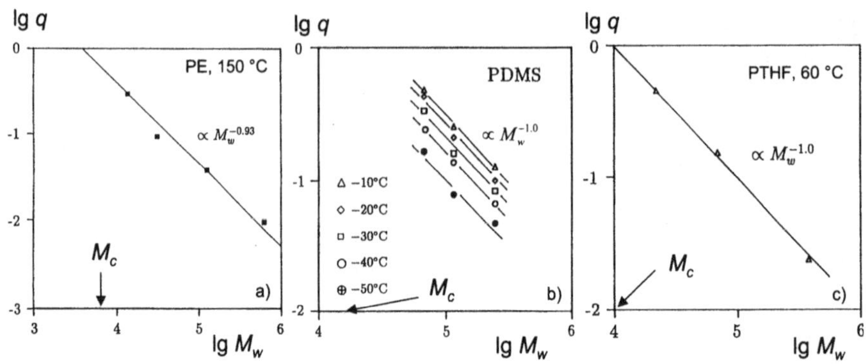

Fig. 23a–c. Relative amplitude of the slowly decaying tail of the transverse relaxation curve measured in **a** linear polyethylene (PE), **b** polydimethylsiloxane (PDMS), and **c** polytetrahydrofuran (PTHF) as a function of the molecular mass [36]. The data have been recorded with standard Hahn echo and Carr/Purcell/Meiboom/Gill pulse sequences. The values of the critical molecular weight are indicated on the abscissa axes. Extrapolating the data for $M_w \rightarrow M_c$ suggests $q \rightarrow 1$ as predicted by the whip model (Eq. 167)

the relative amplitude q. Figure 23 shows results for a series of different polymers as a function of the molecular mass. The data contradict the CLF model represented by Eq. 166, whereas they are largely in accordance with Eq. 167 predicted for the whip model. Moreover, extrapolating the data to $q=1$ suggests abscissa values very close to the critical molecular weight as suggested by Eq. 167. This, and the fact that the chain-end block relaxation time T_2^l of PDMS turned out to be essentially independent of the molecular weight in contrast to the central block relaxation function $g(t)$ [36, 143], strongly support the whip model further.

Figure 24 shows a schematic illustration of the scenario that chain-end dynamics is underlying. Any chain fold that happens to be formed near a chain end consists of a more mobile section (drawn in light gray) that can diffuse back and forth on a curvilinear path along the chain contour between the chain end and the tip of the fold. If this chain section moves towards the chain end, the mass of the moving entity is *increasing* by "unrolling" the fold (the position of which is assumed to be static for simplicity). If the chain section moves towards the fold tip, its mass is *decreasing* by "rolling up" the fold. The friction obstructing this sort of curvilinear displacement is proportional to the length ξ of the moving entity, so that the curvilinear diffusion coefficient of the mobile chain section will be proportional to ξ^{-1}. That is, displacements directed towards the fold tip are favored since ξ is reduced, whereas displacements towards the chain end are slowed down because of the increase of ξ associated with this direction of motion. This mechanism results in a motion resembling that of the cord of a cracking whip. A formalism describing this peculiar whip-like motion due to ter-

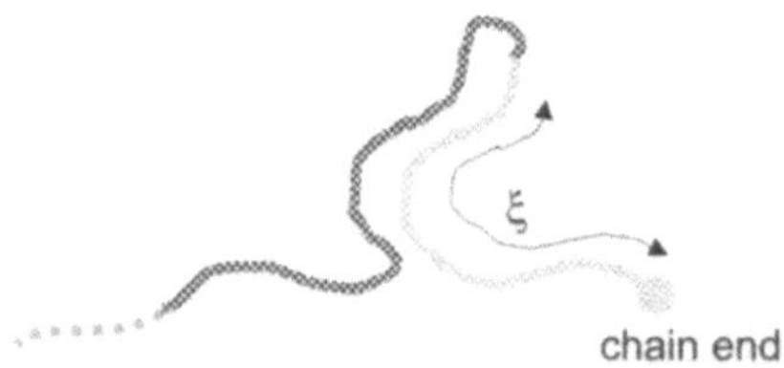

Fig. 24. Illustration of the whip model of chain-end dynamics. A chain fold that happens to be near a chain end consists of a relatively immobile strand towards the middle of the chain (*left*; drawn in *dark gray*) and a chain section on the right (drawn in *light gray*) that can quickly migrate back and forth. Its curvilinear length ξ between the chain end and the tip of the fold is decreased when it is displaced towards the tip of the fold and is increased when it moves in the chain-end direction. As a consequence, the curvilinear diffusion coefficient is increased or decreased, respectively, depending on the displacement direction. Therefore, motions towards the tip of the fold are favored, and the fold tends to unroll not unlike the cord of a cracking whip

minal chain folds close to chain ends has been described [36]. Recent ^{13}C spin–lattice relaxation measurements in the rotating frame corroborate the different dynamics of chain-end blocks [129].

4.3 Free-Volume and Void Effects

The whip effect on chain ends may be considered to be the molecular origin of the free volume which is known to be connected with chain ends [145, 146]. Since the relative number of segments near chain ends decreases with increasing molecular weight, free-volume effects show up predominantly at low molecular weights and in particular below the critical value M_c. The consequence is an additional chain length dependence of segment motions particularly for $M_w<M_c$. That is, the molecular weight dependence intrinsic to chain modes is superimposed by that due to free-volume variations. This effect is accounted for in the empirical Williams/Landel/Ferry (WLF) formalism that can be used to correct data for free-volume effects [52, 147–150].

The free volume and its distribution in heterogeneous samples can be directly probed by the enhancement of spin–lattice relaxation by molecular oxygen dissolved in the sample. Molecular oxygen is electron paramagnetic and, hence, an efficient relaxation agent. The local spin–lattice relaxation rate therefore monitors the local oxygen concentration. Exposing the sample to an oxygen atmosphere at an elevated pressure prompts oxygen molecules to diffuse into areas of reduced density. The concentration reached in equilibrium is a matter of the available free volume. A corresponding formalism can be found in Ref. [151]. In order to avoid leveling of the local relaxation rates by flip-flop spin diffusion, a perdeuterated semicrystalline polyethyl-

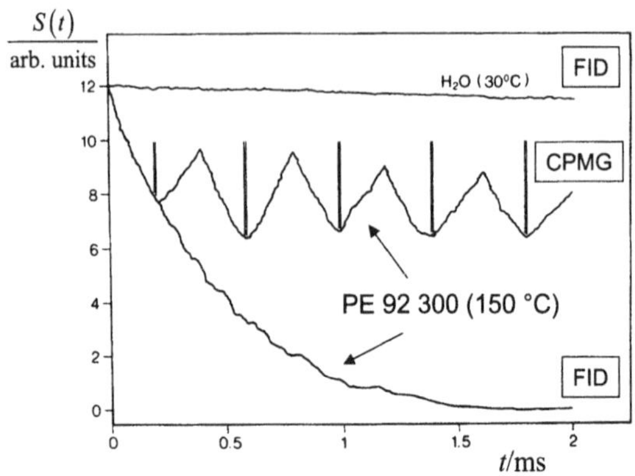

Fig. 25. Free-induction decay (FID) after a 90° pulse at t=0 and a series of Carr/Purcell/Meiboom/Gill (CPMG) spin echoes in linear polyethylene (M_w=92,300) at 150 °C. The 180° pulses of the CPMG sequence are indicated by *vertical lines*. The homogeneity of the external magnetic flux density is demonstrated by an FID of water at 30 °C. The data demonstrate the existence of strong internal field gradients arising from voids [152]

ene sample leaving only a few isolated undeuterated ethylene groups was examined [128]. It was shown that all free volume accessible by oxygen must be attributed to the amorphous part of the sample.

As a consequence of the existence of free volume related to chain ends, the density in polymer melts is expected to be heterogeneous even if the polymer chain length is practically monodisperse. In NMR, density inhomogeneities reveal themselves via the magnetic susceptibility distribution in the form of magnetic field gradients. The corresponding (inhomogeneous) broadening or the corresponding shortening of the free-induction decays in polyethylene were shown to reach two orders of magnitude [152]. Figure 25 demonstrates the effect of such internal field gradients on the free-induction decay (FID). The decay time to $1/e$ of the initial signal intensity is plotted in Fig. 26 as a function of the molecular weight. Interestingly a characteristic molecular weight similar to M_{BC} for the proper (effective) transverse relaxation time (see Fig. 19) shows up in the form of a crossover from a power law decay to a plateau at high molecular weights. This is an indication that component C acts as a very efficient motional narrowing mechanism as discussed above.

4.4 Evidence for Rouse Dynamics ($M<M_c$)

Field-cycling NMR relaxometry is an excellent monitor of chain modes. As illustrated in Fig. 5, it specifically and predominantly probes component B. The other components, A and C, matter only at low temperatures/high frequencies and high temperatures/low molecular weights, respectively. That is,

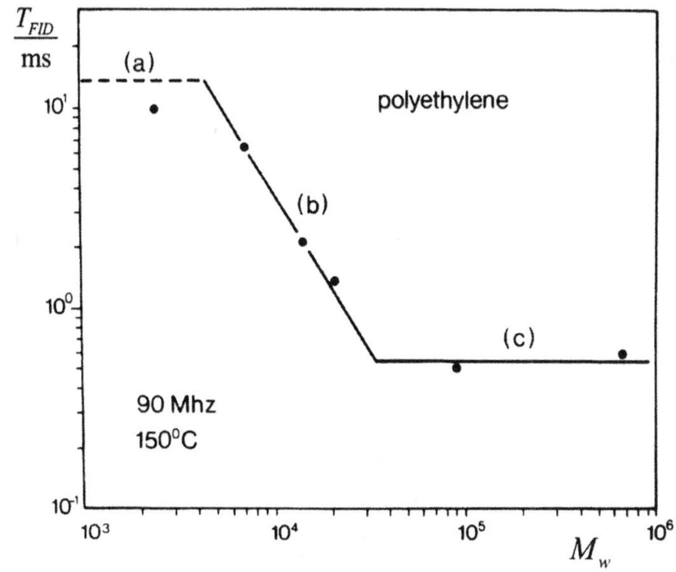

Fig. 26. Decay times T_{FID} to $1/e$ of the initial value of the free-induction signals in melts of linear polyethylene at 150 °C as a function of molecular mass. The *horizontal broken line* (a) indicates the instrumental homogeneity limitation determined with a water sample at 30 °C (see Fig. 25). The motional averaging region (b) is characterized by a strong molecular mass dependence equivalent to that of the proper (effective) transverse relaxation time (see Fig. 19d). It is therefore attributed to the combined influence of components B and C. The plateau (c) reached at high molecular masses reflects the unaveraged effect of internal field inhomogeneities corresponding to a full width at half maximum of the resonance line of about 7 ppm. This value can be estimated on the basis of the magnetic susceptibility of polyethylene relative to that of empty voids [152]

field-cycling NMR relaxometry is particularly suitable to examine component B and to directly test predictions of the model theories.

Rouse dynamics is expected to apply to molecular weights below the critical value where entanglement effects are not yet effective. Experimental data sets for the proton spin–lattice relaxation dispersion [47, 49, 153] are shown in Fig. 27 in comparison to the theoretical frequency dependence predicted by Eq. 64. Very interestingly, the values for the segment fluctuation time τ_s fitted to the T_1 data coincide with those derived from the T_1 minima (see Fig. 14) corrected for the temperature of the field-cycling measurements. That is, the two independent determination methods lead to consistent results.

The logarithmic Rouse formula given by Eq. 64 is valid for $\omega \ll \tau_s^{-1}$. Deviations of the theoretical curves from the experimental data points are therefore only expected at frequencies close to the "minimum condition" $\omega\tau_s \approx 1$. where simultaneously component A starts to become perceptible (compare Fig. 16). This explains why the PDMS data fit better to the model than poly-

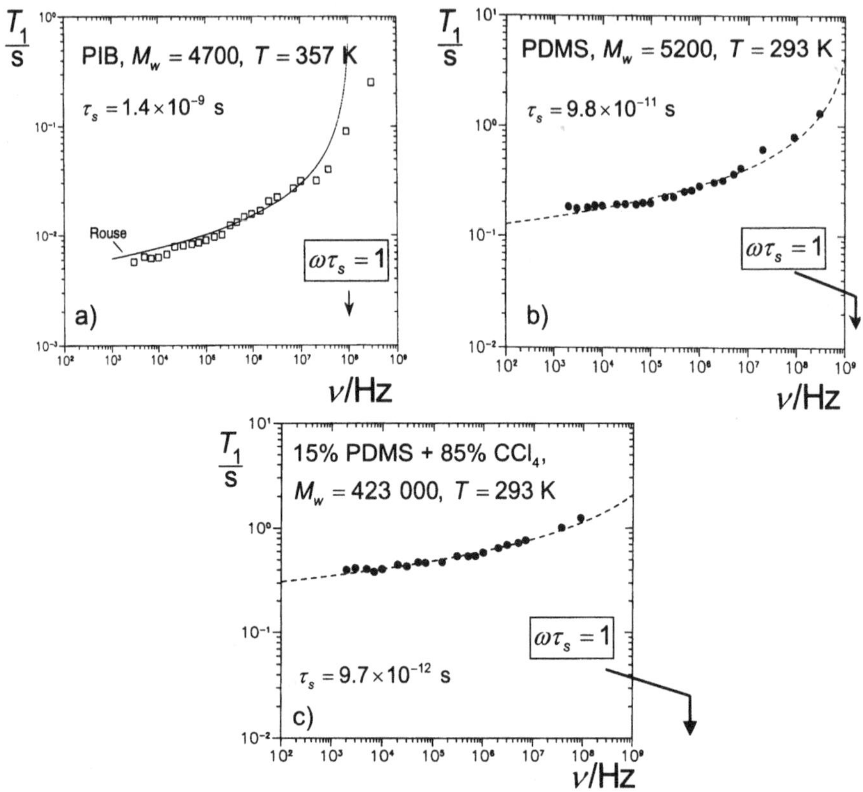

Fig. 27a–c. Proton spin–lattice relaxation dispersion under conditions where Rouse dynamics is expected to apply. The theoretical curves have been calculated with the aid of Eq. 64. The validity of this model is restricted to $\omega << \tau_s^{-1}$. The positions on the frequency axes where the condition $\omega\tau_s=1$ applies are indicated by *arrows* for the segment fluctuation time τ_s fitted to the experimental data. The τ_s values are in accord with those derived from the T_1 minimum data (see Fig. 14) where applicable. **a** Polyisobutylene (M_w=4,700 < M_c≈15,000), melt at 357 K [49]. **b** Polydimethylsiloxane (M_w=5,200 < M_c≈20,000), melt at 293 K [47, 125, 153]. **c** Solution of 15% polydimethylsiloxane (M_w=423,000) in CCl_4 at 293 K [47, 125, 153]. The critical molecular weight of solutions is modified according to Eq. 162

isobutylene (PIB) with a segment fluctuation time one order of magnitude longer than that of PDMS.

Entanglement effects can be reduced by dissolving the polymer even if its molecular weight is well above the critical value in the melt. This is demonstrated in Fig. 27c for PDMS dissolved in CCl_4. The data can again be well described by the Rouse model.

The conclusion from these findings is that the Rouse model is perfectly corroborated by NMR relaxometry experiments provided that entanglement effects are excluded or not effective on the time/frequency scale of the experiment. On the other hand, Rouse dynamics was also predicted for entangled dynamics in limit $(I)_{DE}$ of the tube/reptation model and for the short-time limits of the renormalized Rouse models (see Tables 1, 2 and 3). However, no such behavior was ever found for polymer melts on the basis of NMR relaxometry. The conclusion is that the short-time/high-frequency dynamics of entangled-polymer melts is dominated by component A (including side-group motions) rather than by component B in the Rouse limit. For polymer melts, there is no evidence for an intermittent Rouse regime under entanglement conditions. On the other hand, entangled-polymer solutions where the critical molecular weight is shifted to correspondingly higher values (see Eq. 162) obviously do show such an intermittent Rouse limit as demonstrated in Figs. 27c and 31d.

4.5 The Three Regimes of Spin–Lattice Relaxation Dispersion in Entangled Polymer Melts, Solutions, and Networks ($M<M_c$)

Figure 5 schematically shows the experimental window probed with field-cycling NMR relaxometry in combination with conventional pulse techniques for the determination of spin–lattice relaxation times. In the following, we specifically refer to molecular weights and temperatures large enough so that the relaxation data recorded exclusively refer to component B. Very large molecular weights also exclude any perceptible influence of the whip effect and of free volume connected with chain-end blocks. That is, segments can be assumed to form a dynamically homogeneous phase on the time scale of spin–lattice relaxation experiments (an exception is reported below in context with mesogenic order). Under such conditions, field-cycling NMR relaxometry [2, 18–21] turned out to be an unmatched tool for the elucidation of dynamics in the chain-mode regime.

Figures 28, 29, 30, 31, 32, 33, and 34 show a series of typical spin–lattice relaxation dispersion curves. The technique has been applied to melts, solutions, and networks of numerous polymer species. As experimental parameters, the temperature, the molecular weight, the concentration, and the cross-link density were varied. For control and comparison, the studies are partly supplemented by rotating-frame spin–lattice relaxation data and, of course, by high-field data of the ordinary spin–lattice relaxation time. Furthermore, the deuteron spin–lattice relaxation was employed for identifying the role different spin interactions are playing for relaxation dispersion.

Remarkably, a series of distinct and apparently universal NMR relaxation dispersion regimes can be identified from the data in Figs. 28 to 34 for the diverse polymer species. We consider the time/frequency window $\tau_t << (t,\omega^{-1}) << \tau_s$ (see Fig. 5). The terminal chain relaxation time, τ_t, is indicated by a crossover to a low-frequency T_1 plateau (see Fig. 16); the seg-

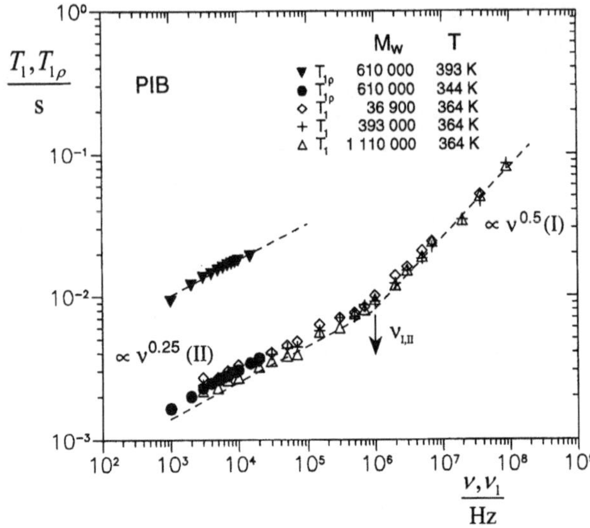

Fig. 28. Proton spin-lattice relaxation times in the laboratory system (T_I) and in the rotating frame ($T_{I\rho}$) of polyisobutylene (PIB) melts as a function of the frequency (ν, Larmor frequency in the laboratory frame; $\nu_1 = \gamma B_1/(2\pi)$, rotating-frame nutation frequency) [125]. The data refer to the molecular weight independent chain-mode regimes I (high-mode-number limit) and II (low-mode-number limit) [49]. The *arrow* indicates the crossover frequency $\nu_{I,II}$ between regions I and II

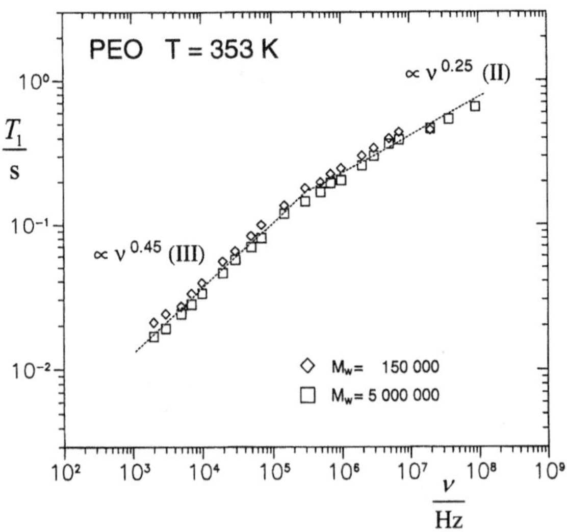

Fig. 29. Proton spin-lattice relaxation time of polyethyleneoxide (PEO) melts as a function of the frequency [49]. The data refer to the molecular weight independent chain-mode regime II (low-mode-number limit) and regime III influenced by intersegment dipolar interactions [154]

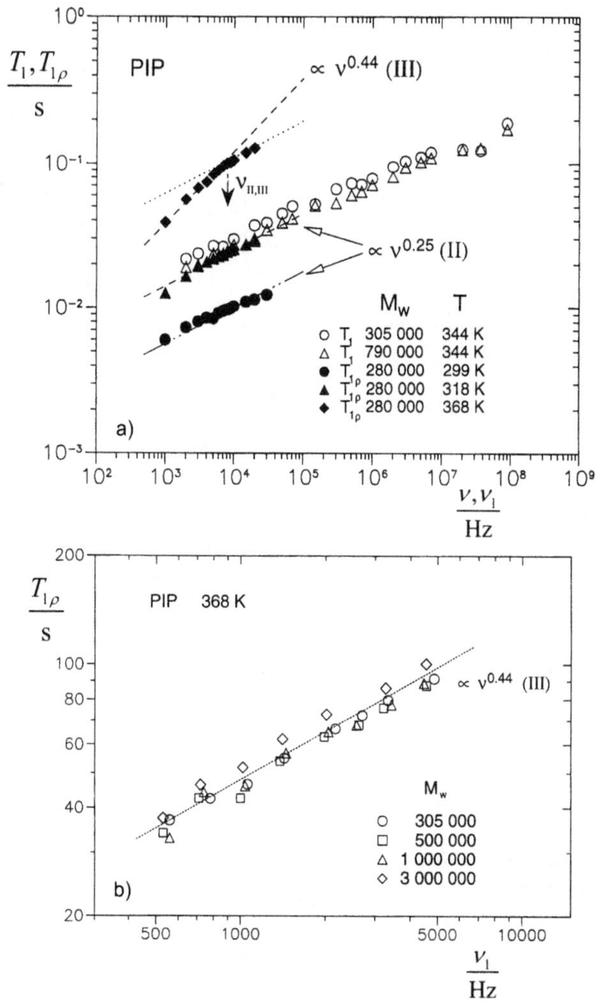

Fig. 30a, b. Proton spin-lattice relaxation times in the laboratory system (T_1) and in the rotating frame ($T_{1\rho}$) of polyisoprene (PIP) melts as a function of the frequency (ν, Larmor frequency in the laboratory frame; $\nu_1=\gamma B_1/(2\pi)$, rotating-frame nutation frequency) [125]. The data refer to the molecular weight independent chain-mode regime II (low-mode-number limit) and regime III influenced by intersegment dipolar interactions [154]. **a** Comparison of T_1 and $T_{1\rho}$ data. The *arrow* is to indicate the crossover frequency $\nu_{I,II}$ between regions I and II. **b** $T_{1\rho}$ data in detail

ment fluctuation time, τ_s, can be determined from T_1 minima (see Figs. 14 and 15). Under such conditions, there is clear experimental evidence for three distinct proton dispersion regimes for entangled polymers, $M_w >> M_c$:

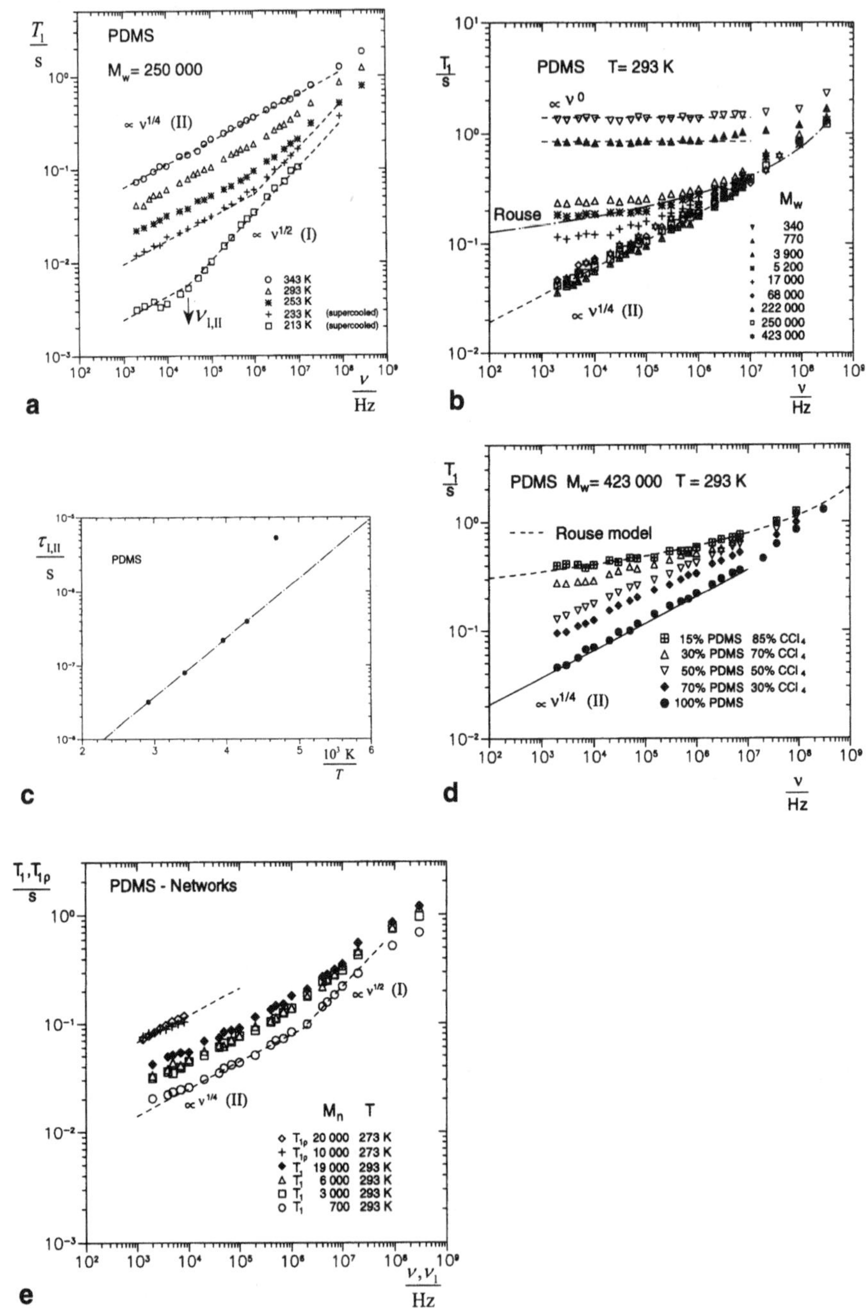

Fig. 31a–e. Proton spin-lattice relaxation times in the laboratory system (T_1) and in the rotating frame ($T_{1\rho}$) of polydimethylsiloxane (PDMS) melts and solutions as a function of the frequency (ν, Larmor frequency in the laboratory frame; $\nu_1=\gamma B_1/(2\pi)$, rotating-frame nutation frequency) [125]. The melt data for $M_w>M_c\approx 24{,}000$ refer to the molecular weight independent chain-mode regimes I (high-mode-number limit) and II (low-mode-

$$\uparrow^{\omega}\downarrow_{T}\begin{cases} T_1 \propto M_w^0 \omega^{0.5\pm0.05} & \text{(region I; "high mode number limit")} \\ T_1 \propto M_w^0 \omega^{0.25\pm0.05} & \text{(region II; "low mode number limit")} \\ T_1 \propto M_w^0 \omega^{0.45\pm0.05} & \text{(region III; "inter - segment interaction limit")} \end{cases} \qquad (168)$$

These power laws are in accordance with data for the frequency dependence of the spin–lattice relaxation time in the rotating frame, $T_{1\rho}$, in a range as far as accessible by this technique (see Figs. 28 and 30–32).

The three proton relaxation dispersion regions I (fast motions), II, and III (slow motions) appear in that sequence from high to low frequencies or from low to high temperatures (see the up and down arrows in Eq. 168. The frequency window of the field-cycling NMR relaxometry technique (Fig. 5) is often not broad enough to cover all three regions at once. Mobile polymers like polyisobutylene (Fig. 28), polydimethylsiloxane (Fig. 31), and polydiethylsiloxane in the isotropic phase (Fig. 32) are subject to regions I and II under the experimental conditions. Less mobile polymers like polyethyleneoxide (Fig. 29) and polyisoprene (Fig. 30) tend to reveal regions II and III in the instrumental frequency window.

Nevertheless it is possible to demonstrate that all three proton relaxation dispersion regions are intrinsic to entangled polymer dynamics. The temperature/frequency range accessible with NMR relaxometry in polybutadiene (Fig. 33) permits the recording of all three regions one after the other in the same sample. At low temperatures, regions I and II dominate; at elevated temperatures region III comes into play. The shift of the crossover time between regions I and II,

$$\tau_{I,II} = 1/\left(2\pi\nu_{I,II}\right), \qquad (169)$$

number limit) [49]. **a** Melts of PDMS 250,000 at different temperatures. The crossover between regimes I and II at $\nu_{I,II}$ is shifted to higher frequencies with increasing temperature. **b** Melts of PDMS at 293 K for different molecular weights. For $M_w<M_c$ the chain-mode regimes I and II characteristic for entangled dynamics are absent and are replaced by motions not subject to entanglements. The theoretical T_1 dispersion curve expected for the Rouse model is shown for comparison. **c** Temperature dependence of the "crossover time constant" $\tau_{I,II}=1/(2\pi\nu_{I,II})$ evaluated from the plot in Fig. 31a. The *line* represents the Arrhenius law $\tau_{I,II}=1.1\times10^{-10}$ s $\exp\{(15.8 \text{ kJ mol}^{-1})/RT\}$. The deviation of the data point at 213 K indicates the influence of the supercooled state at this temperature. **d** Solutions of PDMS 423,000 in CCl_4 at 293 K for different concentrations. For low concentrations the chain-mode regimes I and II characteristic for entangled dynamics fade more and more. The theoretical T_1 dispersion curve expected for the Rouse model is shown for comparison. **e** Cross-linked PDMS for different cross-link densities specified by the mesh molecular weight M_n (number average)

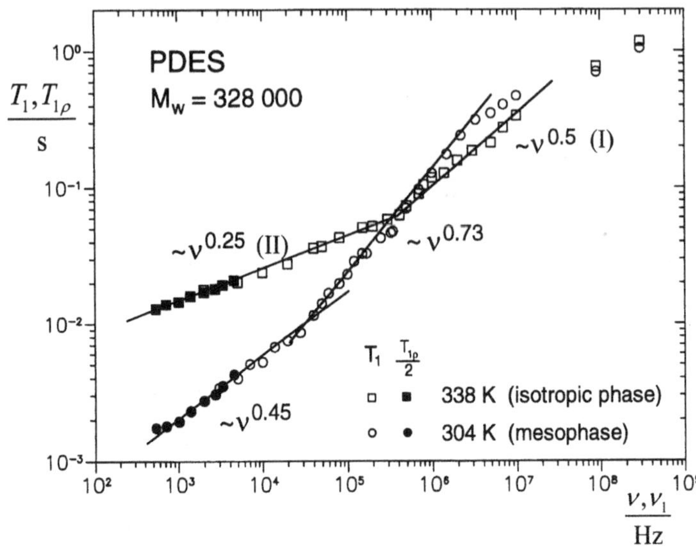

Fig. 32. Proton spin–lattice relaxation times in the laboratory system (T_1) and in the rotating frame ($T_{1\rho}$) of polydiethylsiloxane (PDES) melts in the isotropic and mesomorphic phases as a function of the frequency (ν, Larmor frequency in the laboratory frame; $\nu_1=\gamma B_1/(2\pi)$, rotating-frame nutation frequency) [155]. The data of the isotropic melt phase refer to the molecular weight independent chain-mode regimes I (high-mode-number limit) and II (low-mode-number limit). In the (ordered) mesomorphic phase the dispersion slopes are much steeper and the crossover frequency is shifted to a lower value. At very high frequencies, the influence of component A (including side chain motions) becomes gradually visible

for polydimethylsiloxane is plotted in Fig. 31c as a function of the reciprocal temperature. The Arrhenius-like behavior corroborates that the chain-mode regimes I and II are subject to thermal activation.

These three proton spin-lattice relaxation dispersion regions must not be identified with the Doi/Edwards limits $(I)_{DE}$, $(II)_{DE}$, and $(III)_{DE}$ which are predicted for dynamic ranges with significantly different frequency and molecular weight dependences (see Table 1). No Rouse-like dynamics corresponding to limit $(I)_{DE}$ can be identified in the experimental data sets shown for entangled polymer melts. In the frame of the tube/reptation model, there is no such thing as a frequency dependence like region II, $T_1 \propto M_w^0 \omega^{0.25\pm0.05}$. The square root molecular weight dependence predicted in that model for limit $(III)_{DE}$ does not occur either. However, it will be shown that Doi/Edwards predictions can indeed be verified in perfect detail if chains are confined to artificial tubes prepared in a solid polymer matrix [95]. On the other hand, the high- and low-mode-number limits discussed in the frame of the renormalized Rouse models (Tables 2 and 3) provide a perfect explana-

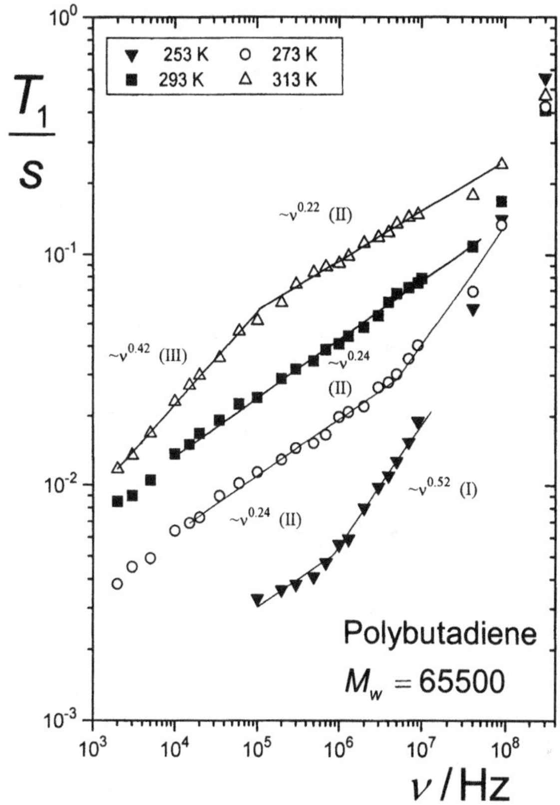

Fig. 33. Proton spin-lattice relaxation time of a melt of linear polybutadiene (M_w= 65,500) at different temperatures as a function of the frequency [154]. By varying the temperature, all three dispersion regimes show up in the experimental frequency window one by one (see Fig. 5)

tion of dispersion regions I and II, whereas region III can be shown to be due to intersegment dipolar interactions.

4.5.1 High- and Low-Mode-Number Limits (Dispersion Regions I and II)

The appearance of dispersion regions I and II in experiments is an exultant confirmation of the high- and low-mode-number, short-time limits predicted by the twice renormalized Rouse models (Tables 2 and 3). The exponents of the power laws suggested by the experimental data even match the theoretical predictions almost perfectly [47, 49]. Nevertheless, the good coincidence of the numerical values of these exponents is not considered to be the decisive finding backing up the renormalized Rouse models. The problem is that the theoretical exponents are slightly affected by the renormalization

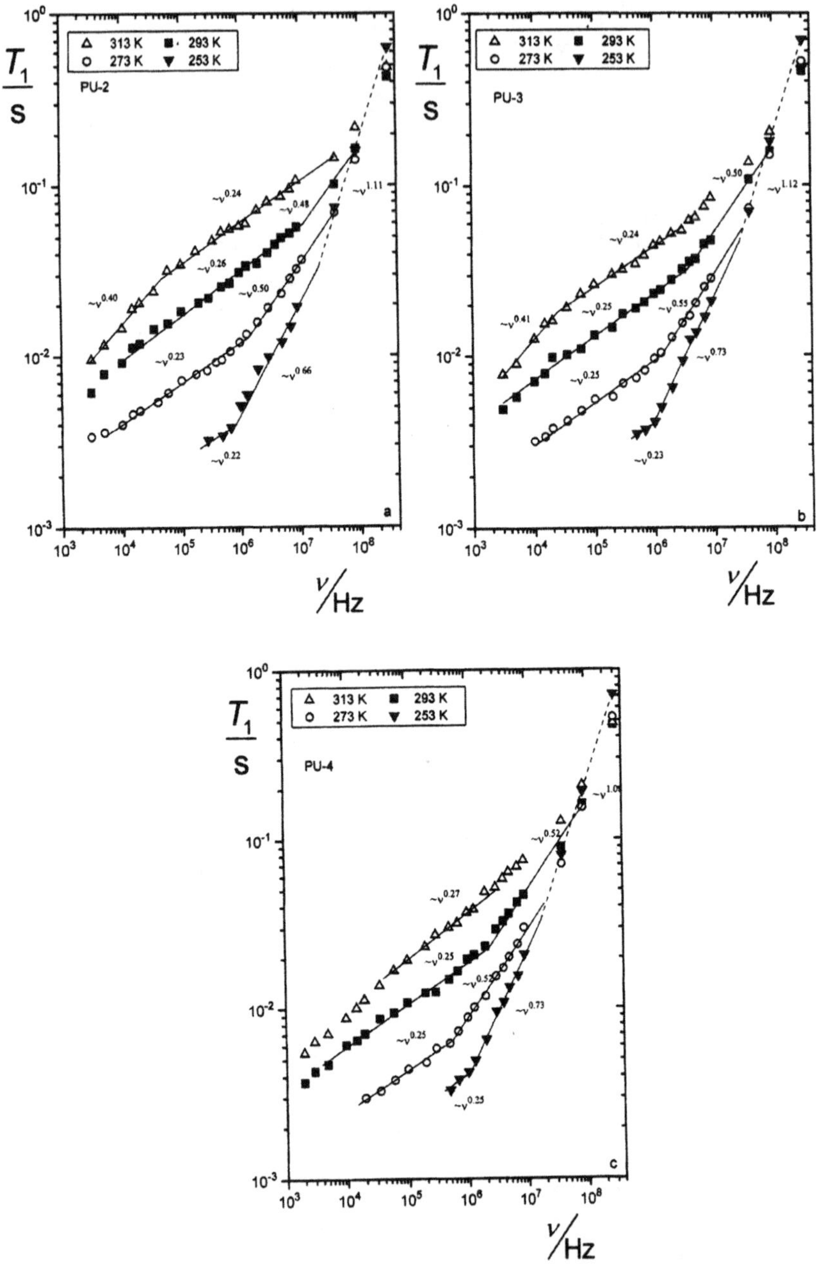

Fig. 34a–c. Proton spin-lattice relaxation time of thermoreversible networks of polybutadiene at different temperatures as a function of the frequency [131]. Linear PB (M_w=51,000) was cross-linked by addition of 4-phenyl-1,2,4-triazoline-3,5-dione (phenylurazole, PU). The crossover frequencies between regimes I, II, and III are shifted depending on the cross-link density. At the lowest temperatures the dispersion slopes tend to be steeper than in ordinary melts (see Eq. 168). **a** 19 phenylurazole groups per chain. **b** 28 phenylurazole groups per chain. **c** 37 phenylurazole groups per chain

ansatz that cannot be traced back to elementary principles and is subject to some ambiguity. The important result is rather that (i) two such limits become visible and that (ii) the exponents change almost exactly as predicted for the crossover between the high- and low-mode-number limits.

The distinction of two such limits with the right variation of the power law exponents appears to be a result reflecting the analytical structure of the generalized Langevin equation, which is thus proven to represent the essential features of entangled chain dynamics in the short-time limit, $t<<\tau_t$. The generalized Langevin equation obviously contains crucial elements determining entangled-chain dynamics.

4.5.2
Intra- and InterSegment Spin Interactions (Dispersion Region III)

Intrasegment dipolar interaction fluctuates as a consequence of segment reorientations and conformational changes. If the interacting nuclear dipoles are residing on different segments or even different chains, variations of the internuclear vector tend to take much longer because they are the consequence of displacements of the dipole hosting segments by self-diffusion relative to each other. Any spin–lattice relaxation dispersion affected by intersegment dipolar interactions is therefore expected at very low frequencies. This can be tested by comparing proton with deuteron spin–lattice relaxation dispersion of the same polymer species. Deuteron relaxation is dominated by (intrasegment) quadrupole coupling with the local electric field gradient, whereas proton relaxation can be determined by intra- as well as intersegment dipolar interactions [2, 154].

Figure 35 shows such comparisons for polyethyleneoxide and polybutadiene melts. The crossover between dispersion regions II and III observed with proton NMR obviously disappears in the deuteron studies. Region III does not occur with deuterons. The different dispersion slope measured with deuteron NMR was also demonstrated with samples of lower molecular weights in Figs. 36 and 45b. That is, dispersion region III must not be considered as a limit specific for polymer theories. Rather it appears that is mainly an effect intrinsic to NMR relaxation by dipolar interaction.

A schematic representation of the situation one is dealing with in this context is shown in Fig. 37. We consider the representative segments # k on chain # α and # l on chain # β. The internuclear vector, $\vec{r}_{kl}(t)$, fluctuates because of self-diffusive displacements $\vec{R}_{rel}(t)$ of segment # l relative to segment # k. That is, the origin of the reference frame is considered to be fixed at segment # k.

The average correlation function for the intersegment dipolar interaction can be expressed as [154]

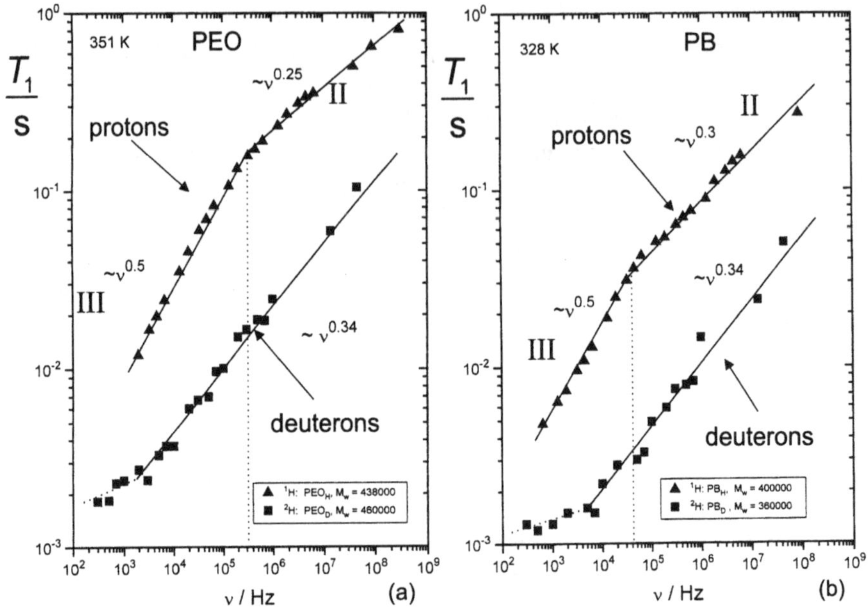

Fig. 35a, b. Proton and deuteron spin–lattice relaxation times of polyethyleneoxide (PEO) and polybutadiene (PB) melts as a function of the frequency [154]. In addition to intrasegment dipolar coupling, proton relaxation is also subject to intersegment dipolar couplings leading to the dispersion regime III specific for this sort of relaxation mechanism. Deuteron relaxation is predominantly due to quadrupole interaction which is of an intrasegment nature. A crossover from regime II to regime III therefore does not occur with deuteron resonance. **a** Polyethyleneoxide (for further deuteron relaxation data see Figs. 36 and 45b). **b** Polybutadiene

$$G_{\text{inter}}^{(m)}(t) = \rho_s \int g(r_0) G_{kl}^{m}(t) d^3 r_0 \quad (m=1,2), \tag{170}$$

where

$$G_{kl}^{(m)}(t) = \left\langle \frac{Y_{2,m}(\vartheta_0,\varphi_0)Y_{2,-m}(\vartheta_t,\varphi_t)}{r_0^3 r_t^3} \right\rangle \Big/ G_{kl}^{(m)}(0) \tag{171}$$

(see Eq. 28), $g(r_0)$ is the radial segment pair correlation function, and ρ_s is the spin number density. The variables r_0 and r_t in Eqs. 170 and 171 stand for r_{kl} (0) and r_{kl} (t), respectively. That is, $r_0 \equiv r_{kl}$ (0) and $r_t \equiv r_{kl}$ (t). For the analytical treatment the radial segment pair correlation function may be approximated by (see Fig. 38)

$$g(r_0) \approx \begin{cases} 0 \text{ if } r_0 \leqslant \sigma \\ 1 \text{ otherwise} \end{cases}. \tag{172}$$

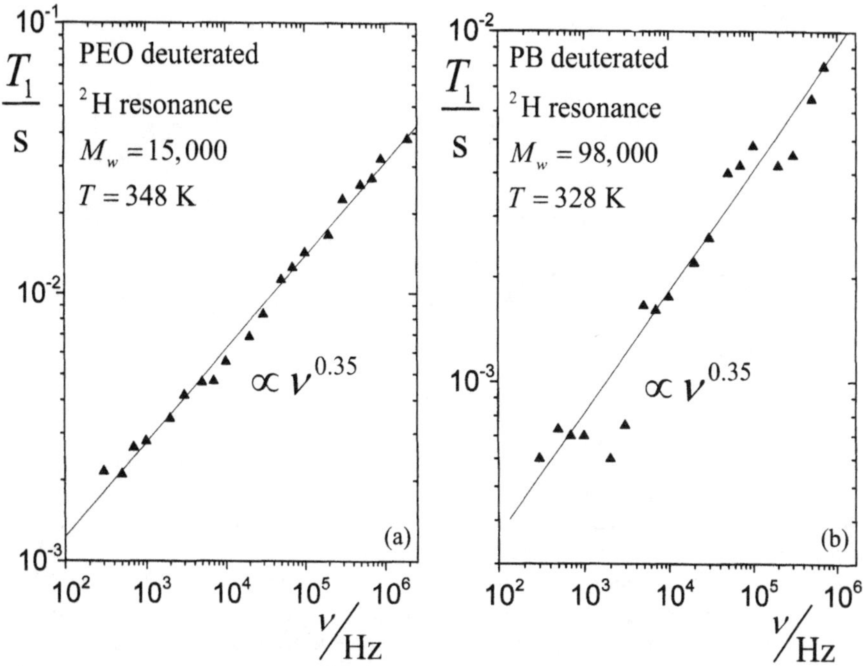

Fig. 36a, b. Deuteron spin–lattice relaxation times of deuterated polyethyleneoxide (PEO) (**a**) and polybutadiene (PB) (**b**) as a function of the frequency [156]

The correlation function in Eq. 171 can be approximated by the following consideration. For relative displacements $R(t)$ (see Fig. 37) much larger than the initial internuclear distance $r_0 \equiv r_{kl}(0)$, the distance and polar angle variation leads to total loss of any correlation to the initial distance vector, whereas the correlation is completely retained in the opposite limit. That is,

$$G_{kl}^{(m)}(t) \to 0 \text{ for } R_{rel}(t) >> r_0 \tag{173}$$

and

$$G_{kl}^{(m)}(t) \approx 1 \text{ for } R_{rel}(t) << r_0. \tag{174}$$

An ansatz accounting for these limits is

$$G_{kl}^{(m)}(t) \approx \left\langle \left| \frac{Y_{2,m}(0)}{r_0^3} \right|^2 \right\rangle P(t) = \frac{1}{r_0^6} P(t), \tag{175}$$

where $P(t)$ is the probability that segment # l is in a spherical volume $\propto r_0^3$ around its initial position $\vec{r}_0$ relative to segment # k. This probability is specified by the following limits:

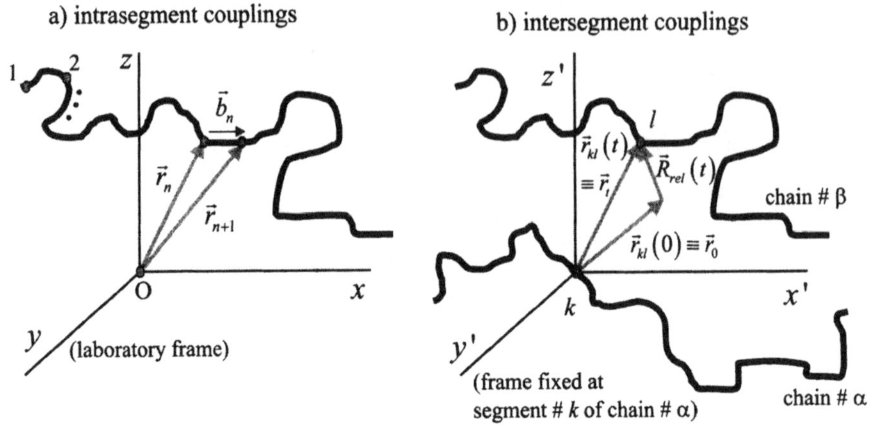

Fig. 37a, b. Intra- and intersegment spin couplings. **a** In the frame of the experiments referred to in this article, intrasegment spin couplings can be dominated by dipole–dipole interactions (^{1}H, ^{13}C), quadrupole couplings (^{2}H), and anisotropic chemical shift interactions (^{13}C). They fluctuate due to segment reorientations relative to the laboratory frame. For component B, the relevant segment orientation is represented by the chain tangent vector $\vec{b}_n$ at segment # n. **b** Intersegment interactions (segments on the same or different chains) are exclusively of a dipolar nature. Spin–lattice relaxation on these grounds originate from reorientation and length variation of the distance vector $\vec{r}_{kl}$ between segments (in the scheme denoted as # k and #l). These fluctuations are caused by translational displacements $\vec{R}_{rel}$ of one segment relative to the other. This is described best in a frame fixed on one of the segments of the interacting pair. Since self-diffusion is a relatively slow process, spin–lattice relaxation by intersegment interactions becomes relevant only at relatively low frequencies

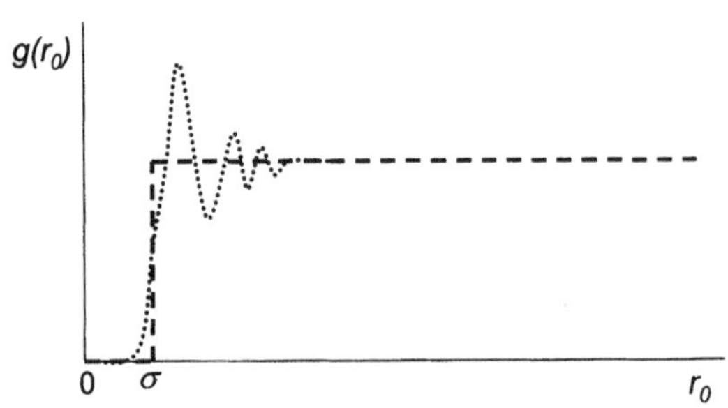

Fig. 38. Schematic representation of the radial segment pair correlation function. The "real" function characterized by a peak series of decaying amplitudes at the nearest coordination shells (*dotted line*). For the analytical treatment, this course is approximated by a step function (*broken line*). σ is the nearest-neighbour distance of segments

$$P(t) \to 1 \qquad \text{for } \langle R_{rel}^2(t) \rangle << r_0^2,$$

$$P(t) \to \frac{r_0^3}{\langle R_{rel}^2(t)\rangle^{3/2}} \quad \text{for } \langle R_{rel}^2(t)\rangle >> r_0^2. \tag{176}$$

The expression

$$P(t) \approx \frac{r_0^3}{\left[r_0^2 + \langle R_{rel}^2(t)\rangle\right]^{3/2}} \tag{177}$$

complies to both limits given in Eq. 176 and will therefore be taken as an approximation. Inserting Eqs. 172, 175, and 177 in Eq. 170 gives

$$G_{\text{inter}}^{(m)}(t) \approx \rho_s \frac{\ln\left\{2\langle R_{lab}^2(t)\rangle/\sigma\right\}}{\langle R_{lab}^2(t)\rangle^{3/2}}, \tag{178}$$

where the mean squared displacement in the laboratory frame is related to the mean squared displacement in the frame fixed at segment # k by $\langle R_{lab}^2(t)\rangle = \frac{1}{2}\langle R_{rel}^2(t)\rangle$. The logarithmic term in Eq. 178 varies slowly with time so that we may approximate further

$$G_{\text{inter}}^{(m)}(t) \approx \frac{\rho_s}{\langle R_{lab}^2(t)\rangle^{3/2}}. \tag{179}$$

The time dependence of the mean squared segment displacement in the laboratory frame was derived on the relevant time scale as low-mode-number, short-time limits of the renormalized and twice renormalized Rouse models as $\langle R_{lab}^2(t)\rangle \propto t^{2/5}$ and $\langle R_{lab}^2(t)\rangle \propto t^{1/3}$ (limit $(\text{II})_{\text{TRR}}$), respectively (see Tables 2 and 3). Inserting these power laws in Eq. 179, one obtains after Fourier transformation according to Eqs. 29 and 30

$$\frac{1}{T_1} = \frac{1}{T_1^{\text{intra}}} + \frac{1}{T_1^{\text{inter}}}, \tag{180}$$

where T_1^{intra} refers to the expressions given in Tables 2 and 3 for the low-mode-number, short-time limits, and T_1^{inter} scales as

$$T_1^{\text{inter}} \propto \nu^{0.4\cdots0.5}. \tag{181}$$

That is, dispersion region III is explained by the dominance of this intersegment contribution, whereas regions I and II are dominated by intrasegment spin interactions.

All three dispersion regions occurring in the frame of component B of chain dynamics can be described in a consistent way based on the same elementary theory. Region III very remarkably is determined by the effect of a certain relaxation mechanism in combination with segment self-diffusion properties as independent phenomena. The consistency of the interpretation of the three proton dispersion regions I, II, and III with the aid of the renormalized Rouse theory is striking. Quantitative evaluations of proton spin-lattice relaxation times on this basis can be found elsewhere [157].

The slope of the deuteron spin-lattice relaxation dispersion curves representing intrasegment interactions is characterized by the exponent 0.34 (Figs. 35, 36, and 42b). This is somewhat larger than the exponent 0.25 found in the average for the proton relaxation dispersion in region II (Eq. 168) interpreted above also to be due to intrasegment interactions alone. This discrepancy may be due to some influence of intersegment dipolar interactions on the proton relaxation dispersion already in region II.

Actually, taking the first nearest neighbor peak (first segment coordination shell) of the radial segment pair correlation function (see Fig. 38 and compare Ref. [69]) into account in a numerical evaluation permitted Herman recently to predict a corresponding effect [158]. One may conclude that dispersion region II of proton relaxation represents a superposition of intrasegment and short-range intersegment relaxation mechanisms, whereas the proton dispersion region III is dominated by long-range intersegment interactions. Here "short" and "long-range" are meant relative to the radius of the first segment coordination shell (Fig. 38).

4.5.3 Mesomorphic Phases

Linear polydiethylsiloxane (PDES) and its higher homologues form ordered mesophases in temperature intervals between the solid and the isotropic melt phases [159, 160]. This mesophase is to be juxtaposed to monomeric [161–166] or main-chain polymeric nematic liquid crystals [167]. The results to be described in the following demonstrate the sensitivity of NMR techniques to microstructural changes.

Figure 32 shows spin-lattice relaxation dispersion data for PDES (having -CH_2CH_3 as side groups) both in the isotropic and in the mesomorphic phases [155]. The dispersion of the isotropic melt is governed by the same empirical power laws for regions I and II as stated before (Eq. 168) and in particular as measured in PDMS melts (side groups: -CH_3).

Two power law regimes were also found in the mesophase, but with larger exponents and a crossover frequency shifted by a factor of about 10 (after correction for the different temperatures) to a lower value. In summary, the power laws observed in PDES in broad frequency ranges are (see Fig. 32 and Ref. [155])

$$T_1 \propto \begin{cases} M_w^0 \omega^{0.50\pm0.05} & \text{(region I "isotropic phase")} \\ M_w^0 \omega^{0.73\pm0.05} & \text{(region I "mesophase")} \\ M_w^0 \omega^{0.25\pm0.05} & \text{(region II "isotropic phase")} \\ M_w^0 \omega^{0.45\pm0.05} & \text{(region II "mesophase")} \end{cases} \qquad (182)$$

The slopes in region II for both the isotropic and the mesomorphic phases are in accordance with data for the spin–lattice relaxation time in the rotating frame, $T_{1\rho}$.

These results for the ordered mesophase are to be compared with the spin–lattice relaxation dispersion expected for nematic liquid crystals where a power law

$$T_1 \propto \nu^{0.5} \quad \text{(ODF)} \tag{183}$$

is predicted for order director fluctuations (ODF) at low frequencies [161–167]. With nematic compounds this law and the collective fluctuations in the ordered state it represents should show up irrespective of whether it is a monomeric or polymeric substance.

This is in contrast to the mesophase of PDES. The monomers of this compound do not show any order because they do not contain mesogenic groups. That is, the order observed in the PDES mesophase must have a thermodynamic background different from that of nematic phases.

It is concluded that the modified power laws for spin–lattice relaxation dispersion in the mesophase reflect a modified behavior of chain modes rather than collective fluctuations of ensembles of molecules in ordered domains. This conclusion is corroborated by the identical frequency dispersion of the spin–lattice relaxation times T_1 and $T_{1\rho}$ in the laboratory and in the rotating frames, respectively. If the order in the PDES mesophase would be of a nematic nature and the fluctuations causing dispersion region II consequently would be of the ODF type, the frequency dispersion of $T_{1\rho}$ should vanish while that of T_1 is retained [22].

In this context, a study of the dipolar correlation effect in the mesophase of PDES is of particular interest [168]. Using this technique, it was concluded that the sample consists of defect-enriched areas and ordered domains. The two dynamic states are connected with segments of different mobility, and they fluctuate temporally and spatially. The exchange times between the two mobility states range between 0.1 and 1.0 s in the mesomorphic temperature range.

The exchange process between the two environments is slow compared with the transverse relaxation rates so that the two signal components can clearly be distinguished. This is in contrast to spin–lattice relaxation where flip-flop spin diffusion as an additional exchange mechanism mediates averaging over all heterogeneities. Spin–lattice relaxation curves therefore decay mono-exponentially in a wide range [155], whereas the transverse magnetization is subject to a superposition of virtually two exponential functions [168]. As concerns the different power law regimes of spin–lattice relaxation given in Eq. 182, it is not yet known to which of the two segment phases they must be attributed or to what degree exchange averaging matters in this respect.

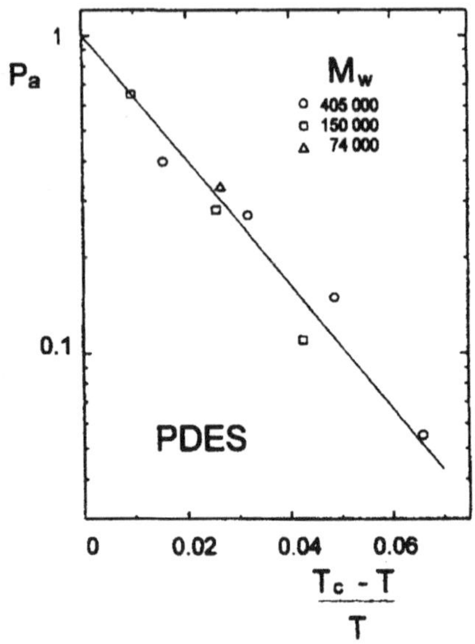

Fig. 39. Fraction of the more mobile segments in the mesophase of polydiethylsiloxane (PDES) as a function of the reduced temperature. The *solid line* represents Eq. 184

The dipolar correlation effect is based on stimulated-echo signals [2] and therefore contains elements both of transverse and longitudinal relaxation. Flip-flop spin diffusion as well as material transport may contribute to the exchange mechanism identified this way. An interesting result of the dipolar correlation effect study is the temperature dependence of the fraction of the more mobile segments which was shown to obey the empirical law (see Fig. 39) [168]

$$P_a = \exp\left\{45\frac{T - T_c}{T}\right\}, \tag{184}$$

where T_c is the (molecular weight dependent) isotropization temperature. Approaching T_c from below increases the fraction of the more mobile segments exponentially as a function of the reduced temperature.

Figure 40 shows an attempt to visualize how the areas of different mobility could arise based on "collapsed" coils. The "gain" in chain mobility in the ordered mesophase is attributed to conformations containing many chain folds compensating the dynamic restrictions entailing the almost nematic-like order in this phase (compare the discussion in context with chain-end dynamics and the statistical physics treatment of chain folds in Ref. [124]).

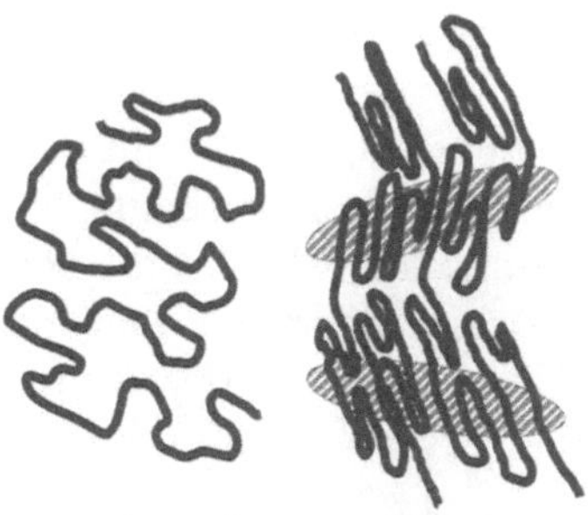

Fig. 40. Attempt to schematically represent the ordered mesophase of PDES juxtaposed to the isotropic phase. The "collapsed" coil structure assumed in the mesophase entails chain folds that provide additional motional degrees of freedom and is therefore thermodynamically stabilized in a certain temperature range (compare the statistical physics treatment in Ref. [124]). The *shaded ellipses* indicate somewhat closer packed, more ordered and less mobile areas in contrast to the more mobile, more distorted chain conformations in regions where many folds accumulate

4.5.4 Solutions

The features of entangled polymer dynamics fade upon dilution by a low-molecular solvent. In other words, the critical molecular weight characteristic for the crossover between Rouse-like and entangled dynamics is growing with the degree of dilution (see Eq. 162). NMR relaxation measurements turned out to be particularly suitable for studies of the dilution process. The crossover from the melt to the dilute solution was already demonstrated with the T_2/T_1 ratio in Fig. 20. It can also be visualized with the aid of spin–lattice relaxation dispersion of PDMS dissolved in CCl_4 as a proton-free solvent. Figure 31d shows corresponding data.

Compared to the region II dispersion occurring as one of the characteristics of melts of entangled polymers the spin–lattice relaxation dispersion becomes flatter upon dilution. With a PDMS content of 15% in CCl_4 the data can be described by the Rouse model as the incarnation of free-chain dynamics (see Fig. 31d). This may appear surprising since the critical molecular weight for this concentration according to Eq. 162 is below the weight-average molecular weight of the polymer examined, M_w=423,000, so that entangled dynamics should still apply.

On the other hand, all polymer dynamics models unspecifically predict some short-time/high-frequency limit where Rouse-like behavior should dominate (see Tables 1–3). In melts of entangled polymers such a regime could never be identified by spin–lattice relaxation dispersion because component B in the form of the high-mode-number limit (region I) appears to overlap with the local segment fluctuation regime (manifesting itself as component A). That is, there is no intermediate time or frequency window left in entangled melts where Rouse dynamics could develop in full.

Such a dynamic gap obviously arises in solutions if the polymer concentration is low enough. Component A representing local segment fluctuations becomes accelerated upon dilution, whereas the entanglement effect on the chain modes is shifted to correspondingly longer length and time scales. That is, a very important theoretical requirement is satisfied in this way. We will come back to the topological constraint problem later in the review.

4.5.5 Networks

Permanent or thermoreversible cross-links mediate the opposite effect on chain dynamics compared with dilution by a solvent: instead of releasing topological constraints by dilution, additional hindrances to chain modes are established by the network. With respect to NMR measurands relatively large cross-link densities are needed to affect chain modes visible in the experimental time/frequency window, as demonstrated with proton spin–lattice relaxation dispersion of polyethylene cross-linked by 10-Mrad irradiation with electron beams [123] and with styrene–butadiene rubbers [29]. However, there is a very strong effect on the dipolar correlation effect which probes much slower motions and can therefore be used favorably for the determination of the cross-link density [29, 176, 177].

In the plot shown in Fig. 31e, spin–lattice relaxation dispersion curves of permanently cross-linked PDMS are shown. With decreasing mesh length, the chain modes appear to be shifted to lower frequencies. This is indicated by lower values of the relaxation times while the dispersion slopes of regions I and II are retained. On the other hand, the effect on the crossover frequency is minor.

The fact that the dispersion regions I, II, and III of proton spin–lattice relaxation retain their qualitative appearance in the presence of (in this case thermoreversible) cross-links is demonstrated in Fig. 34. The influence of fluctuating cross-links on the dispersion curves can be explained by an effectively modified monomeric friction coefficient [131]. A result of particular interest is the shift of the crossover frequencies between dispersion regions I, II, and III. The shifts can be explained by the ratios

$$\frac{\nu_{I,II}}{\tilde{\nu}_{I,II}} = \frac{\nu_{II,III}}{\tilde{\nu}_{II,III}} = \frac{\tilde{\tau}_s}{\tau_s}, \tag{185}$$

where the quantities with tilde refer to networks. The friction effect on segment fluctuations equally slows down chain modes at the lowest frequencies.

4.6
Translational Segment Diffusion and Spin Diffusion

The NMR diffusometry techniques described in this article are frequently and suitably used for tests of the limits for the mean squared segment displacement predicted by polymer theories. The time and length scales probed by the techniques can be adjusted by the corresponding pulse intervals and the field gradient strength (the "wave number"), respectively. In this way, one gets access to the center-of-mass diffusion limit as well as to the anomalous-segment diffusion regime. The root mean squared displacements examined in the former case are larger than the coil dimension, whereas the latter limit is probed if the displacements are shorter than the coil extension.

4.6.1
Center-of-Mass Diffusion

NMR diffusometry is suitable for studies of the center-of-mass diffusion coefficient provided that the diffusion time can be chosen longer than the terminal relaxation time. In terms of the tube/reptation model, this is limit $(IV)_{DE}$ defined by $t >> t_d$. Eq. 75 suggests $D \propto M_w^{-2.0}$ which appears to have been verified several times in experiment [8, 148–150, 169–171]. Figure 41 shows polydimethylsiloxane melt data as a typical example [8]. Data for other polymer melt examples reported in the literature refer to polystyrene and polyethylene at different temperatures [169, 148, 149], and polyethyleneoxide [11, 170].

It appears that all those data perfectly corroborate the validity of limit $(IV)_{DE}$ of the tube/reptation model. However, there are four reasons why the molecular weight dependence observed in experiments may be modified by factors other than the tube constraints:

- The condition $t >> t_d$ is difficult to achieve with high molecular weights so that the anomalous segment displacement regime may gain an increasing influence from low to high molecular weights, as was clearly demonstrated at least with the two highest molecular weights in the plot shown in Fig. 41 [8]. The consequence is an apparently *weaker* molecular weight dependence. The crossover regime between the center-of-mass and the anomalous segment displacement regimes was extensively studied [172] for polystyrene solutions. It appears that the center-of-mass diffusion coefficient cannot be evaluated reliably by the field-gradient technique without considering the crossover to the anomalous diffusion regime at the same time.
- Free volume connected with chain-end blocks influences the molecular weight dependence of chain dynamics in the vicinity of M_c. An apparently *steeper* molecular weight dependence results on these grounds particularly below M_c. The failure of NMR techniques to directly render the molecular

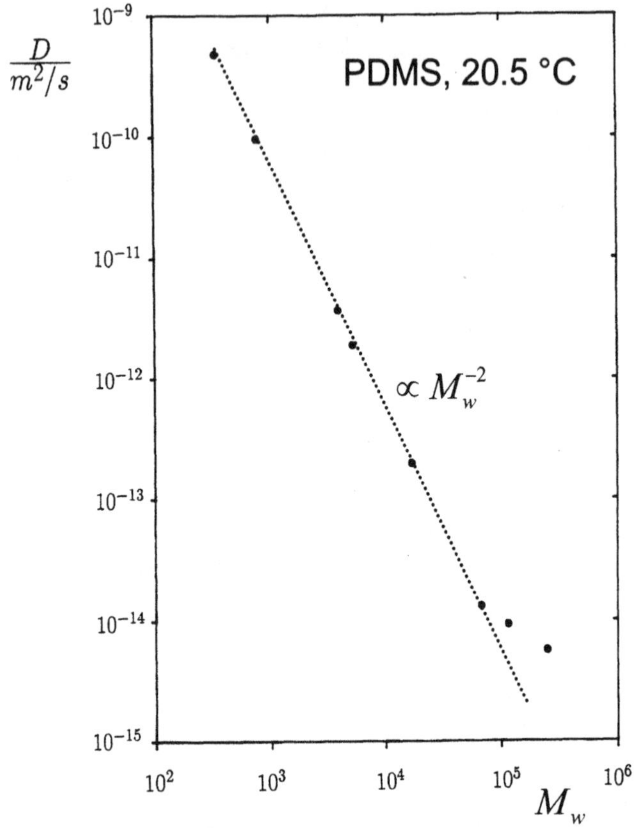

Fig. 41. Center-of-mass diffusion coefficient in polydimethylsiloxane (PDMS) melts at 20.5 °C [8] as a function of the molecular weight. This self-diffusion coefficient is relevant for root mean square displacements exceeding the random-coil dimension. The experimental condition is that the echo attenuation is recorded with small field-gradient strengths and correspondingly long diffusion times. The *straight line* represents the law $D \propto M_w^{-2}$ which seemingly coincides with the prediction by the tube/reptation model for limit $(IV)_{DE}$ (see Eq. 75). The deviation from the M_w^{-2} dependence above M_w=100,000 is due to the fact that in this case the diffusion time could not be chosen long enough to ensure center-of-mass diffusion. Actually we are here already in the anomalous segment diffusion regime. For M_w=250,000, for instance, the root mean square displacement reached along the gradient direction in a diffusion time of 30 ms is only 20 nm compared with a radius of gyration of 25 nm

weight dependence $D \propto M_w^{-1.0}$ predicted for the Rouse model (see Eq. 59) was explained in this way [148, 149].

- At very high molecular weights, when the diffusion coefficient due to flip-flop spin diffusion of the order 10^{-15} m^2/s is getting comparable to the ordinary Brownian self-diffusion coefficient, a *flatter* chain length dependence of the effective diffusion coefficient is expected [10].

- If very broad, the molecular weight distribution may influence the (average) center-of-mass diffusion coefficient evaluated from echo attenuation curves [169, 175].

In view of the contour-length fluctuation mechanism suggested by Doi as an explanation of the fractional exponent of the power law for the zero-shear viscosity [142], $\eta \propto M_w^{3.4}$, a molecular weight dependence somewhat stronger than predicted by Eq. 75 was predicted also for the center-of-mass diffusion coefficient [169]. More recent studies mainly carried out with concentrated solutions even suggest power laws with universally stronger molecular weight dependences than predicted by the tube/reptation model [172–174]:

$$D \propto M_w^{-2.4\pm0.1} \phi^{-1.8\pm0.2},$$

$$\eta \propto M_w^{3.4\pm0.1} \phi^{3.8\pm0.2}, \tag{186}$$

where ϕ is the volume fraction of the polymer.

4.6.2 Anomalous Segment Diffusion

Because of the relatively short displacement time or length scales typically probed by NMR diffusometry, it is particularly well suited to detect anomalies in the segment displacement behavior expected on a time scale shorter than the terminal relaxation time, that is for root mean squared displacements shorter than the random-coil dimension. All models discussed above unanimously predict such anomalies (see Tables 1–3). Therefore, considering exponents of anomalous mean squared displacement laws alone does not provide decisive answers. In order to obtain a consistent and objective picture, it is rather crucial to make sure that (i) the absolute values of the mean squared segment displacement or the time-dependent diffusion coefficient are compatible with the theory, (ii) the dependence on other experimental parameters such as the molecular weight are correctly rendered, and (iii) the values of the limiting time constants are not at variance with those derived from other techniques.

For instance, the segmental friction coefficient or the mean squared end-to-end distance can be determined directly from viscoelasticity or neutron scattering data. Estimating segment diffusion with the aid of the tube/reptation model based on such data leads to unrealistic predictions. The predicted segment displacements tend to be much too large [9, 12, 178]. Figure 42 shows polyethyleneoxide data as an example. One identifies a time dependence $D(t) \propto t^{-0.5}$ at short diffusion times in contrast to the law $D(t) \propto t^{-3/4}$ predicted for $\tau_e << t << \tau_R$ as limit $(\mathrm{II})_{\mathrm{DE}}$ (see Table 1) apart from an inconsistent molecular weight dependence [12, 172]. The incompatibility becomes particularly obvious if the echo attenuation curves are evaluated in full

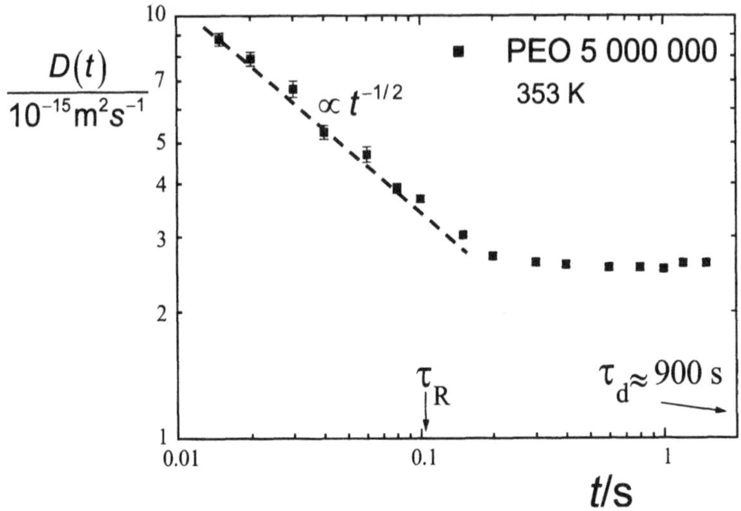

Fig. 42. Time-dependent diffusion coefficient measured in a polyethyleneoxide melt (M_w=5,000,000) at 353 K as a function of the diffusion time [12]. The data were evaluated according to Eq. 27. The Rouse relaxation time τ_R and the tube disengagement time τ_d predicted by the tube/reptation model based on the neutron scattering value of the ratio $\langle R^2_{ee}\rangle/M_w$=1.01×10^{-20} m^{2}mol/g [179, 180] are indicated. The *broken line* represents the power law $D(t)\propto t^{-1/2}$ which, according to the tube/reptation model, should appear in the time interval *above* τ_R as limit (III)$_{DE}$ and not below (see Table 1). The time range where this power law appears and the value for τ_R estimated on the basis of the tube/reptation model are not consistent with each other. The plateau of the experimental data for $t >> 0.1$ s is due to flip-flop spin diffusion which physically limits the detection of molecular displacements by NMR diffusometry

(compare Fig. 44) using the formalism specifically derived for the tube/reptation model (see Eqs. 76–79). In the frame of the experimental accuracy, it is not possible to describe the whole family of echo attenuation curves on the basis of the unmodified predictions in the full diffusion time range for all molecular weights at the same time [9, 12, 178], even if flip-flop spin diffusion (see the following section) is taken into account.

On the other hand, if a static "tube" really exists as anticipated in the tube/reptation model, the predictions for anomalous segment diffusion are adequate, of course, and can be verified in experiment. This will be demonstrated by considering linear polymers confined in nanopores.

4.6.3 Flip-Flop Spin Diffusion

Flip-flop spin diffusion is a phenomenon specific for NMR. It does not matter with other techniques. Spin diffusion based on Zeeman energy conserving flip-flop transitions of dipolar coupled spins, that is interchange of spin-up and spin-down states, is normally negligible compared to molecular self-diffusion in liquids. However, with viscous systems with little motional averaging of dipolar coupling and with the aid of very strong field gradients, NMR diffusometry is able to detect this immaterial transport mechanism competing with molecular diffusion. Spin echoes can be attenuated on this basis just as with any other incoherent displacement process [181].

Employing the strong fringe-field gradient of a superconducting magnet, it was possible to demonstrate the influence of flip-flop spin diffusion on the echo attenuation of polymer melts [10, 12]. The experiments were carried out by diluting the polymer under investigation isotopically. That is, undeuterated polymers were dissolved in a perdeuterated matrix of polymers of equivalent molecular weight. The reduction of intermolecular dipolar interactions diminishes the proton spin flip-flop rate as expected.

The effect is also evident in Fig. 42 referring to a polyethyleneoxide melt with an extremely large molecular weight, M_w=5,000,000. The time-dependent diffusion coefficient was evaluated in the low wave number approximation, Eq. 27, according to $\langle R^2(t)\rangle=6D(t)t$. The time-independent plateau appearing at times $t >> 0.1$ s was shown to be due to flip-flop spin diffusion [10, 12].

As a consequence, the proper evaluation of segment diffusion at long diffusion times turned out to be more complicated than often anticipated. However, even with flip-flop spin diffusion theoretically taken into account, it is not possible to fit the formulas predicted by the tube/reptation model to the experimental data with respect to both the time and molecular weight dependences in a consistent way and without assuming unrealistic parameters [12].

5
Chain Dynamics in Pores ("Artificial Tubes")

The motivation to study the dynamics of polymer chains confined in nanoporous materials with more or less rigid pore walls is twofold. Firstly, there may be important technological applications requiring knowledge on the dynamic behavior of polymers under such conditions [182]. The second reason making this field intriguing for polymer science is the possibility to study chain dynamics under topological model constraints in the form of artificial tubes. In the following we will focus on this latter point.

The tube introduced in the frame of the Doi/Edwards reptation model for the treatment of bulk systems of entangled polymers is a fictitious one. As outlined above, the laws predicted on this basis (see Table 1) can provide only a rather crude picture failing to account for numerous experimental findings quantitatively as well as qualitatively. It may therefore be helpful to study chain dynamics in tube-like pores of a physically real nature.

The diffusion and relaxation behavior of diverse oligomers and polymers has been studied in nanoporous silica glasses [183]. Both measuring techniques provide clear evidence for modified chain dynamics compared with the bulk. However, the problem with this sort of system is that interactions of the polymers with the pore walls play an important role and must be distinguished from the geometrical confinement effect. A corresponding analysis was possible by comparison of polymer data with data obtained with short oligomers of the same chemical species. In contrast to the polymers, the oligomers were assumed to be subject to adsorption but not to modifications of the chain modes by geometrical confinement. Subtraction of the oligomer relaxation rates from those measured with polymers in the pores suggests a spin–lattice relaxation dispersion reduced for the influence of geometrical confinements on the chain-mode distribution. Actually a power law dispersion reproducing that predicted for limit $(\mathrm{II})_{\mathrm{DE}}$ of the tube/reptation model (see Table 1) could be elucidated this way.

A more direct verification of tube/reptation features was possible with systems where the solid matrix and the mobile polymer chains confined to nanopores are of similar organic chemical composition. In the experiments referred to in the following, linear polymers were confined in a solid, that is strongly cross-linked, polymer environment. Under such conditions, the geometry effect can be expected to dominate whereas the wall adsorption phenomenon is of negligible influence.

This sort of system was prepared in the form of so-called semiinterpenetrating networks. Preparation details are described in Ref. [184]. The matrix

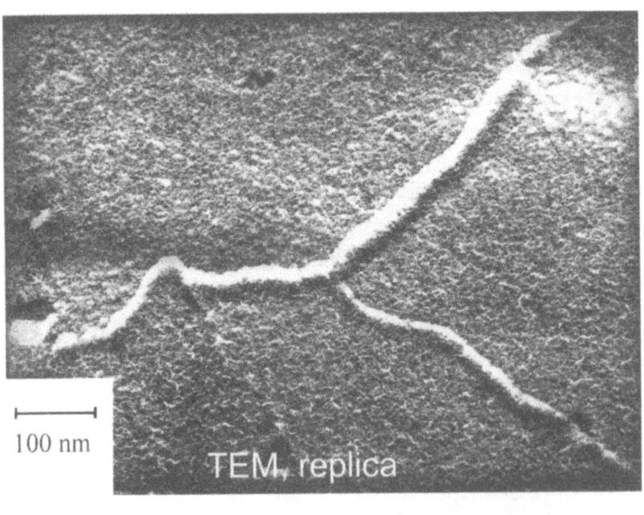

Fig. 43. Transmission electron micrograph of a replica of a freeze-fractured surface of polyethyleneoxide M_w=6,000 in PHEMA [11]. The pore channels have a width of about 10 nm

consisted of cross-linked polyhydroxyethyl methacrylate (PHEMA). Linear polyethyleneoxide (PEO) was incorporated in the 10-nm pores. The molecular weights of the PEO samples used in experiment were chosen large enough to ensure that the root mean squared random-coil diameter in bulk exceeds the pore diameter. Figure 43 shows an electron micrograph of pore channels in this material having a width of about 10 nm.

5.1 Diffusion in Pores

Segment diffusion in pores suggests itself as a typical model scenario representing the premisses of the tube/reptation model. NMR diffusometry is suitable to probe the time or length scales of the Doi/Edwards limits $(II)_{DE}$, $(III)_{DE}$ and beyond. Since the wall adsorption effect in the PHEMA system is expected to be negligible, one can therefore expect that the reptation features of the anomalous segment diffusion regime, especially with respect to limit $(III)_{DE}$, are faithfully rendered by the experiments.

Figure 44 shows typical echo attenuation curves recorded with PEO in a solid PHEMA matrix [186]. The good coincidence of the theoretical curves calculated on the basis of Eqs. 76–79 for all accessible values of the experimental parameters *k* and *t* for different molecular weights [11] corroborates the validity of the tube/reptation model for linear polymers confined to arti-

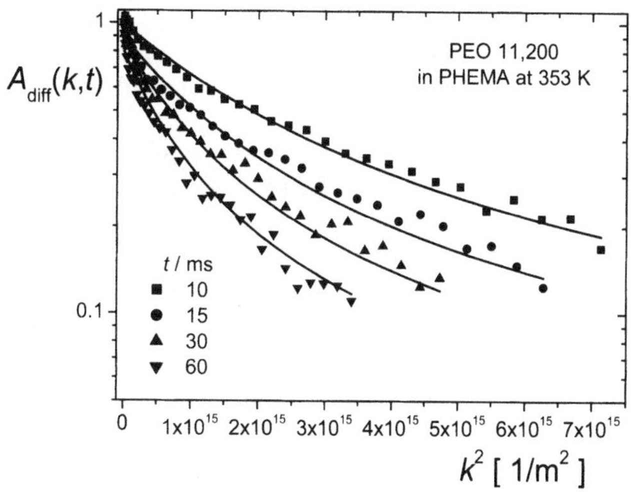

Fig. 44. Echo attenuation curves for polyethyleneoxide, M_w=11,200, confined to PHEMA pores at 80 °C as a function of the squared wave number, k^2, for different diffusion times. The *solid lines* represent a fit of Eqs. 76–79 to the experimental data. The tube diameter is found to be a=(9±1) nm as the only fitting parameter. Other parameter values such as $N=M_w/853$, $D_0=9.66\times10^{-10}$ m^2/s, and $b=8.38\times10^{-10}$ m were taken from the literature [180, 187]

ficial nanopores in a solid matrix. Very remarkably the tube diameter a (as the only free parameter) fitted to the diffusion data coincides with the pore diameter determined with other techniques such as electron microscopy, X-ray scattering or NMR spin diffusion [184].

5.2 Spin–Lattice Relaxation Dispersion in Pores

Field-gradient NMR diffusometry experiments tend to be largely dominated by the time and length scales of limit $(\mathrm{III})_{\mathrm{DE}}$. On the other hand, the much shorter time scale of limit $(\mathrm{II})_{\mathrm{DE}}$ can specifically be examined by NMR relaxometry, which under suitable conditions can indirectly probe translational diffusion of polymer segments. The laws predicted in Table 1 for anomalous segment diffusion on the one hand and spin–lattice relaxation on the other are based on the very same translational segment displacement mechanisms. They stipulate each other compellingly. Spin–lattice relaxation at low frequencies, as they can be covered with the aid of field-cycling NMR relaxometry, is therefore suitable to extend the range of field-gradient NMR diffusometry toward much shorter times.

The total proton frequency range that can be probed by NMR spin–lattice relaxation techniques is 10^3 Hz $< \nu <$ 10^9 Hz. For the present application, deuteron resonance selectively applied to perdeuterated polymers confined to the pores is superior to proton resonance which, unlike the situation in field-gradient experiments, is affected by signals from the matrix and flip-flop spin diffusion across the matrix. The deuteron frequency range is shifted by a factor of 0.15 to lower frequencies. This frequency window largely matches the time scale of chain modes of polymers with medium molecular masses.

The exceptional frequency and molecular weight dependence for the spin–lattice relaxation time expected for limit $(\mathrm{II})_{\mathrm{DE}}$,

$$T_1 \propto M^0 \omega^{3/4} \quad (\omega\tau_s \ll 1 \ll \omega\tau_e) \tag{187}$$

(see Table 1), is particularly indicative for reptational segment diffusion. Figure 45a shows the frequency dependence of the deuteron spin–lattice relaxation time of perdeuterated PEO of different molecular weights in a solid PHEMA matrix [95,185]. In the frequency regime, where the standard Bloch/Wangsness/Redfield relaxation theory is applicable, the law given in Eq. 187 is reproduced as

$$T_1 \propto M_w^{0\pm0.05} \nu^{0.75\pm0.02}. \tag{188}$$

The experimental deuteron frequency range in which this frequency dependence was observed is 5×10^5 Hz $< \nu <$ 6×10^7 Hz. De Gennes' prediction for limit $(\mathrm{II})_{\mathrm{DE}}$ has thus been verified for polymers confined in artificial

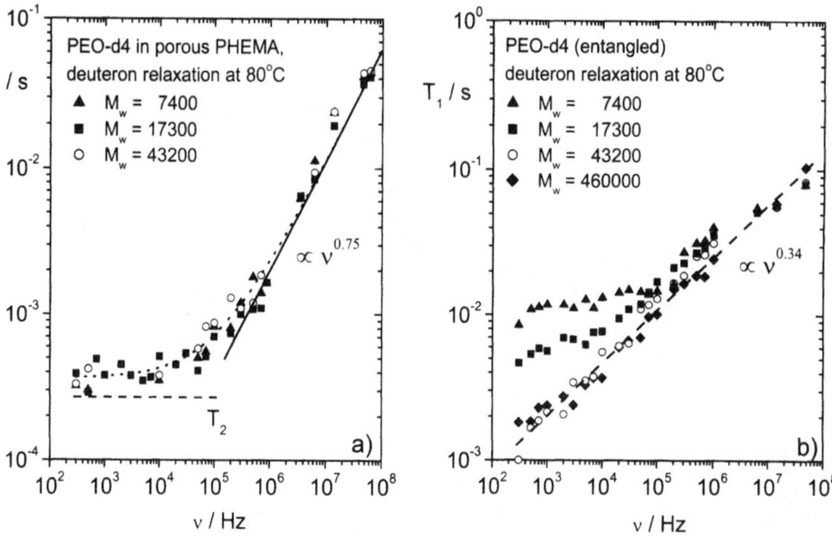

Fig. 45a, b. Frequency dependence of the deuteron spin-lattice relaxation time of perdeuterated PEO confined in 10-nm pores of solid PHEMA at 80 °C (**a**) and in bulk melts (**b**) [95, 185]. The dispersion of the confined polymers verifies the law $T_1 \propto M_w^0 \omega^{0.75}$at high frequencies as predicted for limit $(II)_{DE}$ of the tube/reptation model (see Table 1). The low-frequency plateau observed with the confined polymers indicates that the correlation function implies components decaying more slowly than the magnetization relaxation curves, so that the Bloch/Wangsness/Redfield relaxation theory [2] is no longer valid in this regime. The plateau value corresponds to the transverse relaxation time, T_2, for deuterons extrapolated from the high-field value measured at 9.4 T

tubes in full accordance with the theory. This is in contrast to the bulk behavior of the very same polymers studied with deuteron resonance again. The data in Fig. 45b reproduce features observed with proton and deuteron resonance with bulk melts of other polymer species as discussed above (see Figs. 35 and 36). Polymer chains in bulk melts must obviously have additional degrees of freedom not existing under pore confinements.

5.3 Theoretical Crossover from Rouse to Reptation Dynamics

The theoretical background of the confinement effect in (artificial) tubes was recently examined in detail with the aid of an analytical theory as well as with Monte Carlo simulations [70]. The analytical treatment referred to a polymer chain confined to a harmonic radial tube potential. The computer simulation mimicked the dynamics of a modified Stockmayer chain in a tube with “hard” pore walls. In both treatments, the characteristic laws of the tube/reptation model were reproduced. Moreover, the crossover from reptation (tube diameter equal to a few Kuhn segment lengths) to Rouse dy-

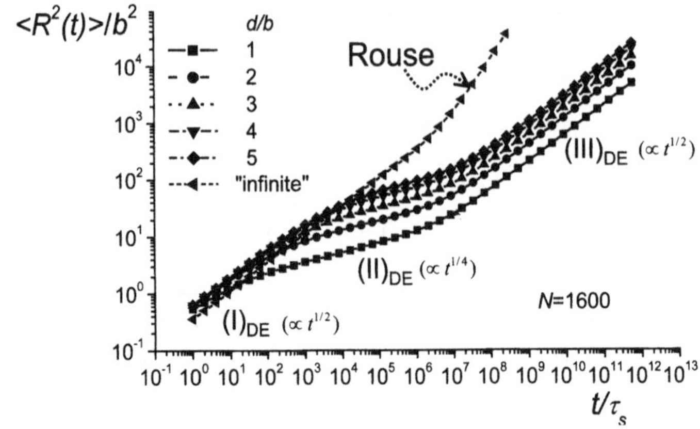

Fig. 46. Theoretical mean squared segment displacement of a chain confined in a randomly coiled tube versus time according to the harmonic radial potential theory [70]. The tube diameter d is given in multiples of the Kuhn segment length b. The crossover tendency to free, unconfined Rouse chain dynamics with increasing tube diameter is obvious. The mean squared displacement is given in units b^2, the diffusion time t in units of the segmental fluctuation time τ_s. The chain length was assumed to be N=1,600 Kuhn segments. The three "anomalous" Doi/Edwards limits (see Table 1) are reproduced with finite tube diameters

namics (tube diameter "infinite") was demonstrated by varying the tube diameter.

Figure 46 shows the time dependence of the mean squared segment displacement as predicted by the harmonic radial potential theory [70]. The three anomalous diffusion limits, $(I)_{DE}$, $(II)_{DE}$, and $(III)_{DE}$, of the tube/reptation model are well reproduced. Note the extended width of the transition regimes between these limits, which should be kept in mind when discussing experimental data with respect to a crossover between different dynamic limits. Increasing the effective tube diameter is accompanied by the gradual transition to Rouse-like dynamics of an unconfined chain (where entanglement effects are not considered).

The spin-lattice relaxation dispersion was derived with the same sort of theory. Figure 47 shows data obtained under the same conditions as assumed for the mean squared displacement data in Fig. 46. Again the dispersion most specific for the tube/reptation model, namely $T_1 \propto \omega^{0.75}$ which is predicted for limit $(II)_{DE}$, was perfectly reproduced at low frequencies.

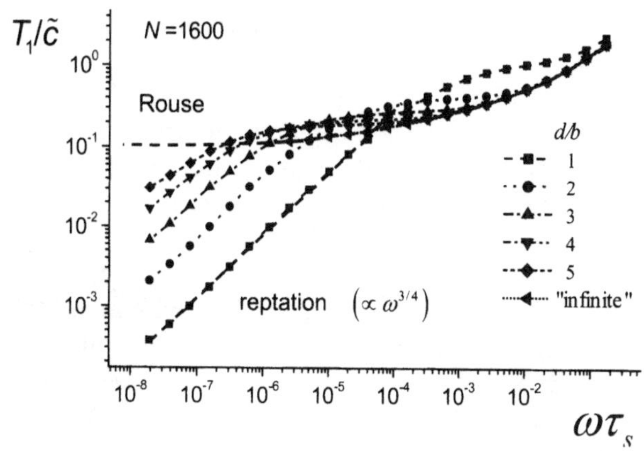

Fig. 47. Spin–lattice relaxation dispersion for a chain of N=1,600 Kuhn segments (of length b) confined to a randomly coiled tube with a harmonic radial potential with varying effective diameters $\dot{d}$. The data were calculated with the aid of the harmonic radial potential theory [70]. $\tilde{c}$ is a constant. At low frequencies the curves visualize the crossover from Rouse dynamics depending on the effective tube diameter. The latter case is described by a T_1 dispersion proportional to $\omega^{3/4}$ characteristic for limit $(II)_{DE}$ of the tube/reptation model

6 Concluding Remarks

NMR may be considered as a particularly versatile class of investigation methods suitable for polymer dynamics studies. In particular, field-cycling NMR relaxometry employed in parallel with field-gradient NMR diffusometry is an unmatched combination of methods for the elucidation of features of chain dynamics. A great wealth of clearly documented phenomena for different measurands, experimental parameters, and different polymer species in different environments and under different thermodynamic conditions were discovered, and appear to be suitable for direct comparisons with theoretical model predictions. One can say that the experimental data available so far already render a closed picture from the empirical point of view.

Transverse and longitudinal NMR relaxometry of ^1H and ^{13}C nuclei was shown to be governed by three components of chain dynamics. With respect to tests of theoretical model concepts, the most important one is component B which reflects the influence of chain modes and can be probed with the field-cycling technique.

Field-gradient NMR diffusometry is suitable for recording translational displacement properties by self-diffusion. This in particular refers to the anomalous segment displacement regime. Center-of-mass diffusion is also accessible this way but must be handled with some care because of the experimental limitation of the maximum diffusion time. Furthermore the influence of flip-flop spin diffusion must be taken into account when examining long polymers.

Three basic model theories have been considered: the *Rouse model*, the *tube/reptation concept*, and the *renormalized Rouse formalism*. Depending on the sample system and the experimental conduct, characteristic features of all these theories have been shown. Some spectacular predictions of theories were verified, others were ruled out.

The frequency dependence of the spin–lattice relaxation time predicted by the *Rouse model* was verified with polymer melts for $M_w<M_c$ in accordance with the segmental fluctuation time determined from the T_1 minimum condition. Melts with $M_w>>M_c$ did not show any such behavior in the short-time limit although predicted so by all three model theories. On the other hand, dilution of the polymer by a low-molecular solvent increases M_c and diminishes the entanglement effect. Under such conditions, Rouse-like behavior was observed.

The *tube/reptation model* of entangled polymer dynamics suggests a number of peculiar anomalous-diffusion limits that were consistently and quantitatively verified in artificial tubes of a solid porous matrix with the aid of field-cycling NMR relaxometry and field-gradient NMR diffusometry. This clear corroboration of a model prediction is however in contrast to the dynamics concluded for bulk melts. In that case neither the prediected T_1 dispersion nor the anomalous segment diffusion behavior was found with respect to the frequency and molecular weight dependences. The assumption of static fictitious tubes as a replacement scheme for the entangling matrix effect is obviously too crude for bulk melts, and does not render the features of polymer dynamics correctly.

Based on the generalized Langevin equation, the *renormalized Rouse models* suggest dynamic high- and low-mode-number limits as an implicit structural feature of this equation of motion. This is a stand-alone prediction of paramount importance independent of any absolute values of power law exponents that arise and are measured in the formalism and in experiment, respectively. The two limits manifesting themselves as power law spin–lattice relaxation dispersions were clearly identified in bulk melts of entangled polymers of diverse chemical species.

Apart from these two relaxation dispersion regimes, which are mainly based on intrasegment spin interactions (and possibly nearest neighbor intersegment contributions), a third dispersion region specific for dipolar-coupled nuclei exists at low frequencies. This was shown to be due to (long-distance) dipolar interactions being relevant for protons but not for deuterons.

A most important finding in general and for establishing of model theories is the fact that segment dynamics is affected by microstructural details already at relatively short times. The local fluctuations of a Kuhn segment typically occur on a time scale shorter than 10^{-10} s and on a length scale of up to 1 nm. It was shown that factors such as chain entanglements, cross-linking or mesomorphic order quite amazingly, and contrary to most model expectations, have an influence on segment dynamics already rather close to these time and length scales. This finding became evident from the relaxometry studies reported and was also observed in the literature [129]. On the other hand, chain modes with low mode numbers and even global, molecular weight dependent motions corresponding to chain relaxation times beyond the Rouse relaxation time do show up at the predicted time and length scales.

Acknowledgments We thank Elmar Fischer, Alexei Denissov, Margarita Kroutieva, Hans Wiringer, and Birgit Körner for discussions and excellent assistance in the preparation phase of this article. Grants by the DFG, the Volkswagen-Stiftung, CRDF, and RFBR are gratefully acknowledged.

References

1. Doi M, Edwards SF (1986) The theory of polymer dynamics. Clarendon, Oxford
2. Kimmich R (1997) NMR tomography, diffusometry, relaxometry. Springer, Berlin Heidelberg New York
3. Fatkullin N, Kimmich R (1995) Phys Rev E 52:3273
4. Torrey HC (1956) Phys Rev 104:563
5. Stejskal EO, Tanner JE (1965) J Chem Phys 42:288
6. Tanner JE, Stejskal EO (1968) J Chem Phys 49:1768
7. Saarinen TR, Johnson CS (1988) J Magn Reson 78:257
8. Kimmich R, Unrath W, Schnur G, Rommel E (1991) J Magn Reson 91:136
9. Fischer E, Kimmich R, Fatkullin N (1996) J Chem Phys 104:9174
10. Fischer E, Kimmich R, Fatkullin N (1997) J Chem Phys 106:9883
11. Fischer E, Kimmich R, Beginn U, Möller M, Fatkullin N (1999) Phys Rev E 59:4079
12. Fischer E, Kimmich R, Fatkullin N, Yatsenko G (2000) Phys Rev E 62:775
13. Fujara F, Geil B, Sillescu H, Fleischer G (1992) Z Phys B 88:195
14. Kimmich R, Fischer E (1994) J Magn Reson A 106:229
15. Kimmich R, Fischer E, Callaghan P, Fatkullin N (1995) J Magn Reson A 117:53
16. Grinberg F, Kimmich R (1995) J Chem Phys 103:365
17. Abragam A (1961) The principles of nuclear magnetism. Clarendon, Oxford
18. Kimmich R (1980) Bull Magn Reson 1:195
19. Noack F (1986) Progr NMR Spectr 18:171
20. Kimmich R (1996) Molecular motions: T_1 frequency dispersion in biological systems. In: Grant DM, Harris RK (eds) Encyclopedia of nuclear magnetic resonance, vol 5. Wiley, Chichester, pp 3083–3088
21. Seitter R, Kimmich R (1999) Magnetic resonance: relaxometers. In: Lindon J, Trantar G, Holmes J (eds) Encyclopedia of spectroscopy and spectrometry. Academic, London, pp 2000–2008
22. Anoardo E, Grinberg F, Vilfan M, Kimmich R (2004) Chem Phys 297:99
23. Cohen-Addad JP (1993) Progr NMR Spectr 25:1
24. Guillermo A, Cohen-Addad JP, Bytchenkoff D (2000) Chem Phys 113:5098
25. Cohen-Addad JP, Guillermo A (2000) Phys Rev Lett 85:3432

26. Brereton MG (1990) Macromolecules 23:1119
27. Ball RC, Callaghan PT, Samulski ET (1997) J Chem Phys 106:7352
28. Callaghan PT, Samulski ET (1998) Macromolecules 31:3693
29. Fischer E, Grinberg F, Kimmich R, Hafner S (1998) J Chem Phys 109:846
30. Maxwell RS, Balazs B (2002) J Chem Phys 116:10492
31. Graf R, Heuer A, Spiess HW (1998) Phys Rev Lett 80:5738
32. Dollase T, Graf R, Heuer A, Spiess HW (2001) Macromolecules 34:298
33. Schnur G, Kimmich R (1988) Chem Phys Lett 144:333
34. Köpf M, Schnur G, Kimmich R (1988) J Polym Sci Polym Lett 26:319
35. Kimmich R, Schnur G, Köpf M (1988) Progr NMR Spectr 20:385
36. Kimmich R, Köpf M, Callaghan P (1991) J Polym Sci Polym Phys 29:1025
37. Volkenstein MV (1963) Configurational statistics of polymer chains. Interscience, New York
38. Birshtein TM, Pticin OB (1966) Conformation of macromolecules. Interscience, New York
39. Flory PG (1969) Statistical mechanics of chain molecules. Interscience, New York
40. Yamakava H (1971) Modern theory of polymer solutions. Harper and Row, New York
41. de Gennes PG (1979) Scaling concepts in polymer physics. Cornell University Press, Ithaka
42. Grosberg AYu, Khokhlov AR (1994) Statistical physics of macromolecules. AIP, New York
43. Gotlib YuYa, Darinskii AA, Svetlov YuE (1986) Physical kinetics of macromolecules. Khimia, Leningrad (in Russian)
44. Grosberg AYu, Khokhlov AR (1997) Giant molecules: here, there, and everywhere... Academic, San Diego
45. Kargin VA and Slonimskii GL (1949) Journal Fizhimii (USSR) 23:563
46. Rouse PJ (1953) J Chem Phys 21:1272
47. Fatkullin N, Kimmich R, Weber HW (1993) Phys Rev E 47:4600
48. Kimmich R, Fatkullin N, Weber HW, Stapf S (1994) J Non-Cryst Solids 172–174: 689
49. Fatkullin N, Kimmich R (1994) J Chem Phys 101:822
50. Khazanovich TN (1963) Polymer Sci USSR 4:727
51. Ullman R (1965) J Chem Phys 43:3161
52. Ferry JD (1980) Viscoelastic properties of polymers. Wiley, New York
53. Edwards SF (1967) Proc Phys Soc 92:9
54. de Gennes PG (1971) J Chem Phys 55: 572
55. Doi M, Edwards SF (1978) J Chem Soc Faraday Trans II 74:1789
56. Klein J (1978) Macromolecules 11:852
57. Doi MJ (1983) Poly Sci Poly Phys 21:667
58. Guenza M (2002) Phys Rev Lett 88:025901
59. Rubinstein M (1987) Phys Rev Lett 59:1946
60. de Cloizeaux (1988) J Europhys Lett 5:437
61. Graessley WW (1982) Adv Polym Sci 47:1
62. Pearson D (1987) Rubber Chem Tech 60:437
63. Marruci G (1985) J Polym Sci Polym Phys 23:159
64. Watanabe H, Tirrel M (1989) Macromolecules 22:927
65. Doi M, Pearson D, Kornfield J, Fuller G (1989) Macromolecules 22:1488
66. Fixman M (1991) J Chem Phys 95:1410
67. Gao J, Weiner JH (1992) Macromolecules 25:1348
68. Gao J, Weiner JH (1994) Science 266:748
69. Loriot G, Weiner JH (1998) J Polym Sci Polym Phys 36:143

70. Denissov A, Kroutieva M, Fatkullin N, Kimmich R (2002) J Chem Phys 116:5217
71. Kimmich R (1975) Polymer 16:851
72. Kimmich R (2002) Chem Phys 284:253
73. Ardelean I, Kimmich R (2003) Ann Rep NMR Spectr 49:43
74. Fatkullin N, Kimmich R (1999) JETP Lett 69:762
75. Fatkullin N, Kimmich R (1999) Macromol Symp 146:103
76. Padding JT, Briels WJ (2001) J Chem Phys 115:2446
77. Fatkullin N (2002) J Non-Cryst Solids 307–310:824
78. Skolnick J, Kolinski A (1990) Adv Chem Phys 78:223
79. Kremer K, Grest GS (1990) J Chem Phys 92:5057
80. Kremer K, Grest GS (1992) J Chem Soc Faraday Trans 88:1707
81. Pakula T (1996) Recent Res Devel Polym Sci 1:101
82. Binder K, Paul W (1997) J Polym Sci Polym Phys 35:1
83. Pakula T (1998) Computational Theor Polym Sci 8:21
84. Shaffer JS (1995) J Chem Phys 103:761
85. Binder K, Paul W (1997) J Polym Sci Polym Phys 35:1
86. Khalatur PG (1996) Computer simulations in polymer systems. In: Kuchanov SI (ed) Mathematical methods in contemporary chemistry. Gordon and Breach, New York
87. Padding JT, Briels WJ (2002) J Chem Phys 117:925
88. Paul W (2002) Chem Phys 284:59
89. Hermann MF (1990) J Chem Phys 92:2043
90. Hermann MF, Panajotova B, Lorenz KT (1996) J Chem Phys 105:1153
91. Douglas JF, Hubbard JB (1991) Macromolecules 24:3163
92. Chatterjee AP, Loring RF (1995) J Chem Phys 103:4711
93. Ngai KL, Phillies GDJ (1996) J Chem Phys 105:8385
94. Ngai KL (1999) Macromol Symp 146:117
95. Kimmich R, Seitter R-O, Beginn U, Möller M, Fatkullin N (1999) Chem Phys Letters 307:147
96. Zwanzig R (1974) J Chem Phys 60:2717
97. Bixon M, Zwanzig R (1978) J Chem Phys 68:1890
98. Schweizer KS (1989) J Chem Phys 91:5802
99. Schweizer KS (1989) J Chem Phys 91:5822
100. Schweizer KS (1993) Physica Scripta T49:99
101. Fuchs M, Schweizer KS (1997) J Chem Phys 106:347
102. Guenza M, Schweizer KS (1997) J Chem Phys 106:7391
103. Schweizer KS, Fuchs M, Szamel G et al. (1997) Macromol Theory Simul 6:1037
104. Guenza M (1999) J Chem Phys 110:7574
105. Fatkullin N, Kimmich R, Kroutieva M (2000) JETP 91:150
106. Genz U (1994) Macromolecules 27:3501
107. Genz U, Vilgis TA (1994) J Chem Phys 101:7101
108. Genz U, Vilgis TA (1994) J Chem Phys 101:7111
109. Zwanzig R (1960) J Chem Phys 33:1336
110. Zwanzig R (1961) Phys Rev 124:985
111. Mori H (1965) Progr Theor Phys 33:423
112. Mori H (1965) Progr Theor Phys 34:765
113. Hansen JP, McDonald (1991) Theory of simple liquids, 2nd edn. Academic, London
114. Baluccani U, Zoppi M (1994) Dynamics of the liquid state. Clarendon, Oxford
115. Berne BJ, Pecora R (1976) In: Berne BJ (ed) Dynamical light scattering. Wiley, New York
116. Edwards SF, Grant JW (1973) J Phys A Math Nucl Gen 6:1169

117. Edwards SF, Grant JW (1973) J Phys A Math Nucl Gen 6:1189
118. Ronca G (1983) J Chem Phys 79:1031
119. Fixman MF (1988) J Chem Phys 89:3892
120. Hess W (1988) Macromolecules 21:2620
121. Rostiashvili VG (1990) Sov Phys JETP 70:563
122. Pokrovskii VN (1992) Sov Phys Usp 35:384
123. Kimmich R, Bachus R (1982) Coll Polym Sci 260:911
124. Kimmich R, Köpf M (1989) Progr Coll Polym Sci 80:8
125. Weber HW, Kimmich R (1993) Macromolecules 26:2597
126. Gotlib YY, Torchinskii IA, Shevelev VA (2001) Polym Sci A 43:1066
127. Gotlib YY, Torchinskii IA, Shevelev VA (2002) Polym Sci A 44:206
128. Voigt G, Kimmich R (1980) Polymer 21:1001
129. Qui XH, Ediger MD (2002) Macromolecules 35:1691
130. Allen G, Connor TM, Pursey H (1963) Trans Faraday Soc 59:1525
131. Kimmich R, Gille K, Fatkullin N, Seitter R, Hafner S, Müller M (1997) J Chem Phys 107:5973
132. Koch H, Bachus R, Kimmich R (1980) Polymer 21:1009
133. Schneider H, Hiller W (1990) J Polym Sci Polym Phys 28:1001
134. Singer A, Hiller W (1985) Polym Bull 14:469
135. Manelis GB, Erofeev LN, Provotorov BN, Khitrin AK (1989) Sov Sci Rev B Chem 14:1
136. Kuhn W, Barth P, Hafner S, Simon G, Schneider H (1994) Macromolecules 27:5773
137. Cohen-Addad JP, Guillermo A (1984) J Polym Sci Polym Phys 22:931
138. Kimmich R (1984) Polymer 25:187
139. Kimmich R, Rosskopf E, Schnur G, Spohn KH (1985) Macromolecules 18:810
140. Weber HW, Kimmich R, Köpf M, Ramik T, Oeser R (1992) Progr Coll Polym Sci 90:104
141. Köpf M, Schnur G, Kimmich R (1988) Macromolecules 21:3340
142. Doi M (1981) J Polym Sci Polym Lett 19:165
143. Köpf M (1991) PhD thesis, Universität Ulm
144. Kimmich R (1985) Helv Phys Acta 58:102
145. Doolittle AK (1951) J Appl Phys 22:1471
146. Fox TG, Gratch S, Lashaek S (1956) In: Eirich FR (ed) Rheology, vol.1. Academic, New York, p 431
147. Bailey RT, North AM, Pethrick RA (1981) Molecular motion in high polymers. Clarendon, Oxford
148. Fleischer G (1984) Polym Bull 11:75
149. Fleischer G (1987) Coll Polym Sci 265:89
150. von Meerwall E, Ferguson RD (1982) J Polym Sci Polym Phys 20:1037
151. Kimmich R, Peters A, Spohn K-H (1981) J Membrane Sci 9:313
152. Bachus R and Kimmich R (1983) Polym Communications 24:317
153. Kimmich R, Weber HW (1993) J Chem Phys 98:5847
154. Kimmich R, Fatkullin N, Seitter R-O, Gille K (1998) J Chem Phys 108:2173
155. Kimmich R, Stapf S, Möller M, Out R, Seitter R-O (1994) Macromolecules 27:1505
156. Link T, Kimmich R (2000) unpublished
157. Fenchenko KV (1997) Polym Sci B 39:146
158. Herman, MF (2000) J Chem Phys 112:3040
159. Godovsky YK, Papkov VS (1989) Adv Polym Sci 88:129
160. Kögler G, Loufakis K, Möller M (1990) Polymer 31:1538
161. Pincus P (1969) Solid State Commun 7:415
162. Blinc R, Hogenboom D, O'Reilly D, Peterson E (1969) Phys Rev Lett 23:969

163. Doane JW, Johnson DL (1970) Chem Phys Lett 6:291
164. Ukleja P, Pirs J, Doane JW (1976) Phys Rev A 14:414
165. Dong RY (1983) Isr J Chem 23:370
166. Vilfan M, Kogoj M, Blinc R (1987) J Chem Phys 86:1055
167. Zeuner U, Dippel T, Noack F, Müller K, Mayer C, Heaton N, Kothe G (1992) J Chem Phys 97:3794
168. Grinberg F, Kimmich R, Möller M, Molenberg A (1996) J Chem Phys 105:9657
169. Bachus R, Kimmich R (1983) Polymer 24:964
170. Maklakov AI, Skirda VD, Fatkullin NF (1990) Self diffusion in polymer melts and solutions. In: Cheremisinoff NP (ed) Encyclopedia of fluid mechanics, vol 9: polymer flow engineering. Gulf, Houston
171. Manz B, Callaghan PT (1997) Macromolecules 30:3309
172. Komlosh ME, Callaghan PT (1998) J Chem Phys 102:1648
173. Tao H, Lodge TP, von Meerwall ED (2000) Macromolecules 33:1747
174. Lodge TP (1999) Phys Rev Letters 83:3218
175. Fleischer G (1985) Polymer 26:1677
176. Grinberg F, Garbarczyk M, Kuhn W (1999) J Chem Phys 111:11222
177. Garbarczyk M, Grinberg F, Nestle N, Kuhn W (2001) J Polym Sci Polym Phys 39:2207
178. Rommel E, Kimmich R, Spülbeck M, Fatkullin NF (1993) Progr Colloid Polym Sci 93:155
179. Kugler J, Fischer EW (1983) Makromol Chemie 184:2325
180. Smith GD, Yoon DY, Jaffe RL, Colby RH, Krishnamoorti R, Fetters LJ (1996) Macromolecules 29:3462
181. Fatkullin NF, Yatsenko GA, Kimmich R, Fischer E (1998) J Exp Theor Phys 87:294
182. de Gennes PG (1999) Advan Polym Sci 138:91
183. Stapf S, Kimmich R (1995) Macromolecules 29:1638
184. Beginn U, Fischer E, Pieper T, Mellinger F, Kimmich R, Möller M (2000) J Polym Sci Polym Chem 38:2041 (note that a misprint occurred in that paper with the length scale specified in the electron micrograph given at Fig. 8)
185. Kimmich R, Fatkullin N, Seitter R-O, Fischer E, Beginn U, Möller M (1999) Macromol Symp 146:109
186. Fischer E, Beginn U, Fatkullin N, Kimmich R (2004) Macromolecules 37:3277
187. Graessley WW, Edwards SF (1981) Polymer 22:1329

Editor: Sam Edwards Received: February 2003

APS (2004) 170:115–167
DOI 10.1007/b95533

Emerging Technologies for the 3D Analysis of Polymer Structures

Hiroshi Jinnai[1] · Yukihiro Nishikawa[1] · Takayuki Ikehara[2] · Toshio Nishi[3] (✉)

[1] Department of Polymer Science and Engineering, Kyoto Institute of Technology, 606-8585 Kyoto, Japan
hjinnai@kit.ac.jp
[2] Department of Applied Chemistry, Faculty of Engineering, Kanagawa University, 3-27-1 Rokkakubashi, Kanagawa-ku, 221-8686 Yokohama, Japan
[3] Department of Organic and Polymeric Materials, School of Science and Engineering, Tokyo Institute of Technology, 2-12-1 Ohokayama, Meguro-ku, 152-8551 Tokyo, Japan

This review covers recent progress of studies using new methods for analyzing polymer structures, with emphasis on three-dimensional (3D) microscopy, e.g., laser scanning confocal microscopy (LSCM) and transmission electron microtomography (TEMT). These methods have been applied to morphological studies of structures in polymer alloys over the past few years. We review some of the studies dealing with 3D structures of binary polymer blends and block copolymers that are self-assembled during the phase separation processes and/or phase transition. Typical sizes of such self-assembled structures are of the order of micro- and nanometers, respectively. A great deal of new structural information, which has never been obtained from conventional microscopy or various scattering methods, can be evaluated *directly* from the 3D volume data. The structural information obtained from 3D microscopy offers essential insight into the physics of nonlinear, nonequilibrium phenomena and the statistical mechanics of long-chain molecules.

Keywords Polymer alloy · 3D microscopy · 3D digital image analysis · Bicontinuous structures

Abbreviations and Symbols

$A_{tri}(i;j)$	Area of i-th triangles (generated by MCA) sharing the j-th POI
CMC	Constant mean curvature
CT	Computerized tomography
d	Signed displacement of the parallel surface
D	Thickness of (blend) specimen
D_{CT}	Diameter (size) of an object reconstructed in CT
DPB	Deuterated polybutadiene
f	Copolymer composition
f_S	Styrene volume fraction in microphase-separated structure of SIS block copolymer
E	Number of edges of faces subdividing the surface
F	Number of faces subdividing the surface
$F(x)$	Scaled structure factor
$f(x,y)$	Intersection of an object
H	Mean curvature
$\tilde{H}$	Scaled mean curvature
$\langle H \rangle$	Area-averaged mean curvature
$I(q,t)$	Scattering intensity in TRLS experiment
$I(x,y,z)$	Intensity of fluorescence from a particular point in a focal plane (x–y plane) at a depth z
K	Gaussian curvature
$\tilde{K}$	Scaled Gaussian curvature
$\langle K \rangle$	Area-averaged Gaussian curvature
L	Crystallographic unit cell edge
LS	Light scattering

LSCM	Laser scanning confocal microscopy
MCA	Marching cubes algorithm
N	Number of pixels along the edge of the image
N_{CT}	Number of projections in CT
N_{j}	Coordination number at a junction
NA	Numerical aperture of an objective
N_{tri}	Total number of triangles comprising interface generated by MCA
n	Normal vectors at vertices of the polygon generated by MCA
n_{tri}	Normal vector of each triangle generated by MCA
p	Positional vector of a surface
PB	Polybutadiene
$p(r,\varphi)$	Projection at an angle φ
$P(H,K)$	Joint probability density of surface curvatures
$P_H(H)$	Marginal probability densities of the mean curvature
$\tilde{P}_H(\tilde{H};t)$	Scaled marginal probability densities of the mean curvature
p_i	Coordinates of the i-th vertices of the triangle generated by MCA
$P_K(K)$	Marginal probability density of the Gaussian curvature
$\tilde{P}_K(\tilde{K};t)$	Scaled marginal probability density of the Gaussian curvature
POI	Point of interest at a surface
PSM	Parallel surface method
q	Wave number
$q_{\mathrm{m}}(t)$	Wave number at which $I(q,t)$ shows a maximum
q_{c}	Crossover wave number from q^{-6} to q^{-4} with increasing q in $I(q,t)$ of spinodally decomposing polymer mixtures
r	Resolution of reconstructed image in CT
RI	Roughness index
R_{m}	Area-averaged mean radius in Tomita's scattering theory
$R_{\mathrm{m,LS}}$	Mean radius estimated from LS based on Tomita's scattering theory (Eqs. 28 and 29)
$R_{\mathrm{m,LSCM}}$	Mean radius estimated from LSCM based on Tomita's scattering theory (Eq. 29)
R_{g}	Radius of gyration of polymers
S	(Total) interfacial area
SCFT	Self-consistent field theory
SD	Spinodal decomposition
SFM	Sectioning and fitting method
S_i	Area of i-th neighboring triangles generated by MCA
SIS	Poly(styrene-*b*-isoprene-*b*-styrene) triblock copolymer
$S(q)$	Scattering structure factor
t	Phase-separation time
TDGL	Time-dependent Ginzburg–Landau equation
TEM	Transmission electron microscopy
TEMT	Transmission electron microtomography
u and v	Curvilinear coordinates of a surface
V	Number of vertices of faces subdividing the surface
V_{s}	Volume of the whole system
z	Depth along the optical axis of LSCM
χ	Euler–Poincaré characteristic
$\delta A(j)$	Local area assigned to j-th POI

ΔF	Full width at half maximum (fwhm) of optical sections in LSCM observations
ΔH	Class interval of H
ΔK	Class interval of K
ΔZ	Increment along depth direction of LSCM
$\langle \eta(t)^2 \rangle$	Mean squared refractive index fluctuations
θ	Scattering angle in TRLS experiments
κ_i	Principal curvatures (i=1 and 2)
λ	Wavelength of laser of LSCM
λ_{LS}	Wavelength of laser in TRLS experiments
$\Lambda_m(t)$	Length characterizing the global size of SD bicontinuous structures
σ_H	Standard deviation of the mean curvature distribution
$\sum(t)$	Surface area per unit volume
ϕ	Volume fraction of one of the phases in polymer mixture
ϕ_{PB}	Volume fraction of the PB-rich phase in DPB/PB mixture
φ	Angle of the projection slice in computerized tomography (CT)

1 Introduction

Polymer materials are ubiquitous in our daily life. Such materials often consist of more than one species of polymer and thus become multicomponent systems, such as polymer blends [1,2] and block copolymers [3]. Due to repulsive interaction between the constituent polymers, the multicomponent polymer materials involve "phase separation".

Studies to characterize the phase-separated structures inside the materials have been growing intensively over the past three decades. Academic interest in the complex fluids (to which polymeric systems belong) as well as ceaseless industrial need for developing new materials activated such studies. In academia, for example, pattern formation and self-assembling processes of binary polymer blends are one of the most fascinating research themes in nonlinear, nonequilibrium phenomena. Block copolymers likewise self-organize into (equilibrium) phase-separated *nano*structures (termed "microphase-separated structures" in contrast to the "macrophase-separated structures" in the polymer blend). Since it is required for the block copolymers to place their junctions at the interface and thus the microdomain-forming blocks must uniformly fill space in the most entropically favored manner, the block copolymer morphologies are often highly organized. Studies of the block copolymer morphologies belong to the statistical mechanics of long-chain molecules. In industry, such phase-separated polymer systems have been found important to achieve superior physical properties. Hence, the structure–property relationship in multicomponent polymeric materials is of significant importance, basic studies on which eventually render new designs of polymer materials satisfying the diverse requirements of industry.

Up to now, morphological studies of the multicomponent polymeric materials have been carried out by various microscopic and scattering methods. Optical microscopes, transmission electron microscopes (TEM), scanning electron microscopes (SEM), and atomic force microscopes (AFM) are commercially available and widely used. The biggest advantage to using such microscopes is that they offer intuitive real-space representation of morphologies by showing *images*. However, when it comes to "measurements", especially in a quantitative way, the microscopes sometimes give only limited pieces of structural information with inferior statistical accuracy compared with scattering methods. The scattering techniques give statistically averaged structural information over the scattering volume, which is normally a lot bigger than the area of the microscopes. Although accurate, the scattering methods do not *directly* provide intuitive insight into the morphologies. Moreover, in general, the scattering methods may require knowledge of the morphologies beforehand to obtain structural parameters. In other words, for the complete characterization of the morphologies, one may need to know them first by the microscope and subsequently evaluate statistically the structural parameters by the scattering on the basis of the morphologies. Thus, the two methods have the pros and cons and are complementary to each other. According to the size of the morphologies, one may use an optical microscope together with light scattering (LS) for micron-scale morphologies and TEM (or SEM, AFM) with small-angle neutron (or X-ray) scattering for nanometer-scale structures.

One of the biggest problems one might encounter in the course of the above conventional protocol is that the identification of complex structures by the microscopes is often inconclusive and hence interpretation of the scattering data would not be straightforward. The microscopes used so far are able to take two-dimensional (2D) (transmitted or surface) images of three-dimensional (3D) objects. The more complicated the morphologies are, the less convincing the interpretation of the 2D images usually become. It is thus requisite to develop microscopes that are capable of obtaining 3D images. Hereafter, we call such microscopes *3D microscopes* and the methodology *3D-scopy* or *3D microscopy*. Due to these backgrounds, a couple of different types of 3D microscopes have been developed as described in Sect. 2. We emphasize that 3D microscopy should be accompanied by digital image analysis (see Sect. 3) that extracts useful and new structural information from the 3D volume data.

In the present review, we will try to demonstrate that 3D microscopy has great advantages over the conventional methods to observe and analyze complex polymer structures. This review paper is organized in the following way. In Sect. 2, we briefly describe the basic principles of 3D microscopy that will be used later in this review paper. In Sect. 3, *quantitative* 3D image analysis for the 3D volume data on the basis of the geometry will be introduced. Section 4 is dedicated to the visualization of one of the transient structures observed in the polymer blend during the phase separation called spinodal decomposition (SD). It was found that the two domains form networks that coexist in 3D space (*bicontinuous sructure*). The stability of the interface between the two domains, the effects of preferential wetting and confinement on the SD process were examined using the (differential) geometry. 3D direct observation of the nanoscale structure of block copolymers will be discussed in terms of their stability in Sect. 5. The block copolymer morphology we dealt with exhibited bicontinuity similar to that of the one studied in the polymer blend. The effect of incorporation of the covalent bond (to form block copolymers) on the shape and regularity of the resulting morphologies is of particular interest, which will be discussed in Sect. 6. Finally, concluding remarks are given in Sect. 7.

2
3D Microscopy

Figure 1 shows the spatial resolution of various types of microscopes and scattering methods. The figure includes 3D microscopy such as laser scanning confocal microscopy (LSCM), X-ray computerized tomography (X-ray CT), and transmission electron microtomography (TEMT). Although no experimental results obtained from the X-ray CT will be presented in the present review paper, we note that the microscopy [4] is particularly useful to

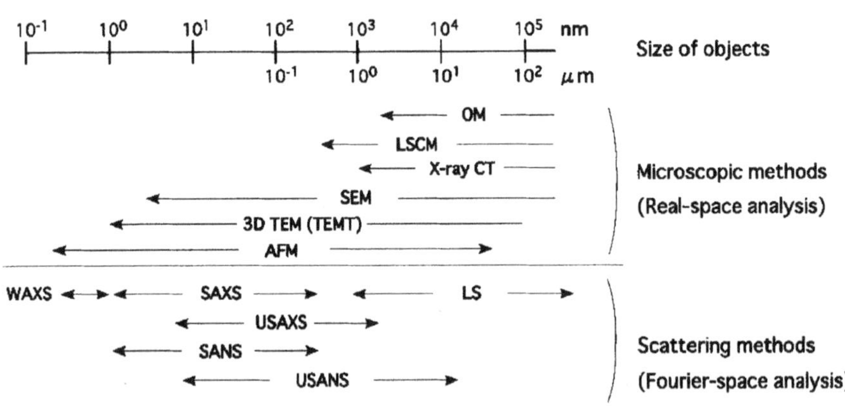

Fig. 1 Schematic illustration showing resolution of both microscopic and scattering methods. Abbreviations are the following: OM (optical microscopy), LSCM (laser scanning confocal microscopy), X-ray CT (X-ray computerized tomography), SEM (scanning electron microscopy), TEMT (transmission electron microtomography), ATF (atomic force microscopy), WAXS (wide-angle X-ray scattering), SAXS (small-angle X-ray scattering), LS (light scattering), USAXS (ultra-small-angle X-ray scattering), SANS (small-angle neutron scattering), and USANS (ultra-small-angle neutron scattering)

observe internal structures of opaque specimens. In the early 1990s, the spatial resolution of the X-ray CT was limited to ca. 100 μm. Recently, a resolution comparable with that of the optical microscope (~1 μm) has been achieved. It should be emphasized that, combining LSCM, X-ray CT, and TEMT, it is possible to establish a 3D microscopy whose resolution is over the wide range from nano- to micrometers. In what follows, we will briefly describe the principles and some applications for LSCM and TEMT.

2.1 Laser Scanning Confocal Microscopy (LSCM)

After LSCM became commercially available in the late 1980s, the instrument generated an excitement among biologists. With LSCM, one can slice incredibly clean thin optical sections out of thick specimens, gain impressive 3D views at high resolution [5, 6].

The focal depth of the conventional optical microscopes is usually deep in comparison to its lateral resolution. This leads to the superposition of the out-of-focus images, and hinders the 3D interpretation of the images. Confocal microscopy has overcome this disadvantage by introducing a pinhole just in front of the detector as shown in Fig. 2. The solid line shows the opti-

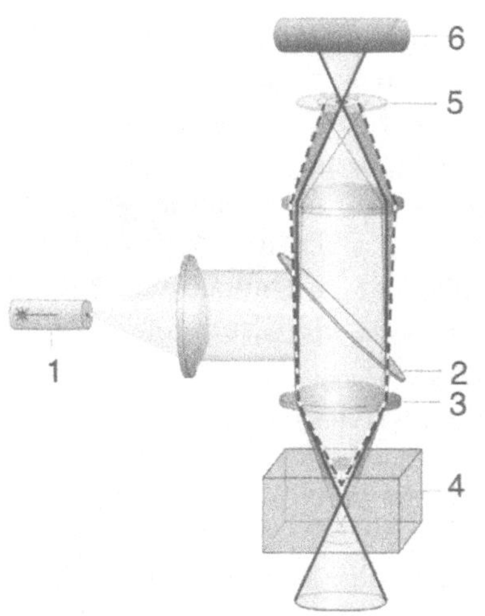

Fig. 2 Optics of confocal microscopy. The incident light emitted from laser (*1*) is reflected by a dichroic mirror (*2*). The reflected light is focused on the specimen (*4*) through objective lens (*3*). At the focus the light may be reflected by the interior structure or the incident laser excites fluorescent molecules attached to the particular site (position) of the specimen. The reflected light or fluorescence from the focal plane refocuses at the confocal pinhole (*5*) and thus it passes the pinhole to reach detector (*6*). The out-of-focus light (fluorescence) is blocked by the confocal pinhole as shown by the *dashed line*

cal path of reflected (or fluorescent) light in the focal plane of the objective lens. The reflected (or fluorescent) light focuses at the pinhole and reaches the detector. Simultaneously, the object may reflect the out-of-focus light (or emit the out-of-focus fluorescence), as shown by the dashed line in Fig. 2. In this case, simple ray tracing shows that the light (or fluorescence) reflected back to the confocal pinhole arrives as a defocused blur, only the central portion of which is detected and contributes to the image. In this way the system eliminates the images which do not lie within the focal region of the lens. The excellent depth resolution thus achieved, which is the full width at half maximum (fwhm) of optical sections, ΔF, is typically 1 μm or may be less, depending on the optical setup. ΔF is a complicated function of an incident laser beam of wavelength, λ, a numerical aperture (NA), magnification of the objective, radius of the confocal pinhole, refractive index of the object, etc. Once this "optical slicing" has been achieved, the objective (or the sample stage) is moved along the optical axis of the microscope to obtain a series of 2D optically sliced images, which is later used to reconstruct a 3D image (see Sect. 3.1 for details).

To our best knowledge, Verhoogt et al. [7] first reported an application of LSCM to a polymer mixture. Although their work was instrument oriented and thus did not provide much insight into the phase-separated structure of the binary mixture itself, the potential of the instrument was clearly demonstrated. Image generation in LSCM of a polymer blend system was discussed by Ribbe et al. [8]. White and Wiltzius [9] discussed a mechanism of the phase separation in an off-critical polymer mixture through the time evolution of the size distribution of spherical domains. LSCM was applied to investigate the phase-separated structure of polymer mixtures at the critical composition, demonstrating that the structure is truly interconnected and periodically arranged, forming a 3D "maze" [10–12]. These studies are considered to take full advantage of 3D reconstruction intrinsic to LSCM. Morphology under a shear flow [13] and that of a ternary polymer mixture [14] were also studied by LSCM.

We note that some other studies rather use the excellent depth resolution of the confocal microscope; surface and bulk structures of the solvent-cast films of a polymer mixture were investigated as a function of distance from the air–polymer surface [15, 16]. LSCM was also applied to investigate and demonstrate the penetration process of a spherulite into another spherulite of a different component in crystalline/crystalline polymer blends [17]. Besides those studies, LSCM was used for structural investigation of gel [18, 19], membrane systems [20, 21].

2.2
Transmission Electron Microtomography (TEMT)

As already mentioned, TEMT is a method to image structures in 3D at nanometer-scale resolution [22]. Unlike confocal microscopy, TEMT uses a series of 2D projections taken by transmission electron microscopy (TEM) that are subsequently used to compute a 3D image of the object on the basis of computerized tomography (CT). TEMT offers a new approach to obtain 3D information of the order of the nanometer scale, and hence was rapidly developed during the past few years first in the biological community. Although the high potential of TEMT as a unique and powerful tool to study various types of structures in material science was recognized [23], it is only recently that the technique became available to the material community [24, 25]. So far, only a few studies [26–30] have used TEMT to investigate the complex 3D structures of polymeric materials, most of which dealt with the microphase-separated structures of block copolymers [see Sect. 5 for an example]. In what follows we will describe the principles of CT and subsequently discuss several experimental limitations in the TEMT measurements.

The CT allows the complete 3D internal structure of an object to be reconstructed from a set of 2D projections at different angular settings, without imposing any *a priori* assumptions or requirements regarding the symmetry of the object. The reconstruction from projections is aided by an understanding of the relationship between an object and its projections in Fourier space: "the central slice theorem" [31] states that the Fourier transform of an object's projection is a central plane in the Fourier transform of the object as shown in Fig. 3. A projection of a 2D object, represented by the function $f(x,y)$, is produced by parallel illumination from the electron source passing through the object, which is tilted at an angle φ about an axis orthogonal to the beam path. $p(r,\varphi)$ denotes the projection in real space. The Fourier transform of $p(r,\varphi)$ is

$$F(q,\varphi) = \int_{-\infty}^{+\infty} p(r,\varphi)e^{-iqr}dr. \qquad (1)$$

Here q is the wave number. The Fourier coefficients, $F(q,\phi)$, are arranged along the line in Fourier space with the same tilting angle ϕ. Gathering 2D projections over as wide a tilting range as possible gives a 3D Fourier transform of the object. Note that the ideal 180° range of tilt is required to give a complete sampling of Fourier space. Taking the inverse Fourier transform of the entire array of coefficients results in the reconstructed volume in real space. In the above 2D object case (in Fig. 3):

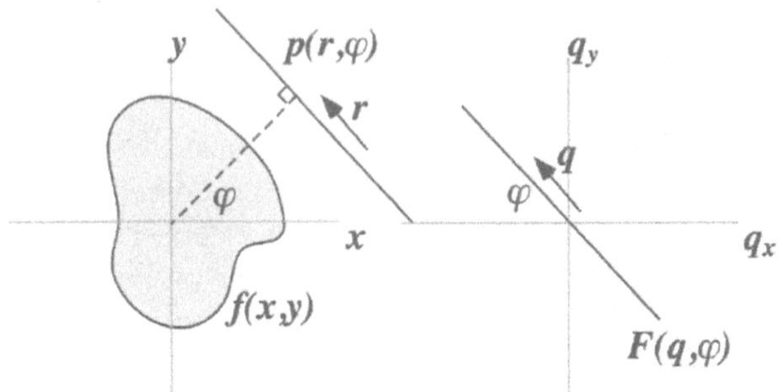

Fig. 3 The discrete form of the central slice theorem in two dimensions. A projection $p(r,\varphi)$ in real space (x,y) at angle φ is a slice $F(q,\varphi)$ at the same angle in Fourier space (q_x,q_y)

$$f(x,y) = \int_0^{2\pi} \int_0^{+\infty} F(q,\varphi) e^{iq(x\cos\varphi + y\sin\varphi)} q\, dq\, d\varphi. \tag{2}$$

Unfortunately the projection data are always sampled at discrete angles leaving regular gaps in Fourier space. The inverse Fourier transform intrinsically requires a continuous function, and therefore azimuthal interpolation is required to fill the gaps in Fourier space. The quality of the reconstruction is significantly affected by the type of interpolation implemented [32]. An alternative and more commonly used algorithm is the radius-weighted back projection technique [33].

In the case of TEMT, the projections at different angles are collected by tilting the specimen with respect to the electron beam in the TEM column. The tilt range achievable in a TEM is restricted by the relative geometries of specimen holder and objective lens since they will contact physically at high tilt angle, e.g., 70°. The missing information due to this limitation becomes a wedge-shaped region of Fourier space, giving rise to a loss of resolution in the reconstructed image especially in the direction parallel to the electron beam. The restoration of the missing structural information has been studied intensively [34]. In addition, the resolution of the reconstructed image is strongly affected by the numbers of projections. The resolution is expressed as [35],

$$r = \pi D_{CT} / N_{CT}, \tag{3}$$

where r is the size of the smallest feature visualized in the reconstruction. D_{CT} and N_{CT} are the diameter (size) of the object or feature being reconstructed and the number of projections, respectively. Since the Fourier coefficients are radially arranged, more information will be lost as q increases, which limits the resolution of the reconstructed images. Equation 3 tells us that the resolution will be compensated by increasing the number of projections gathered; since N_{CT} is increased, the density of the Fourier coefficients at a given q will be more dense. In reality, however, the angular increment in TEMT experiments is also limited to, at smallest, 1° (typically 2°) due to the severe electron beam damage to the polymer specimens.

During tilting the specimen, misalignments in the digitized images are also inevitable due to the imperfect eucentricity of the specimen stage. As a crucial prelude to calculating a 3D reconstruction, alignment of the digitized images is necessary. This is achieved either by least squares tracking of fiducial markers [22], such as small gold particles, or by sequential cross-correlation. Often a combination of both processes is used [36].

3
Digital Image Analysis of 3D Structures

In the preceding section, two types of 3D microscopy were introduced. In some cases, it is sufficient to have 3D images of the structures in order to intuitively grasp their unique features. Actually, many researchers, in material science as well as in the medical and biological fields, seem to be content with 3D "pretty" pictures or the cross-sectional views of their 3D objects. This is merely "observations", not "structural analysis".

Yet 3D microscopy gives much more structural information than either the conventional 2D microscopes or the scattering methods, most of which have never been acquired before. It is expected that the new structural information will clear up the root of unsolved physical problems. In the present section, we will introduce some *quantitative* analytical methods to extract such structural parameters from the 3D digital data.

3.1
3D Reconstruction

In most cases, a 3D image is constructed from a stack of two-dimensional (2D) slices, i.e., a 3D array of image intensities. In the case of fluorescent LSCM, for example, the image intensities correspond to the concentration of fluorescent molecules that are chemically attached to one of the constituents. For quantitative image analysis, it is first necessary to isolate the portions of interest by defining surfaces that separate such particular parts from the other parts of the structures. The surfaces are called "interface".

One can assign the interface by specifying a threshold value for the image intensities. The marching cubes algorithm (MCA) [37] is used to find the interface from the 3D image. In this algorithm, a cube is created from the centers of eight adjacent pixels; four pixels from each of two adjacent slices. The image intensity at the pixels is assigned to the corresponding vertices of the cube. The interface is considered to intersect an edge of this cube if the intensity of one vertex of this edge is larger than the threshold value, and that of another vertex of the same edge is smaller than the threshold value. An intersecting point between the interface and the edge of the cube is determined by a linear interpolation according to the image intensity assigned to the two vertices on the edge. By properly connecting the intersections determined for the cube, the local interface is constructed and represented by a set of triangles. Note that the MCA defines the 14 types of intersection for the interface constructed in the cube by taking rotational and mirror symmetry into consideration (see Fig. 3 in ref. [37]). This gives a polygonal representation of the interface, which eventually reduces to triangles, from the 3D array of data by the MCA. A normal vector of each triangle, n_{tri}, can be estimated by

$$n_{\mathrm{tri}} = \frac{(p_2 - p_1) \times (p_3 - p_2)}{|(p_2 - p_1) \times (p_3 - p_2)|}, \tag{4}$$

where p_1, p_2, and p_3 are the positional vectors of the vertices of the triangle.

The normal vectors thus obtained from the MCA were used to evaluate normal vectors at vertices of the polygon, n. We note that the normal vectors obtained by the MCA, i.e., n_{tri}, were those for the triangles comprising the polygon, not for the vertices. n was estimated by averaging the normal vectors of the neighboring triangles that shared the vertex of interest with those areas as a weighting:

$$n = \frac{\sum_i S_i n_{\mathrm{tri}}^i}{\left|\sum_i S_i n_{\mathrm{tri}}^i\right|}. \tag{5}$$

Here n_{tri}^i and S_i denote, respectively, the normal vector and the area of the i-th neighboring triangle. Equation 5 will be further used to generate a parallel surface to measure surface curvatures (see Sect. 3.2.2 for details).

3.2 Curvature Distribution Measurements

In some systems, the local shape of surfaces is significant in understanding the formation and stability of the morphologies and thus it can be a key structural measure to clarify the underlying physics. Surface curvatures are fundamental parameters characterizing the shape of surfaces in the differential geometry. From the volume data taken by the 3D microscopy, it is possible to determine the surface curvatures *for the first time.*

3.2.1 Surface Curvatures

The curvatures at an arbitrary point on the surface can be expressed by two principal curvatures or the mean and Gaussian curvatures [38]. A 2D curvature is defined for a planar-curved path passing through a point of interest (POI) on the surface. The path is expressed by the intersection between the surface and the plane contains the surface normal vector at the point (see Fig. 4). Paths with different tangential directions on the surface have, in general, different values for the curvatures. Maximum and minimum values among those curvatures at the point of interest are called the principal curvatures, i.e., κ_1 and κ_2, as schematically shown in Figs. 4 and 7. Note that each curvature has a sign: it is positive if the path is convex from one side, negative otherwise, and zero if the path is straight. The mean curvature, H, and the Gaussian curvature, K, are defined as

$$H=\frac{\kappa_1+\kappa_2}{2},\quad K=\kappa_1\kappa_2. \tag{6}$$

They are uniquely determined at the point p. The sign of K classifies the shape of the surface: it is elliptic if $K>0$, parabolic if $K=0$, and hyperbolic if $K<0$.

In what follows we will propose two kinds of computational methods to measure the local curvatures. In Sect. 3.2.2, we will show a method to estimate area-averaged mean, $\langle H\rangle$, and Gaussian, $\langle K\rangle$, curvatures. This method involves measurements of areas of surfaces that are parallel to the interface. In Sect. 3.2.3, an alternative method to measure the distributions of the curvatures that were obtained by calculating the curvatures at many points on

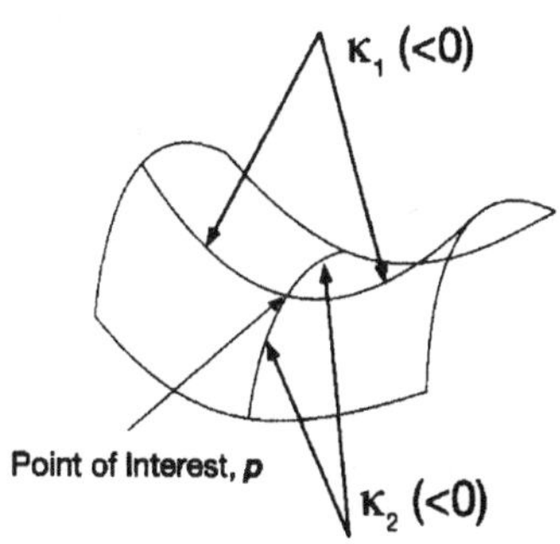

Fig. 4 Definition of the surface curvatures. The *solid line* on the surface is a planar-curved path. The principal curvatures at a point on the surface are defined as the maximum and minimum curvatures. Note that the curvature is signed: the curvature is positive for the case where the center of the radius of the curvature is placed in one side, and negative for the other side

the interface will be demonstrated. Note that both methods are based on the differential geometry.

The accuracy of the measurements depends strongly on the quality of the 3D surface. The roughness of the surfaces is quantified by an index, called "roughness index", RI,

$$\mathrm{RI} \equiv \sqrt{\langle A_\Delta \rangle}\frac{|\langle \kappa_1 \rangle| + |\langle \kappa_2 \rangle|}{2}. \tag{7}$$

Here, $\langle A_\Delta \rangle$ expresses an average area of each triangle, i.e., $\langle A_\Delta \rangle \equiv \sum_i |A^i_\Delta|/N_{\mathrm{tri}}$. A^i_Δ is the area of the i-th triangle and N_{tri} is the total number of triangles. RI gives a measure of the roughness of the interface constructed by many triangles relative to the average radius of curvature. The smaller the value of RI is, the smoother the surface becomes. Note that the square root of the area of a triangle and the average curvature have, respectively, dimensions of length and reciprocal length, and hence the index is a dimensionless quantity.

3.2.2 Parallel Surface Method (PSM)

Here in the present section we will describe the principle of a method to measure area-averaged curvatures, a parallel surface method (PSM). As mentioned in Sect. 3.1, the surface is expressed by many triangles and the normal vector at vertices of each triangle [37]. In the differential geometry, the following relation between the area of the infinitesimal patch at a point p, $da(0,p)$, and that of the parallel patch, $da(d,p)$, holds: [38]

$$da(d,p) = da(0,p)(1 + 2H(p)d + K(p)d^2). \tag{8}$$

Here $H(p)$ and $K(p)$ are the mean and Gaussian curvatures at the point p, respectively. d is a signed displacement of the parallel surface from the original surface ("interface") (see Fig. 5): it is positive if the direction of the displacement points to one side of the surface, and negative otherwise. Sum-

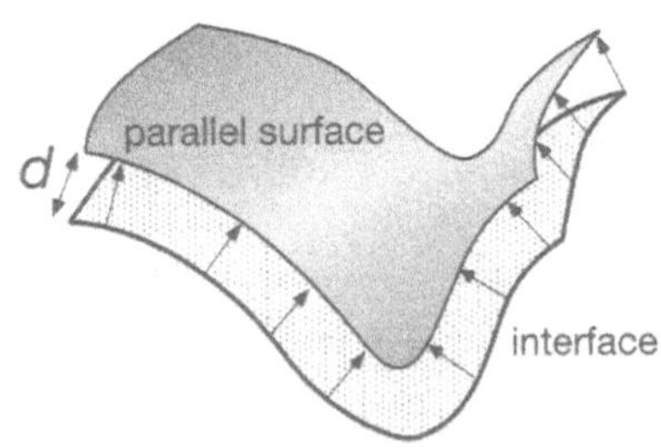

Fig. 5 Parallel surface with the displacement d from the interface

ming Eq. 8 over the whole area of the interface by changing the position of the point of interest p, gives

$$A(d) = A(0)(1 + 2\langle H\rangle d + \langle K\rangle d^2). \tag{9}$$

$A(d) = \int_p da(d,p) \cdot A(0)$ is the total area of the original interface, and $A(d)$ is that of an imaginary surface parallel to the interface ("parallel surface") at the displacement d. The curvatures, $\langle H\rangle$ and $\langle K\rangle$, are expressed as

$$\langle H\rangle = \frac{\int_p H(p)da(0,p)}{\int_p da(0,p)}, \langle K\rangle = \frac{\int_p K(p)da(0,p)}{\int_p da(0,p)}. \tag{10}$$

Thus, $\langle H\rangle$ and $\langle K\rangle$ are the area-averaged quantities.

Following Eq. 9, the parallel surface which will be used for the curvature measurements was computationally created. Displacement vectors with length d at each vertex are created by extrapolating the normal vectors estimated by the MCA at each vertex of the triangles. Connecting the top of the displacement vectors makes the parallel surface, as schematically illustrated in Fig. 5. The area of the parallel surface was obtained by summing up the areas of all the triangles that consist of the parallel surface. The detailed protocol of the PSM and critical test on applicability and precision using various types of model surfaces can be found elsewhere [39]. An example of the determination of area-averaged surface curvatures of the phase-separated structures in a polymer mixture will be given in Sect. 4.3.3.

3.2.3 Sectioning and Fitting Method (SFM)

In Fig. 6, we try to demonstrate the significance of the curvature distribution. Part (a) of the figure shows a flat surface, while a wavy surface is drawn in part (b). Although both surfaces have the same surface geometry in terms of the area-averaged curvatures, i.e., $\langle H\rangle$=0 and $\langle K\rangle$=0, the local geometrical

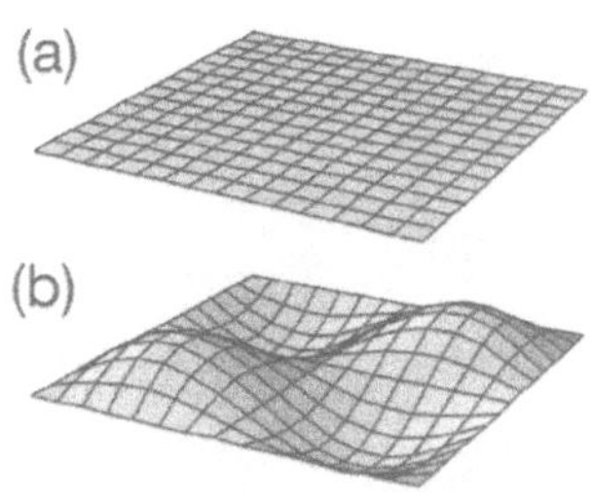

Fig. 6a,b Schematic representation of **a** flat and **b** wavy surfaces. The wavy surface equally has concave and convex parts so that the area-averaged mean, $\langle H\rangle$, and Gaussian, $\langle K\rangle$, curvatures for both surfaces are same: $\langle H\rangle$=$\langle K\rangle$=0

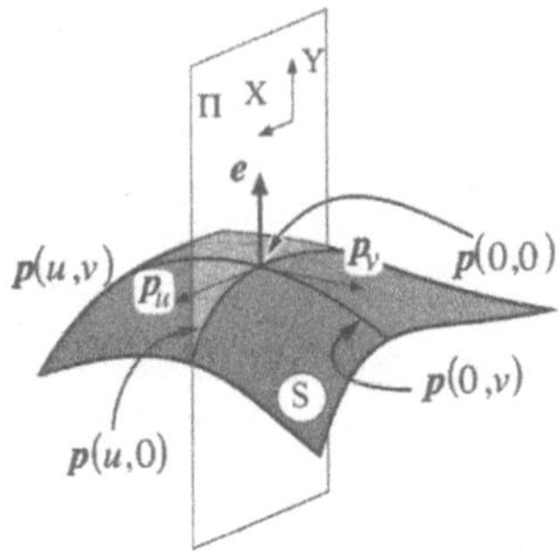

Fig. 7 Schematic diagram of a surface, expressed in a parametric form $p(u,v)$, and a "sectioning plane", Π, which defines $p(u,0)$. $p(0,0)$ is a point of interest (POI) at which the local curvatures are measured. e is the unit normal vector of the surface at the POI

shape is obviously different between the two surfaces. Although for the surface (b) $\langle K\rangle=0$ may not be obvious, it is proved as follows. Let A be the region of the surface concerned, and the boundary of A be ∂A. The surfaces shown in the figures are such examples. According to the Gauss–Bonnet theorem, any surface bounded by a closed curve satisfies the equation $2\pi = \int_A Kda + \oint_{\partial A} \kappa_g ds + \sum_i \varepsilon_i$, where κ_g is the curvature along ∂A and ϵ_i is the exterior angle of the i-th vertex of ∂A if it exists. For the case of the surface (b), κ_g is zero because each edge is a straight line, and there are four vertices of 2π (in radians). Thus the second and third terms on the right-hand side are 0 and 2π, respectively, so that the term $\int_A Kda$ should be zero. $\int_A Kda$ can be written as $\langle K\rangle A$. Because A is a nonzero positive value for any surface, $\langle K\rangle$ is zero for the surface (b)

According to the differential geometry [38], the surface is expressed in a parametric form as $p(u,v)=(x(u,v),y(u,v),z(u,v))$, where p is a positional vector of the surface, and u and v are the curvilinear coordinates on the surface as schematically shown in Fig. 7. The definition of u and v is arbitrary. The first and second fundamental forms of the differential geometry are defined, respectively, as [38]

$$\mathrm{I} = Edudu + 2Fdudv + Gdvdv, \tag{11}$$

and

$$\mathrm{II} = Ldudu + 2Mdudv + Ndvdv. \tag{12}$$

The parameters in Eqs. 11 and 12 are given by

$$E = p_u \cdot p_u, F = p_u \cdot p_v, G = p_v \cdot p_v,$$

$$L = p_{uu} \cdot e, M = p_{uv} \cdot e, N = p_{vv} \cdot e. \tag{13}$$

The subscripts on p designate the partial derivatives. For example, $p_{uv}{\equiv}\partial p/\partial u$ and $p_{uv}{\equiv}\partial^2 p/\partial u\partial v$. e is the unit vector normal to the POI, calculated by the partial derivatives of $p(u,v)$ from $e{=}p_u{\times}p_v/|p_u{\times}p_v|$. Using these parameters, H and K are expressed as

$$H = \frac{EN + GL - 2FM}{2(EG - F^2)}, \tag{14}$$

and

$$K = \frac{LN - M^2}{EG - F^2}. \tag{15}$$

If the surface formula $p(u,v)$ and its derivatives are known, H and K can be determined from Eqs. 14 and 15 at any points on the surface.

Let the POI be $p(0,0)$. The surface is first "sectioned" by a plane including the POI and e. The intersection between the plane and the surface is defined as $p(u,0)$. Parameters E and L in Eq. 13 can be estimated from the partial derivatives of $p(u,0)$ with respect to u at the POI. Subsequently, the surface is cut by another plane that defines $p(0,v)$. Now F, G, and N can be evaluated. The parameter M remains unsolved since determination of the complete functional form of the surface around $p(0,0)$, i.e., $p(u,v)$, turned out to be technically very difficult. Thus, we rearranged Eqs. 14 and 15 to eliminate M:

$$f(i,H,K){\equiv}0 = 4F_i^2\left\{L_iN_i - K\left(E_iG_i - F_i^2\right)\right\} - \left\{E_iN_i + G_iL_i - 2H\left(E_iG_i - F_i^2\right)\right\}^2, \tag{16}$$

where the subscript i denotes the i-th set of the curvilinear coordinates (u,v) defined by the two sectioning planes. The two unknown quantities, H and K, are invariant for any choice of the coordinates, while the other coefficients in Eq. 16 depend on the choice of the curvilinear coordinates, u and v. Therefore, in principle, Eq. 16 with two sets of the parameters $\{E_i,F_i,G_i,L_i,N_i\}$ obtained from two sets of curvilinear coordinates (u,v) give the simultaneous equation for H and K. However, the values of the curvatures obtained by this scheme are, in practice, affected by the choice of the combination of the sectioning planes. Thus, several sets of $f(i,H,K)$ are used to solve the two unknown quantities by using nonlinear regression fitting. A numerical procedure to obtain reliable results can be found elsewhere in detail [41]. We call this method a sectioning-and-fitting method (SFM).

As the SFM provides local surface curvatures at a given point on the surface, the curvature distributions of H and K can be obtained by conducting the above measurements at many points on the surface. A joint probability density of H and K, $P(H,K)$, is calculated as

$$P(H,K) = \frac{1}{\Delta H \Delta K \sum_j \delta A(j)} \times \sum_j \delta A\left(j \middle| H - \frac{\Delta H}{2} \leqslant H(j) < H + \frac{\Delta H}{2} \text{ and } K - \frac{\Delta K}{2} \leqslant K(j) < K + \frac{\Delta K}{2}\right) \tag{17}$$

$H(j)$ and $K(j)$ are the mean and the Gaussian curvatures at the j-th POI. The term $\delta A(j|H-(\Delta H/2)...)$ denotes the area on the j-th POI which satisfies its local curvatures given by $H-(\Delta H/2)\leq H(j)<H+(\Delta H/2)$ and $K-(\Delta K/2)\leq K(j)<K+(\Delta K/2)$. ΔH and ΔK are the class interval of H and K. According to the definition of the joint probability density, the front factor $1/\Delta H\Delta K\sum_j \delta A(j)$ in Eq. 17 was introduced for the normalization condition, $\sum_H \sum_K P(H,K)\Delta H\Delta K = 1$, which can be replaced to $\iint P(H,K)dHdK=1$ if $\Delta H\rightarrow 0$ and $\Delta K\rightarrow 0$. The local area assigned to the j-th POI, $\delta A(j)$, was given by

$$\delta A(j) = \frac{1}{3}\sum_i A_{\text{tri}}(i;j). \tag{18}$$

Here $A_{\text{tri}}(i;j)$ represents the area of i-th triangles sharing the j-th POI. The factor 1/3 is necessary to avoid overcounting the total area by three times, since there are three POI for each triangle. Similar to Eq. 17, the marginal probability densities of the mean curvature, $P_H(H)$, and the Gaussian curvature, $P_K(K)$ are calculated from $P(H,K)$

$$P_H(H) \equiv \frac{1}{\Delta H \sum_j \delta A(j)} \times \sum_j \delta A\left(j \middle| H - \frac{\Delta H}{2} \leqslant H(j) < H + \frac{\Delta H}{2}\right) \tag{19}$$

and

$$P_K(K) \equiv \frac{1}{\Delta K \sum_j \delta A(j)} \times \sum_j \delta A\left(j \middle| K - \frac{\Delta K}{2} \leqslant K(j) < K + \frac{\Delta K}{2}\right) \tag{20}$$

Normalizations of $P_H(H)$ and $P_K(K)$ are $\sum_H P_H(H)\Delta H=1$ and $\sum_K P_K(K)\Delta K=1$, respectively.

3.3 Surface Topology

In the preceding sections, the methods to analyze local structural information, the surface curvatures, are presented. Measurement of interfacial shape is particularly important; for example, block copolymers exhibit several types of complicated nanostructures. This is due principally to immiscibility between the dissimilar constituent sequences. A subtle balance between interfacial tension and molecular packing frustration dictates the local shape of the interface, which ultimately determines the global morphology (see Sect. 5 for details). Likewise, in the case of transient morphologies developed during the phase separation of polymer mixture (see Sect. 4), the onset of morphological formation reflects thermodynamic instability, whereas structural evolution is again driven by an overall reduction in interfacial area and thus the evolution of local interfacial shape should be significant.

Besides the surface curvatures that define the metric relationships of objects, there are some fundamental aspects of shapes which are preserved if the structures are made of "stretchable rubber sheet"—*topology*. Suppose a surface is subdivided into a number of faces, edges, and vertices. A topological characteristic, χ, called the Euler–Poincaré characteristic, depends only on the topology of the facetted surface and is defined:

$$\chi = F - E + V. \tag{21}$$

Here F represents the number of faces; E and V are the number of their edges and vertices, respectively. For example, all polyhedra exhibit χ=2. This can be intuitively understood if one imagines the polyhedra can be deformed into a sphere. In other words, since all polyhedra are topologically equivalent to the sphere, χ is conserved.

Another important topological characteristic is the genus, g, of a surface. It is a measure of connectedness of a surface and is equal to the number of holes or handles in the surface, provided the surface is closed, and is related to χ by

$$g = [2 - \chi]/2. \tag{22}$$

χ of the sphere is 2 and thus g=0. The sphere has zero hole or handles. As one attaches handles (or digs holes) one by one in order to increase connectedness, g increases by 1 and simultaneously χ decreases by 2: the more complicated the structure becomes, the less χ becomes.

Remarkably, the topology is linked to the integral surface curvature by the following simple equation (the *Gauss–Bonnet theorem*):

$$\int_{\text{surface}} K da = 2\pi\chi. \tag{23}$$

This elegant result asserts that no matter how a surface is stretched or squashed, its integral curvature remains unchanged. It relates the local structure of the interface to its global topology.

Both Eqs. 21 and 23 can be used to evaluate χ and subsequently g from Eq. 22. This means that it is now possible to access unexplored but important structural quantities to characterize global features of structures, the connectivity of the structures. It is likely that the connectivity of the network and the mechanical properties of the materials are mutually related.

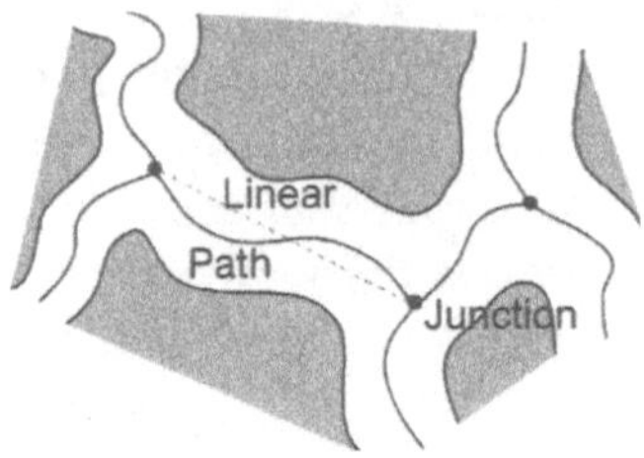

Fig. 8 Schematic illustration of 3D thinning method. The network domain (*white part*) is ultimately "skeletonized" to a skeletal network, from which several topological parameters can be evaluated

Actually, in medical science, the connectivity of trabecular bone attracts increasing attention, as it has been hypothesized that a loss of trabecular elements and consequently a loss in connectivity is a primary reason for decreasing strength and stiffness in osteoporosis [42]. Pursuing a relation between the connectivity and the mechanical properties in polymeric systems may be likewise important [43]. The connectivity is thought to be a very sensitive measure of the occurrence of morphological transition during SD [44].

It is found that both Eqs. 21 and 23 give accurate results if the surface of the objects is very smooth and thus no defects are found in the triangulation process by the MCA. In other words, determination of χ using the number of faces, edges, and vertices was found to be more sensitive to RI than the surface curvature measurements, e.g., PSM and SFM. In most cases, the Gauss–Bonnet theorem offers reliable results.

The above two methods to measure χ (and thus g) are based on surface. There is another way to evaluate the topological measures. It is the "3D thinning method" and is particularly effective to understand the topological features of network-type structures: bicontinuous structures transiently developed in the phase separation process of binary liquid (polymer) mixtures or those formed in the microphase separation of block copolymers [as described in Sects. 4 and 5, respectively]. A 3D thinning algorithm was originally proposed by Toriwaki et al. [45, 46] and modified by us to offer reliable measurements of network-type structures [47, 43]. The algorithm was used to translate the volume image of the complex domain network into a skeletal one, as schematically illustrated in Fig. 8. It involves Euclidean distance transformation [48] to assure that the obtained skeletal networks locate at the center of the phase-separated domains.

The periodic network structures with known coordination number at a junctions, N_j, and distance between adjacent junctions, D_j, e.g., the infinite periodic minimal surfaces (IPMS), were employed to confirm the accuracy of the 3D thinning algorithm. Various types of IPMS, such as the Schoen's

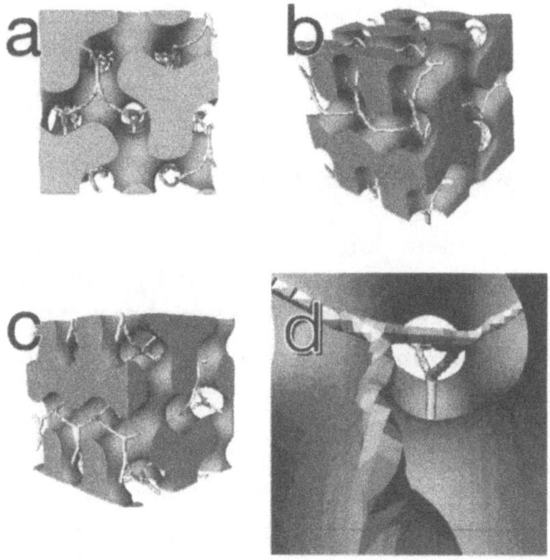

Fig. 9a–d Skeletal network after the 3D thinning of G morphology. **a–c** G morphologies with the skeletal network from different viewpoints. **d** Network strand inside the 3D volume data

gyroid (G) (N_j=3), diamond (N_j=4), I-WP (N_j=8), P (N_j=6), and neovius (N_j=4 and 12), were used [47, 43]. It was found that our 3D thinning algorithm gave accurate measurements for N_j and D_j, as long as (i) there are enough unit cells ($C_{total} \geq 100$, C_{total}: total number of unit cells) in the 3D volume and (ii) the unit cell edge consists of more than 20 pixels.

Figure 9 demonstrates the spatial arrangements of the network strand with respect to the precursor bicontinuous domains, the G morphology. The solid line in each 3D volume represents the network strand. It is clearly observed that the skeletal network computed by the 3D thinning locates at the center of one of the domains. The network strand of the G morphology had three-branched junctions as expected [see Fig. 9d].

The Euler–Poincaré characteristic, χ, and the genus, g, can be estimated from the total number of junctions, N, and total number of branches, B, through the following equation [49]:

$$\chi_{skel} = 2N - B. \tag{24}$$

Here $B \equiv \sum_{i=1}^{N} N_{j,i}$ ($N_{j,i}$ is the coordination number at the i-th junction).

4
Time Evolution of Bicontinuous Structure in Spinodal Interface of a Binary Polymer Mixture

4.1
Basics of Phase Separation in Binary Polymer Mixutures

The dynamics and pattern formation during phase separation processes have been a subject of many experimental and theoretical studies over the past decades as a fascinating example of nonlinear, nonequilibrium phenomena [1, 2]. If a binary mixture is rapidly quenched from the single-phase region to the spinodal region of the phase diagram by changing thermodynamic variables, such as temperature and pressure, the mixture becomes thermodynamically unstable and separates, via spinodal decomposition (SD), into two phases. If the volume fraction of one of the phases is close to 0.5 ("isometric case"), the phase-separated structure is implied to be periodic and bicontinuous with the aid of theories [50], experiments [51, 52], and computer simulations [53–56].

A considerable number of studies have been carried out to investigate the SD processes mainly in the Fourier space using scattering techniques, e.g., light scattering (LS) [57–61] and small-angle neutron scattering [62–66]. As a result of these studies, a great deal of information on the time evolution of the phase-separated structures has been obtained: various characteristic lengths of the structure, e.g., characteristic length of the periodic structure, inverse of interfacial area per unit volume, and interface thickness, have been statistically discussed as a function of the phase-separation time, t. Analysis of the time evolution of such characteristic lengths gave the following notable features of the evolving structures: (i) evolution of concentration fluctuations in the early stage SD can be well described by the linearized theory [67, 68]; and (ii) the structures grow with dynamical self-similarity in the late stage SD, viz., forms of the structures at various times are statistically identical while the characteristic wavelength of the periodic structure increases ("dynamical scaling hypothesis" proposed by Binder and Stauffer) [61, 69–71].

In the late stage of SD, compositions of the coexisting phases have already reached equilibrium values described by the phase diagram [63]. Thus, an excess free energy of the system is localized at the interface between the two phases. Structure growth is driven by *minimizing the interfacial free energy of the system to decrease the interfacial area.* Hence, under such conditions, differential geometrical properties of the interface, e.g., the interface curvatures, and their time evolution are particularly meaningful and essential. With LSCM and 3D structural analysis, direct measurements of such important quantities become possible.

4.2
Direct 3D Visualization of Phase-Separating Bicontinuous Structures

A binary mixture of deuterated polybutadiene (DPB) and polybutadiene (PB) was used to investigate the phase-separation processes of SD. Detailed information about the DPB/PB mixture, e.g., molecular characteristics, phase diagram, and sample preparation, can be found elsewhere [72]. Note that the DPB/PB mixture used here has an upper critical solution temperature (UCST)-type phase diagram and it requires homogenization by mechanical mixing [73] in order to bring the system to the single-phase state. The thickness of the mixture, D, was 200 μm.

The phase-separated structures of the DPB/PB mixtures were observed by LSCM with an incident laser beam wavelength λ of 364 nm. A band pass filter (395–440 nm) installed in front of the detector (photomultiplier) was used to detect only fluorescence from the anthracene molecules that were labeled only to the PB. The PB phase was recognized as a bright phase under the fluorescent LSCM (see Fig. 10). The intensity of fluorescence from a particular point in a focal plane (x–y or lateral plane) at a given depth z, $I(x,y,z)$, was recorded by the detector behind a pinhole ("confocal pinhole") which efficiently excludes out-of-focus light, thus achieving an excellent depth resolution. Here, the z -axis denotes the optical axis of the micro-

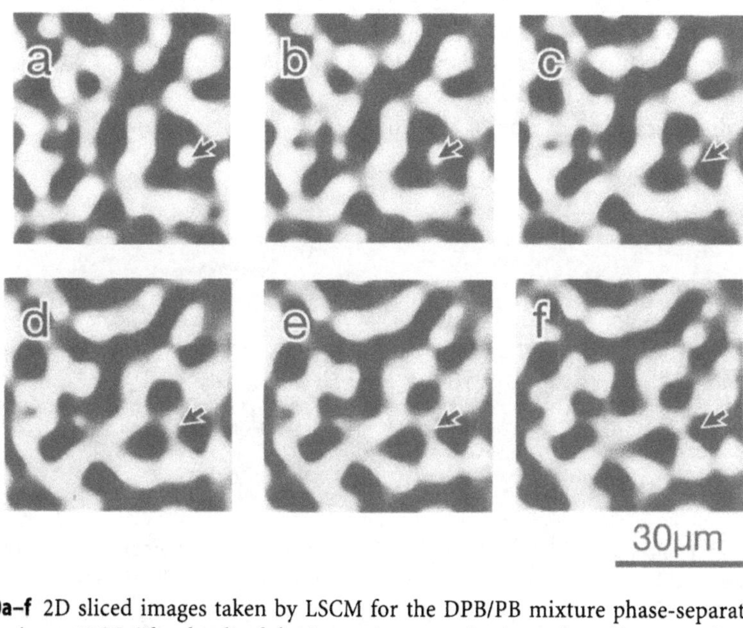

Fig. 10a–f 2D sliced images taken by LSCM for the DPB/PB mixture phase-separated for 1,675 min at 40 °C. The depth of the image increases in the order of image **a**–**f** with the increment of 1.0 μm. The point marked by an *arrow* corresponds to the same pixel position in all images

scope. The laser beam was scanned in the lateral plane, giving rise to a 2D optically sliced image. Thus, a series of 2D sliced images with an increment of ΔZ along the z-axis was obtained, which was used to reconstruct the original volume object.

The mixture was annealed at 40 °C for six different phase-separation times t (t=1,280, 1,675, 2,880, 4,610, 7,365, and 8,610 min, all in the late stage SD). Figure 10 shows a series of 2D sliced LSCM images obtained from the DPB/PB mixture at t=1,675 min. Parts (a) to (f) show 2D slices in the same region of the specimen at different depths. With the excellent depth resolution of LSCM, one can actually investigate how the phase-separated domains are spatially arranged. The bright circular domain marked by an arrow, i.e., a part of the PB phase, in Fig. 10a was surrounded by the dark domain, i.e., DPB phase. It appeared isolated from the neighboring PB domains. As the focal plane got deeper, this part was found to be connected to the neighboring PB domains [see parts (c) and (d) at depths of 2.0 and 3.0 μm relative to part (a)]. This part of the PB domain shown in part (e) at the depth level of 4.0 μm relative to the image in part (a) appeared to be a junction. The marked region in Fig. 10a became a DPB domain at 11 μm below (although not shown). Thus, the two domains are really interconnected forming a 3D maze.

The 2D LSCM slices were "binarized" with the aid of computer image analysis and were stuck together to make a 3D representation of the structures (see Sect. 3.1). The MCA was used to model the interface between the DPB and PB components by contiguous polygons in the reconstruction process. The interfacial area was estimated by summing the area of the polygons, from which the interfacial area per unit volume, $\Sigma(t)$, was estimated. The volume fraction of one of the phases, ϕ, was also measured. The volume

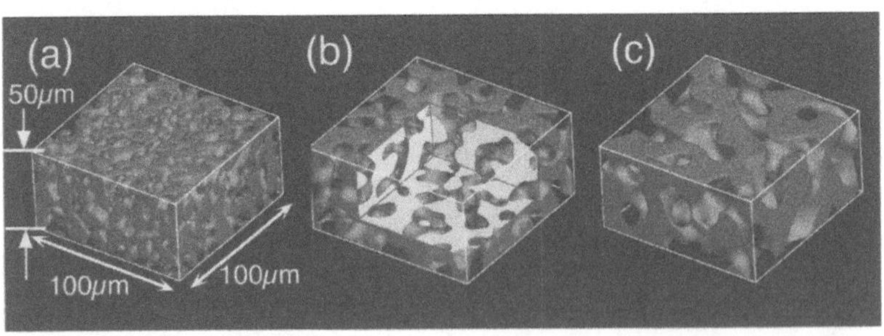

Fig. 11a–c Fluorescence LSCM images showing time evolution of bicontinuous structures for the DPB/PB mixture at critical composition. The image of 100×100×50 μm^3 was obtained at **a** 1,675, **b** 2,880, and **c** 4,860 min after the onset of SD at 40 °C. Parts **a–c** show a "solid model": the images show only the PB-rich phase labeled with anthracene and the other phase (DPB-rich) is left empty. A part of the structure in **b** was removed to show a cross section of the 3D structure

fraction of the DPB-rich phase remained 0.5 over the entire range of our annealing time in the LSCM experiments.

Figure 11 displays the time evolution of the 3D phase-separated structures. The representative phase-separated structures at three different annealing times are shown in the figure, all corresponding to the structures in the late stage SD. The figure clearly demonstrates that the phase-separated structure is bicontinuous and interpenetrating. Such morphology may be called "sponge-like" after the terminology of "sponge" in the field of topology and differential geometry [38] and after the close resemblance of this structure to the sponge-like structure found in a microemulsion at the hydrophile–lipophile balance point [74].

4.3 3D Structural Analysis of Phase-Separating Bicontinuous Structures

In this section, we demonstrate various types of structural analysis from the 3D volume data taken by LSCM, sometimes in conjunction with scattering data taken by time-resolved light scattering (TRLS) [75].

4.3.1 Scaled Structure Factor

The time evolution of the phase-separated structure was also followed in Fourier space for the DPB/PB mixture by using TRLS [75]. In the TRLS experiments, two different wavelengths, i.e., 488 nm (Ar laser) and 632.8 nm (He-Ne laser), were used as incident lasers in order to expand the observable range of wave number, q. Note $q \equiv (4\pi/\lambda_{LS})\sin(\theta/2)$ (λ_{LS}: wavelength of the incident laser, θ: the scattering angle in the medium). The global shape of the phase-separated structures has been studied by the (scattering) structure factor, $S(q,t)$, which can be evaluated not only from the scattering intensity profile but also from the 3D image data obtained by LSCM. In this section, comparison of $S(q,t)$ between TRLS and LSCM will be made.

The time evolution of the scattering intensity, $I(q,t)$, is given by

$$I(q,t) = C\langle\eta(t)^2\rangle\Lambda_m(t)^3 S(q,t) \sim C\langle\eta(t)^2\rangle q_m(t)^{-3} S(q,t), \tag{25}$$

where C is a proportionality constant. $\langle\eta(t)^2\rangle$ is the mean squared refractive index fluctuation proportional to the squared amplitude of concentration fluctuations. $S(q,t)$ is the scattering structure factor at a given t during SD which characterizes the form of the phase-separated structure. $q_m(t)$ is the wave number at which $I(q,t)$ shows a maximum. We now define a scaled structure factor, $F(x,t)$, by

$$F(x,t) \equiv I(x,t) q_m(t^3) \sim C\langle\eta(t)^2\rangle S(x,t), \tag{26}$$

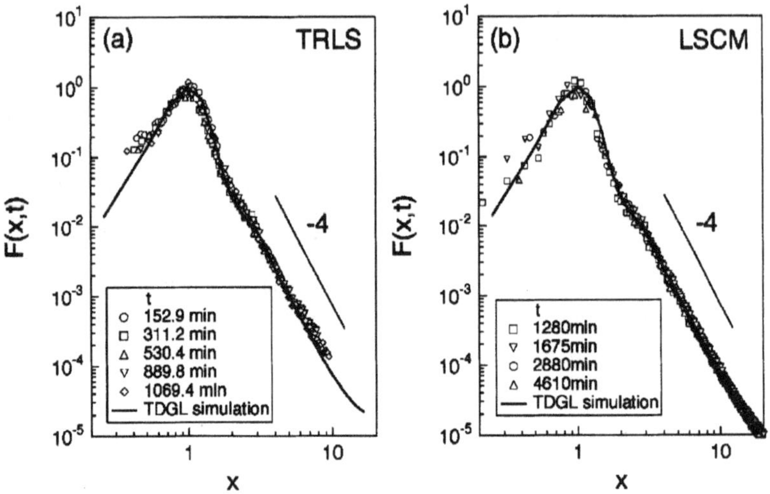

Fig. 12a,b Comparison of scaled structure factors, $F(x,t)$, obtained from LS (**a**) and from LSCM (**b**). $F(x)$ shown by *solid lines* in **a** and **b** are identical and obtained from 3D computer simulation based on the TDGL equation with hydrodynamic interactions

with

$$x \equiv q/q_{\mathrm{m}}(t). \tag{27}$$

In the late stage SD, since $\langle \eta(t)^2 \rangle$ becomes a constant equilibrium value, $F(x,t)$ does not depend on t if $S(q,t)$ becomes a time-independent universal scaling function, $S(x)$, which depends only on the form of the phase-separated structure. Consequently, in the late stage SD, $F(x,t)$ can be rewritten as $F(x)$ ("universal scaled structure factor"), if this criterion is satisfied.

$F(x,t)$ experimentally obtained from TRLS (152.9 min$\leq t \leq$1,069 min) is shown in Fig. 12a. The scaled structure factor from LSCM (1,280 min$\leq t \leq$4,610 min), shown in Fig. 12b, was calculated in the following way: $I(q_x,q_y;t)$ was first obtained from the 3D image by taking the square of the magnitude of its 3D Fourier transformation, where q_i (i=x and y) denotes x and y axes in Fourier space. $|q| \equiv 2\pi j/NL$ (j=0, 1, 2,...,N). $I(q_x,q_y;t)$ thus obtained was then orientationally averaged in the q_x–q_y plane to calculate $I(q,t)$. The abscissa has been scaled by $q_{\mathrm{m}}(t)$ and the ordinate by $I_{\mathrm{m}}(t)$ to obtain $F(x,t)$.

As demonstrated in Fig. 12, $F(x,t)$ fell on to a single master curve both in the TRLS and LSCM experiments, suggesting that the form of the phase-separated structures at various times can be scaled with the single length parameter, i.e., characteristic wavelength of the concentration fluctuations, $\Lambda_{\mathrm{m}}(t)$. The structures grow with self-similarity: the bicontinuous form is maintained over the phase-separation time covered in the experiment. Only the size of the morphology increases with t. $\Lambda_{\mathrm{m}}(t) \equiv 2\pi/q_{\mathrm{m}}(t)$. The solid lines in Fig. 12 show the $F(x)$ obtained from a 3D computer simulation based on the time-dependent Ginzburg-Landau (TDGL) equation incorporating hydrodynamic interactions [56, 76, 77]. $F(x)$ both from TRLS and LSCM were in excellent agreement with the theoretical structure factor over a large intensity scale, as large as five orders of magnitude, and a wide spatial scale, as wide as two orders of magnitude. This result demonstrates that $F(x)$ from the two independent methods (TRLS and LSCM) are *quantitatively* identical, hence assuring that the LSCM captures the real structural entity. Moreover, the TDGL theory gives a physical basis to the experimentally obtained bicontinuous phase-separated structure developed via SD. Furthermore, the unique scaled structure factor manifests the sponge-like real-space structure shown in Fig. 11.

4.3.2 Interfacial Curvature Measurement: Comparison Between LSCM and TRLS

In this section, we show an example of the measurement of the area-averaged curvatures by the PSM (see Sect. 3.2.2) and compare the obtained result with those independently estimated from a scattering measurement, i.e., LS [75]. Note that, unlike the previous section, a poly(styrene-*ran*-butadiene) (SBR) was used instead of DPB as one of the constituents of the polymer mixtures. The phase separation was induced by heating the SBR/PB mixture at 40 °C. Besides the difference in constituent polymers, the SD behavior and underlying physics of the SBR/PB system are essentially the same as those of the DPB/PB system.

As already described, after the LSCM observation and the 3D reconstruction, the interface of the SBR/PB mixture is expressed by many triangles with the normal vectors at vertices. Figure 13a displays the interface between

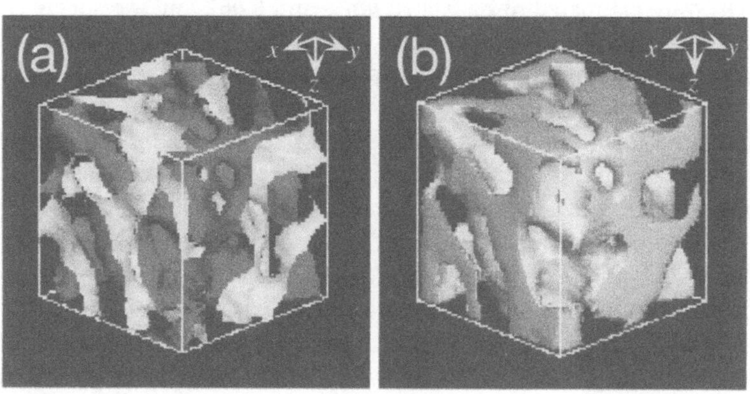

Fig. 13a,b 3D image of the SBR/PB phase-separated structure (box size: 30×30×30 μm^3). **a** Interface between the two coexisting phases.**b** Solid modeling view of the same volume as shown in **a**

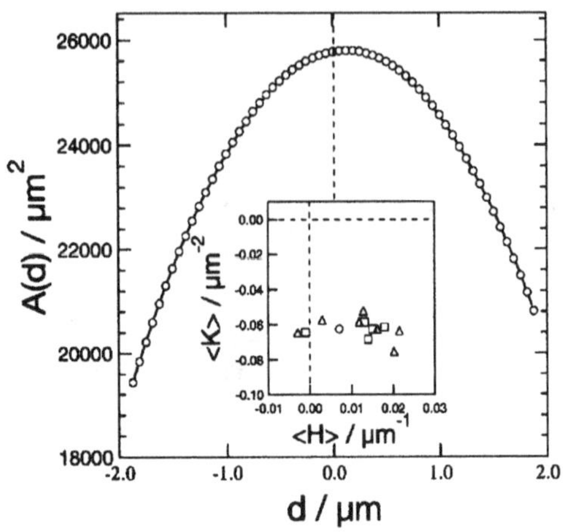

Fig. 14 The area of parallel surface, $A(d)$, plotted against the displacement from the interface, d. The sample volume for the measurement was 62.5×62.5×30 μm^3. The *solid line* shows the best fit of Eq. 9, from which curvatures were determined. In the *inset*, *triangles*, *circle* represent, respectively, the results obtained from a volume of 31.3×31.3×30 μm^3 (△), 46.9×46.9×30 μm^3 (□), and 62.5×62.5×30 μm^3 (○) at the different position of the 3D image

the PB-rich and the SBR-rich phases. One side (the SBR-rich domain) of the interface is gray, while the other side (the PB-rich domain) is white. Figure 13b is the same volume as shown in part (a), where the SBR-rich phase is shown as transparent and the PB-rich phase as solid-like. The phase-separated structure appears to be periodic and bicontinuous, just as the structure presented in Fig. 11.

Figure 14 shows $A(d)$ as a function of d for the SBR/PB mixture. The solid line in the figure represents the best fit of Eq. 9 to the data, from which $\langle H\rangle$ and $\langle K\rangle$ were estimated to be 0.0070 μm^{-1} and 0.062 μm^{-2}, respectively. The negative sign of $\langle K\rangle$ clearly provides convincing evidence that the interface of the phase-separated polymer blend is, *on average*, hyperbolic with effectively zero mean curvature. Note that the results varied with size and the position of the image for the measurements. Such a variation is displayed in the inset of Fig. 14. The averaged principal curvatures can be estimated by solving a quadratic equation, i.e., $x^2-2\langle H\rangle x+\langle K\rangle=0$. $\langle\kappa_1\rangle\cong0.26$ μm^{-1} and $\langle\kappa_2\rangle\cong-0.24$ μm^{-1}. The absolute radii of the curvatures were $|\langle R_1\rangle|\equiv|1/\langle\kappa_1\rangle|\cong3.8$ μm and $|\langle R_2\rangle|\equiv|1/\langle\kappa_2\rangle|\cong4.1$ μm, respectively, the difference of which was close to the lateral resolution of the microscope (i.e., 0.3125 μm/pixel) and may be negligible. The area-averaged mean radius can be defined as $|\langle R\rangle|\cong(|\langle R_1\rangle|+|\langle R_2\rangle|)/2$, which gives $|\langle R\rangle|\cong4.0$ μm in the present case. Figure 15 shows one of the optical slices used to reconstruct the 3D image, in which circles with the radius $|\langle R\rangle|$ are drawn. It demonstrates that many parts of the interface can be fitted by these circles.

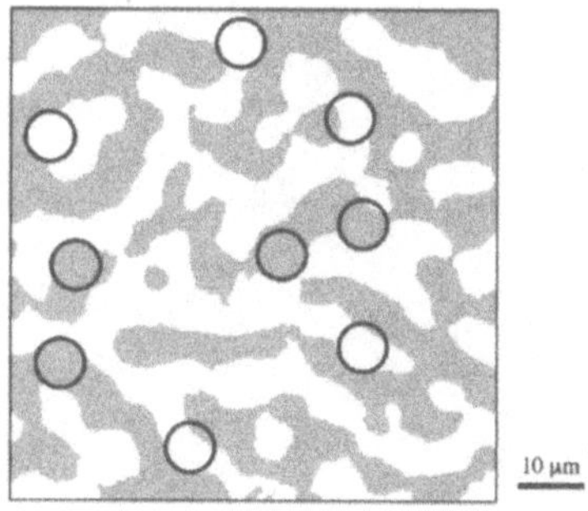

Fig. 15 An optical section taken at a level of 47.5 μm from the coverslip. The *white* phase corresponds to the PB phase, and the *gray* phase to the SBR phase. *Circles* with the radius of 4 μm, which corresponds to $|\langle R\rangle|$ obtained by the PSM, are shown

TRLS (i.e., a reciprocal-space method) has been used to measure the mean radius of curvature of the phase-separated structure [78, 79]. Since we evaluated $|\langle R\rangle|$ from the PSM (i.e., real-space method), it is worthwhile to compare the radius of curvature from both methods. The reciprocal-space method relies on the following Tomita's theory [80].

Tomita discussed a correlation function at a short distance for the late stage phase-separated isotropic system [80], in which semimacroscopic random interfaces were assumed. Fourier transformation of the correlation function gives $S(q)$, whose large q expansion is:

$$S(q)=\frac{2\pi A(0)}{q^4 V_s}\left\{1+\frac{1}{R_m^2 q^2}+\cdots\right\}. \tag{28}$$

V_s is the volume of the whole system. R_m is a kind of area-averaged mean radius ("scattering mean radius") defined by

$$R_m^{-2}=\iint \bar{R}(a)^{-2}da\Big/\iint da. \tag{29}$$

Using $\langle H\rangle$ and $\langle K\rangle$, Eq. 29 can be written as

$$R_m^{-2}=3\langle H^2\rangle/2-2\langle K\rangle/2. \tag{30}$$

Here, Eq. 30 represents the mean square of the sectional curvature averaged around the normal vector. The first term in Eq. 28 is the Porod scattering [81]. The crossover from $q^{-n}(n>6)$ to q^{-4} with increasing q in the structure factor occurs at $q_c R_m \simeq 1$ [78, 79].

Experimentally, Fig. 16 demonstrates that $S(q)$ shows a crossover from $q^{-n}(n\cong 6)$ to q^{-4}. Following Takenaka et al. [78], Hashomoto et al. [82], and Läuger et al. [79], we define q_c at the intersection of these two lines, i.e., $q_c\cong 0.8$ μm^{-1} and hence the mean radius obtained from LS, $R_{m,LS}\cong 1/$

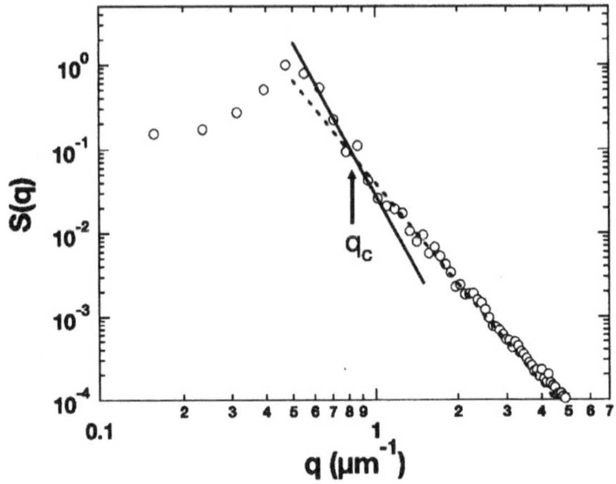

Fig. 16 Structure factor, $S(q)$, vs the wave number, q, obtained from the 3D Fourier transformation of the LSCM image. *Solid and dashed lines* represent q^{-6} and q^{-4}, respectively. The crossover wave number, q_c, is shown (see text for details)

q_c=1.25 μm, which means that the ratio of the radius to the periodic distance of the structure, $R_{m,LS}/\Lambda_m$, is roughly 0.10, as $\Lambda_m=2\pi/q_m$=12.1 μm in this case.

On the other hand, since $\langle H^2\rangle$ was independently evaluated to be 0.022 μm^{-2} from SFM measurements for the SBR/PB blend, the mean radius from LSCM, $R_{m,LSCM}$, was 4.0 μm according to Eq. 30. Moreover, $|\langle R\rangle|\cong$4.0 μm was directly obtained from PSM. Both $R_{m,LSCM}$ and $|\langle R\rangle|$ are in excellent agreement, demonstrating that the framework of Tomita's theory seems to be correct. However, they are bigger than $R_{m,LS}$ by a factor of about 3, which implies that the relation used in LS experiments, i.e., $q_cR_{m,LS}=1$ is not quite true and hence measurement of R_m by TRLS *would not be accurate* after all. Thus, a new relation, such as $q_cR_m\cong3.2$ ($R_m/\Lambda_m\cong0.33$), may be used instead of $q_cR_m=1$ ($R_m/\Lambda_m\cong0.10$).

4.3.3 Curvature Distribution Measurements of Phase-Separating Bicontinuous Structures

As mentioned in Sect. 3.2.3, SFM provides local surface curvatures at a given point on the surface, so the joint probability density, $P(H,K;t)$, at a given t can be obtained. $\iint P(H,K;t)dHdK=1$ was used for normalization. Figure 17 shows $P(H,K;t)$ of the DPB/PB mixture at t=1,675 min, which has a characteristic functional form, elucidating the following characteristics: (i) most of the measured points (up to 93%) lie in the $K<0$ region, demonstrating that the interface is anticlastic—two principal curvatures have opposite signs at a large portion of the interface; (ii) the probability for $K>0$, corresponding to

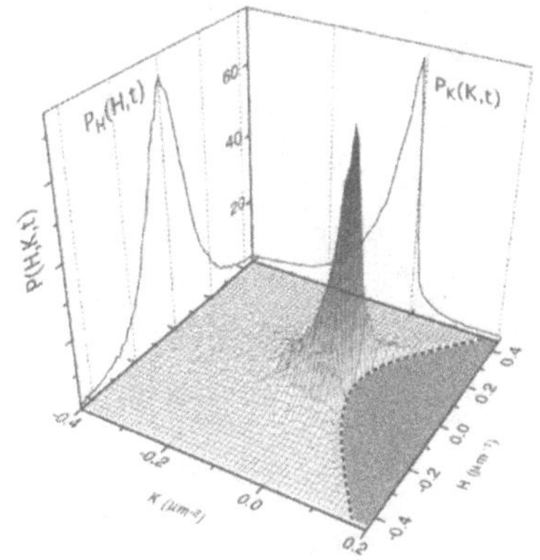

Fig. 17 Bird's-eye view of the joint probability density *P*(*H*,*K*;*t*) for the DPB/PB mixture at *t*=1,675 min. *P*(*H*,*K*;*t*) was obtained by sampling ca. 200×10^3 surface points randomly chosen. Typically, the entire surface consists of about 2×10^6 surface points. We made sure that the number of points used for determining *P*(*H*,*K*;*t*) were enough so that the shape of the distribution became invariant with further sampling. $P_H(H;t)$ and $P_K(K;t)$ are also shown. The *broken parabolic line* shows $K=H^2$

the synclastic surface, is small; and (iii) the observed probability density fulfills a relation of $K\leq H^2$. This relation is required because the principal curvatures, which can be calculated from and via $\kappa_i = H \pm \sqrt{H^2 - K}(i=1,2)$, should be real numbers.

From *P*(*H*,*K*;*t*), the probability densities of the mean curvature, $P_H(H;t)$, and that of the Gaussian curvature, $P_K(K;t)$, are calculated from

$$P_H(H;t) \equiv \int P(H,K;t)dK, \tag{31}$$

and

$$P_K(K;t) \equiv \int P(H,K;t)dH. \tag{32}$$

Figure 18 demonstrates the time evolution of the probability densities. $P_H(H;t)$ is symmetric about H=0 and thus $\langle H \rangle$ stayed close to zero (see Fig. 18a), implying that the system tends to take the most reasonable evolution path with a low free energy in the SD process. $P_K(K;t)$ is distributed mostly in the K<0 region, demonstrating that the interface is hyperbolic regardless of *t*. The full width at half maximum (fwhm) of $P_H(H;t)$ decreased, thus the portion of the distribution with small $|H|$ increased with *t*. Similarly, the portion of small $|K|$ increased as time elapsed [11]. These results suggest that the average radius of curvature of the spinodal interface increased, i.e., the interface became smoother with *t*.

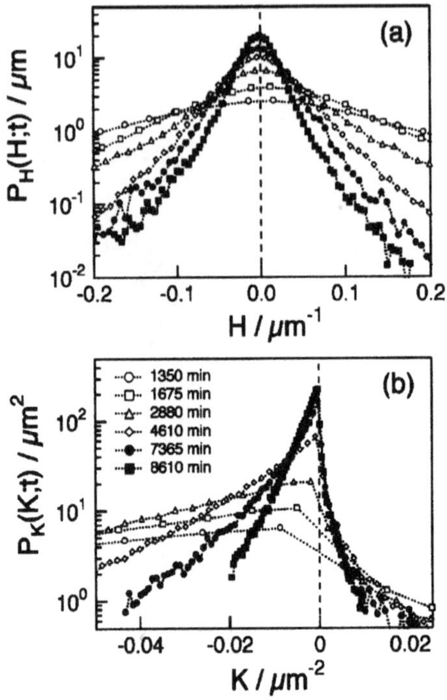

Fig. 18a,b Time evolution of **a** $P_H(H;t)$ and **b** $P_K(K;t)$ for the DPB/PB mixture in the late stage SD

4.4 Scaling Analysis

The examination of the dynamical scaling through $S(x,t)$ has been done many times (see also Sect. 4.3.1 in this review paper); most of the studies showed the self-similar growth of the global structure, which is characterized by the scattering maximum, i.e., a structure corresponding to the length scale of $\Lambda_m(t)$. Although the curvatures and $S(q,t)$ are implicitly related to each other, a general relationship between them is *not at all clear.* Hence, it is impossible to discuss whether or not the local shape of the interface really follows the dynamical scaling law from the scattering data. Meanwhile, the curvatures presented here qualify the local shape of the structure.

In order to test the dynamical scaling law for the local structure, such as the probability densities of the curvatures, the inverse of $\Sigma(t)$ was used as a length characteristic to the bicontinuous structures. The scaled probability densities of H, $\tilde{P}_H(\tilde{H};t)$, and K, $\tilde{P}_K(\tilde{K};t)$, are respectively obtained by

$$\tilde{P}_H(\tilde{H};t) = \Sigma(t)P_H(H;t), \tilde{P}_K(\tilde{K};t)\Sigma(t)^2 P_K(K;t). \tag{33}$$

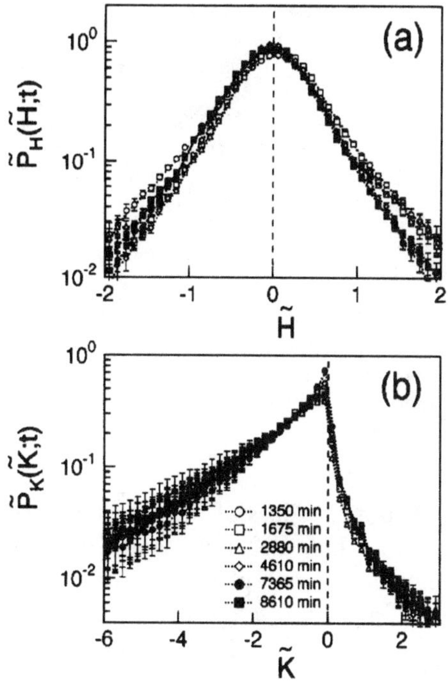

Fig. 19a,b Scaled probability densities (**a**) $\tilde{P}_H(\tilde{H};t)$ and (**b**) $\tilde{P}_K(\tilde{K};t)$ for the DPB/PB mixture in the late stage SD

Here, $\tilde{H} = H\Sigma^{-1}$ and $\tilde{K} = K\Sigma^{-2}$.

The scaled probability densities are calculated using the data presented in Fig. 18 and plotted against t in Fig. 19. The scaled probability densities at all t fell nicely onto single master curves both for H and K, demonstrating that the dynamical scaling law holds for interfacial curvatures within experimental accuracy. The results shown in Fig. 19 stand out by the fact that they demonstrate that not only the global structure but also the local shape of the spinodal interface evolves with the dynamical self-similarity. In addition, the probability densities experimentally obtained in the present study are expected to be universal in a variety of condensed matter systems.

4.5 Effects of Surface Wetting and Confinement on Spinodal Decomposition

Up to here, most studies to date have concentrated on the morphological and temporal features of morphology in the bulk SD, i.e., in the interior of the thick specimen (e.g., D~500 μm). The influences of confinement and spatial dimensionality on the kinetics of the phase separation are interesting problems [83]. In what follows, we will discuss these two effects on SD.

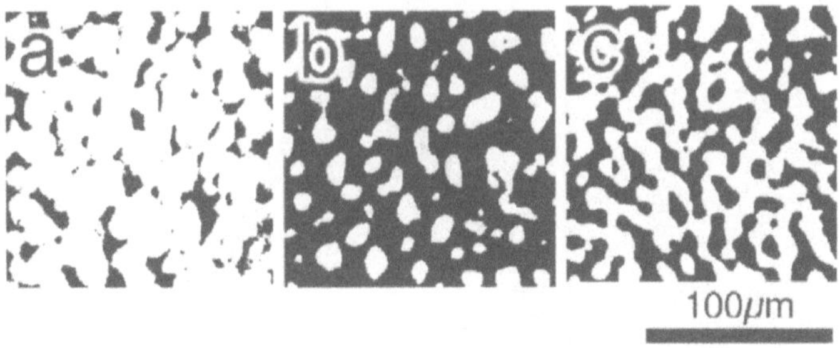

Fig. 20a–c Two-dimensional (2D) optical slices of the DPB/PB mixture undergoing SD at 40 °C for 2,846 min. Binarized images at three different depths from the glass surface, *z*, are presented: **a** z=0 (at glass surface), **b** z~4 μm, **c** z=25 μm. *White* and *black* regions of the 2D images correspond, respectively, to PB-rich and DPB-rich phases. *Bar* corresponds to 100 μm

Figure 20 shows 2D optical slices of the DPB/PB mixture undergoing SD at 40 °C for 2,846 min at three different depths, *z*, from the glass surface. The volume fraction of the PB-rich phase, ϕ_{PB}, was found to be nearly 0.8 at the glass surface (see Fig. 20a), while ϕ_{PB} at z~4 μm was only 0.2 (see Fig. 20b). 2D optical slices of the phase-separated structure at this depth looked much like an "island-in-sea" structure (PB-rich domains are islands), being different from that of the conventional bicontinuous structure deeper inside the specimen ($z \cong 25$ μm) (see Fig. 20c). Such drastic variation of ϕ_{PB} is demonstrated in Fig. 21, in which ϕ_{PB} is plotted against *z* for the DPB/PB mixture (t=2,846 min).

The PB-rich phase preferentially wets the glass surface. Within ca. 10 μm from the glass surface (z<10 μm), there is a wetting layer of the PB-rich

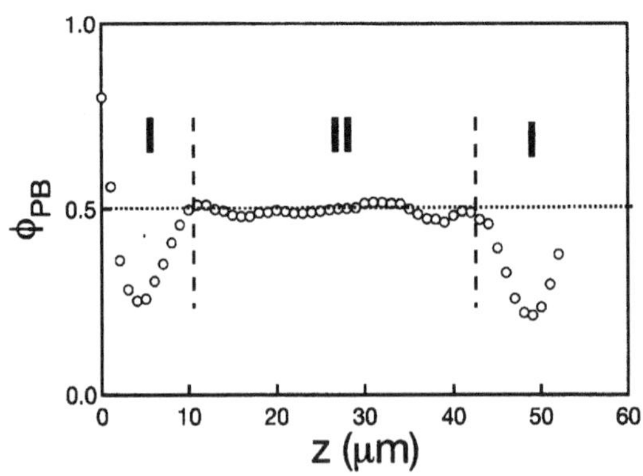

Fig. 21 Depth dependence of volume fraction of PB-rich phase, ϕ_{PB}. Phase-separation time is 2,846 min. The wetting and bulk regions (labeled as *I* and *II*, respectively) were observed

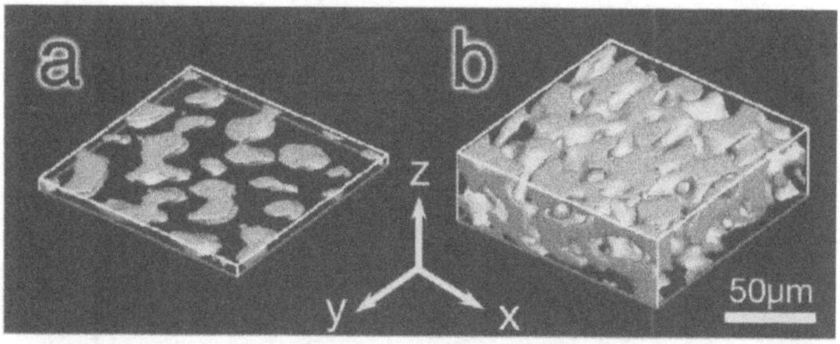

Fig. 22a,b 3D representations of phase-separated structures **a** in the wetting region and **b** in the bulk region. The DPB/PB mixture was annealed for 2,846 min at 40 °C. Images show only PB-rich phase labeled with anthracene and DPB-rich phase is left empty. *Bar* shows 50 μm

phase that is followed by a "depletion layer" where the PB-rich phase becomes a minor phase. Similar composition variation can be observed at the other side of the specimen (43 μm<z<55 μm).

These wetting and depletion layers are similar to those observed in poly(ethylenepropylene) (PEP) and perdeuterated poly(ethylenepropylene) (d-PEP) blend having much smaller thickness, D~9,000 Å [84]. These layers were induced by van der Waals interaction from the surface. Namely, breaking translational and rotational symmetry by the preferential wetting of one of the components to the surface caused the spinodal waves to grow normal to the surface in the early stage of SD ("surface-directed spinodal decomposition") [85]. Note that the thicknesses of the wetting layers were of the order of the radius of gyration of the polymers, R_g. We emphasize that the wetting and depletion layers found in our DPB/PB mixture extended for as long as 10 μm, much longer than R_g. The existence of the thick wetting and depletion layers cannot be explained by such short-range interaction, e.g., the van der Waals interaction, from surfaces alone. Possible explanation can be found elsewhere [83].

In the middle of the sample deeper than the wetting regions, there exists a region where ϕ_{PB} stayed constant at around 0.5. This composition is consistent with the bulk composition previously obtained in the thicker specimens consisting of the same DPB/PB blend (D=200 μm) (Sect. 4.2) [72]. In this region, the phase-separated structure is free from surface effects and thus exhibits isotropic bulk phase separation ("bulk region").

Figure 22 displays 3D phase-separated structures corresponding to the wetting region [z=3~6 μm, part (a)] and the bulk region [z=14~44 μm, part (b)] for the DPB/PB mixture at t=2,846 min. Note that the phase-separated structures near the upper and lower coverslips were essentially the same at each phase-separation time. The phase-separated structure in the wetting region becomes an anisotropic network structure as mentioned above. In the bulk region, the phase-separated structure is an "isotropic" bicontinuous

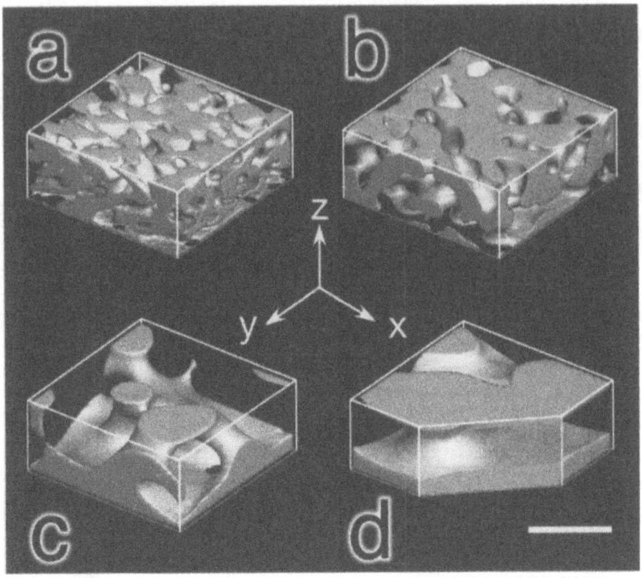

Fig. 23a–d Time evolution of phase-separated structure at **a** t=2,846 min, **b** t=4,310 min, c t=9,823 min, **d** t=21,508 min. *Solid part* of each figure represents PB-rich phase. Bottom plane of each 3D image as shown by *gray* edges represents glass surface. A part of the phase-separated structure in **d** was removed to show a cross section of 3D structure, demonstrating the formation of thick wetting layer and "columnar structure". *Bar* corresponds to 50 μm

network structure, as expected for the polymer mixture with the critical composition.

Demonstrated in Fig. 23 is the time evolution of the phase-separated structure of the DPB/PB mixture (reconstructed from the glass surface (z=0) to the bulk region). Up to t=4,310 min, the bicontinuous network structure was dominant except for the wetting region, while it transformed into the " columnar structure" in which cylindrical domains consisting of PB-rich phase bridge the upper and lower PB wetting layers (see Fig. 23d). As shown below, the occurrence of the topological transition was characterized by the various structural parameters.

The characteristic length of the phase-separated structure, $\Lambda_m(t)$, is a first-choice structural parameter in morphological studies, especially in TRLS experiments. $\Lambda_m(t)$ can be directly calculated from the LSCM 3D volume data (see Sect. 4.3.1) [72]. In the scattering experiments, the incident laser is sent from the film normal (z-axis) and the scattering intensity is measured in the x–y plane, from which $q_m(t)$ and thus $\Lambda_m(t)$ were determined.

Figure 24 shows the time evolution of $q_m(t)$ [and thus $\Lambda_m(t)^{-1}$] in the DPB/PB mixture [D=55 μm, "thin (or confined)" blend]. For comparison,

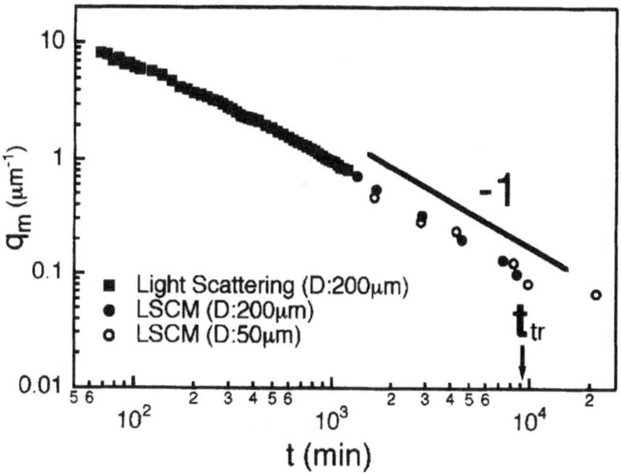

Fig. 24 Time evolution of characteristic wave number, $q_m(t)$, in a double logarithmic plot. $q_m(t)$ was determined from peak position of scattering intensity. *Filled symbols* represent $q_m(t)$ for the DPB/PB "thick" sample (D=200 μm), while *open circles* show $q_m(t)$ for the DPB/PB "thin" sample (D=55 μm). *Circular and square symbols* correspond to the data obtained from light scattering and LSCM, respectively

the time dependence of $q_m(t)$ in a thick DPB/PB blend [D=200 μm, "thick" blend] is also presented. The open circles correspond to $q_m(t)$ estimated for the thin DPB/PB mixture. It is intriguing that the time evolution of $q_m(t)$ in the thin blend traced a similar path to that of the thick DPB/PB mixture. In the late time region, $q_m(t)$ showed power law behavior, i.e., $q_m(t) \sim t^{-1}$ regardless of the thicknesses, being consistent with the experimental observations in liquid mixtures free from viscoelastic effects [56]. The last point of the thin DPB/PB blend, i.e., $q_m(t$=21,508 min), may deviate from the power law behavior. The time evolution of $q_m(t)$ indicates that the growth of the bicontinuous structure in the lateral plane is *not significantly affected* by the confinement.

The time evolution of the scaled structure factor, $F(x,t)$, the interfacial curvature distributions, $P(H,K;t)$, and the interfacial area per unit volume, $\Sigma(t)$, were also examined to study both the global and local features of the DPB/PB phase-separated structure. Depicted in Fig. 25 is the time evolution of $\Sigma(t)$ for the thin and thick DPB/PB mixtures. As reported before [72], a power law behavior, $\Sigma(t) \sim t^{-1}$, was observed for the thick blend, while the thin blend showed the same exponent until t_{tr}, and stayed constant afterwards ($t_{tr} \sim$9,000 min). This experimental observation indicates that the phase-separation kinetics were changed at $t \cong t_{tr}$. Although not shown here, both $F(x,t)$ and $P(H,K;t)$ also exhibited master curves (and thus demonstrated self-similar growth) until a certain phase-separating time, $t < t_{tr}$. Interestingly enough, $\Lambda_m \cong D$ at $t \cong t_{tr}$. In other words, the DPB/PB blend began "feeling" the container wall and the kinetics of the phase separation changed

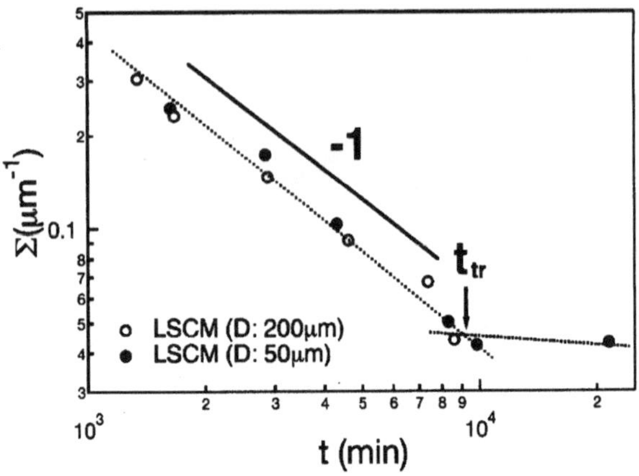

Fig. 25 A double logarithmic plot of interfacial area per unit volume, $\sum(t)$, as a function of *t*. *Open and filled circles* represent data from thick and thin DPB/PB mixtures, respectively

when the characteristic length of the bicontinuous structure reached the thickness of the sample. Together with the LSCM 3D images, at $t \cong t_{tr}$ it was found that the phase-separated structure transformed from the bicontinuous structure to the columnar structure, i.e., tubes bridging the wetting layers.

Note that the local characteristics such as $\Sigma(t)$ and $P(H,K;t)$ were more sensitive measures of the effect of confinement than the global characteristics, $\Lambda_m(t)$ and $S(x,t)$. Therefore, in most of the existing studies that used the global structural parameters alone to characterize the kinetics of SD, the effects had not been evident.

5 Direct Measurement of Interfacial Curvature Distributions in a Bicontinuous Block Copolymer Nanostructure

5.1 Basics of Microphase-Separated Morphologies in Block Copolymers

Block copolymers exhibit periodic nanostructures due to immiscibility between the dissimilar (A and B) sequences [3]. Classical block copolymer nanostructures include spheres of A(B) on a body-centered cubic lattice in a B(A) matrix, cylinders of A(B) on a hexagonal lattice in a B(A) matrix, and coalternating lamellae. Of considerable recent interest are several complex (bicontinuous) nanostructures—the perforated lamellar (PL), gyroid (G), and double-diamond (D) morphologies [86–91]. These nanostructures may develop if the copolymer composition, *f*, falls within a narrow range between the cylindrical and lamellar morphologies. Figure 26 shows an example of

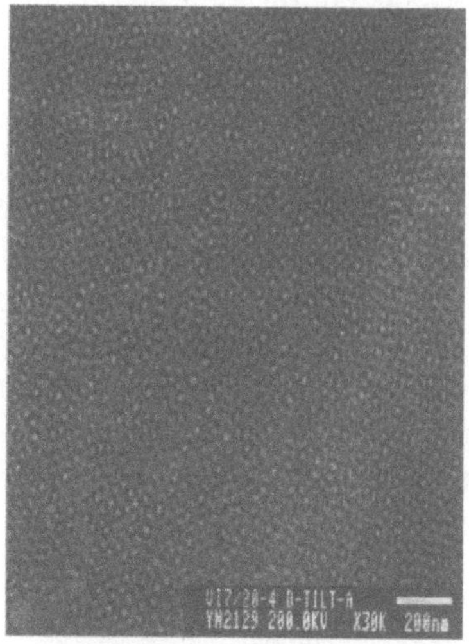

Fig. 26 TEM micrograph of poly(isoprene-*block*-vinylpyridine) diblock copolymer. Due to OsO_4 staining, isoprene-rich regions are electron-opaque and appear dark. The "wagon wheel" projection associated with the [111] axis of G morphology is observed

the G morphology observed by TEM. The characteristic "wagon wheel" projection was observed in the figure. The block copolymer nanostructures presented in Fig. 26, once believed [86] to be D, exemplified by a Schwarz D surface with $Pn3m$ symmetry, have been reclassified [92] on the basis of their SAXS signatures as G, which is represented by the Schoen G surface with $Ia\bar{3}d$ symmetry. Identification of complex nanostructures by TEM is often inconclusive, since they appear identical along several projection axes. In the present section, we demonstrate that with 3D microscopy and image analysis, it will become possible to explore such complicated morphologies which were previously unsolved due to experimental difficulties.

Complex nanostructures similar to the above block copolymer morphology also develop in surfactant and lipid systems due to the formation of surfaces with constant mean curvature (CMC) that minimize contact between immiscible moieties [93]. Since block copolymer nanostructures share common topological features with those of other self-organized systems, the concept of CMC minimal surfaces has been used [87] to explain the stability of complex block copolymer nanostructures. On the basis of self-consistent field theory (SCFT), Matsen and Bates [94, 95] have recently proposed that the area-averaged mean curvature, $\langle H \rangle$, governs the gross morphology (lamellar, bicontinuous, cylindrical or spherical), whereas the standard deviation of the mean curvature distribution, σ_H, determines the delicate stability of the complex nanostructures (G, D or PL). This additional consideration results from packing frustration [96] and implies that, while a surface strives toward CMC, the mean curvature cannot be constant everywhere along the interface since the microdomain-forming blocks must uniformly fill space in the most entropically favored manner. Thus far, neither $\langle H \rangle$ nor σ_H has been measured experimentally despite their apparent importance.

3D visualization of bicontinuous morphologies in block copolymer systems has been achieved [26–27] by TEMT (see Sect. 2.2). This technique affords the real-space structural analysis of complex nanoscale morphologies without a priori symmetry or surface assumptions [97]. Application of numerical methods developed [39, 98] to measure interfacial curvatures from 3D LSCM images of SD polymer blends (see Sects. 3.2.3 and 4.3.3) to a TEMT reconstruction of the G morphology yields the first experimental measurements of interfacial curvature distributions, as well as $\langle H \rangle$ and σ_H, in a complex block copolymer nanostructure.

5.2 Direct 3D Visualization and Curvature Distribution of Bicontinuous Block Copolymer Nanostructure

A poly(styrene-*block*-isoprene-*block*-styrene) (SIS) triblock copolymer was synthesized by living anionic polymerization. Details of the experimental protocols and TEMT measurements are provided elsewhere [22, 26, 99, 100]. Figure 27a shows the reconstructed 3D image of the nanostructure in the SIS triblock copolymer. Volumetric analysis of the reconstruction yields the styrene volume fraction f_S=0.33, in excellent agreement with the known composition of the copolymer (f_S=0.32). The light and dark channel networks evident in Fig. 27a both represent the S microphase. They are shaded differently to demonstrate that the two S channel networks do not intersect. For the sake of clarity in this figure, the I microphase is transparent. A crys-

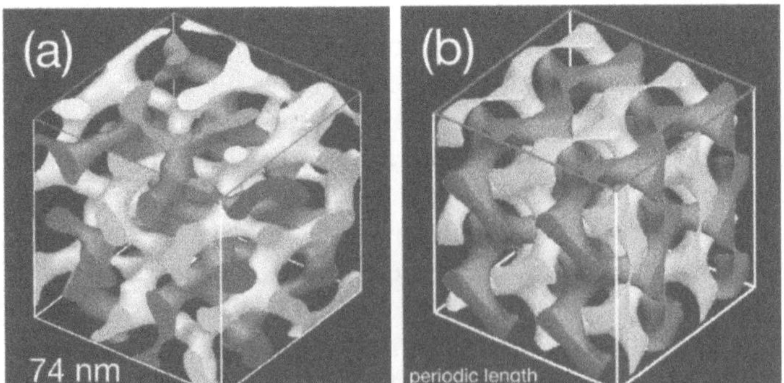

Fig. 27a,b Transmission electron microtomograph of the gyroid morphology in a SIS triblock copolymer (**a**), and the constant thickness model based on Schoen's gyroid surface (**b**). The nonintersecting light and dark channels correspond to the minority microphase (S in the SIS copolymer), while the majority (I) microphase is transparent. The edge of each cube equals twice the periodic length, and the top cross section identifies the (001) plane tallographic analysis of Fig. 27a identifies that the nanostructure is G and reveals that the lattice constants of this nanostructure are a=78 nm, b=71 nm, c=74 nm.

Displayed in Fig. 27b is a model bicontinuous morphology generated from the Schoen G surface. The trigonometric approximation used to generate this surface is given by

$$g(x,y,z) = \cos\frac{2\pi x}{L}\sin\frac{2\pi y}{L} + \cos\frac{2\pi y}{L}\sin\frac{2\pi z}{L} + \cos\frac{2\pi z}{L}\sin\frac{2\pi x}{L}, \tag{34}$$

where L denotes the crystallographic unit cell edge of the gyroid. The periodic minimal surface, which divides space equally, is obtained by setting $g(x,y,z)$ equal to zero. To emulate the microphase-separated nanostructure of the SIS copolymer with f_S=0.33, a new model interface is formed by translating the interface obtained from Eq. 34 along its normals by an equal distance everywhere. Two parallel surfaces with opposite direction but the same displacement, selected so that the volume fraction of the swollen microphase is equal to 0.67, are consequently generated. The parallel surfaces were computed in the same way as described in Sect. 3.2.2. The resultant morphology constitutes an approximate model of the hypothetical CMC interface and, following Hajduk et al. [89], serves as the constant-thickness model. As in Fig. 27a, the nonintersecting light and dark channels shown in Fig. 27b identify the minority microphase, with the majority microphase remaining transparent.

Interfacial curvature distributions, $P(H,K)$, were evaluated from the 3D morphologies in Fig. 27 according to the SFM. The value of RI computed for the TEMT data analyzed here is 0.12. If RI is less than 0.2 in the curvature distribution measurements, a 5% error is expected [98].

5.3 Geometrical and Topological Analysis of Bicontinuous Block Copolymer Nanostructure

Figure 28 shows a surface contour representation of the $P(H,K)$ distribution measured from the G morphology in the SIS triblock copolymer. Included in this figure are the probability densities of the mean and Gaussian curvatures—$P_H(H)$ and $P_K(K)$, respectively—calculated according to Eqs. 19 and 20. The curvature is arbitrarily chosen to be positive if the center of the osculating circle resides within the I microphase. According to Fig. 28, most of the interface (77%) possesses $K<0$, indicating that (i) the two principal curvatures (κ_1 and κ_2) have opposite signs and (ii) most of the interface is hyperbolic. An important characteristic of $P(H,K)$ is that it satisfies the boundary condition $K \leq H^2$ (dashed line). A qualitative conclusion that can be drawn from Fig. 28 is that, since H is not constant everywhere along the in-

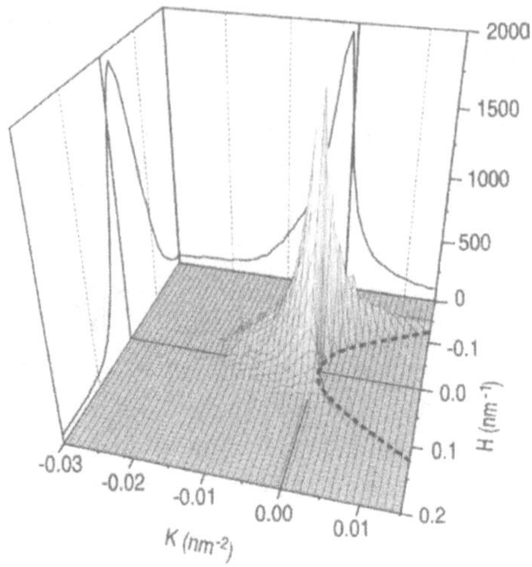

Fig. 28 Surface contour representation of the joint probability density, $P(H,K)$, measured for the G morphology of the SIS triblock copolymer. Marginal probability densities, $P_H(H)$ and $P_K(K)$, are also shown. The *dashed parabolic curve* represents $K=H^2$

terface, the CMC representation does not accurately represent the G morphology. This point is addressed further below.

The interfacial curvature probability densities are displayed in Figs. 29a and 29b for the G morphology in the SIS copolymer and the constant-thickness model surface, respectively. To facilitate comparison, $P_H(H)$ and $P_K(K)$ have been scaled with respect to the interfacial area per unit volume (Σ) in the same fashion as that described earlier (Eq. 33). $P(H,K)$ was scaled in the following way:

$$\tilde{P}(\tilde{H},\tilde{K}) = P(H,K)\Sigma^3. \tag{35}$$

Here, $\tilde{H} = H\Sigma^{-1}$ and $\tilde{K} = K\Sigma^{-2}$, with Σ=0.070 and 0.074 nm^{-1} for the SIS copolymer and constant-thickness model, respectively. Close examination of the scaled probability densities in Fig. 29a reveals that a part of the scaled joint probability density for the SIS G morphology possesses $\tilde{H} < 0$ and $\tilde{K} > 0$, implying that the interface is an elliptic surface curved inward relative to the I microphase. Such interfacial concavity is not evident from $\tilde{P}(\tilde{H},\tilde{K})$ derived from the constant-thickness model of the G morphology in Fig. 29b, in which nearly all (just under 100%) of the measured points possess $\tilde{K} < 0$. Moreover, $\tilde{P}(\tilde{H},\tilde{K})$ of the constant-thickness model exhibits two interesting characteristics. The first is that the measured data are distributed along $\tilde{H} = C_0\tilde{K}$, where the constant C_0 is related to the displacement used to construct the constant-thickness model in Fig. 27b from the Schoen G sur-

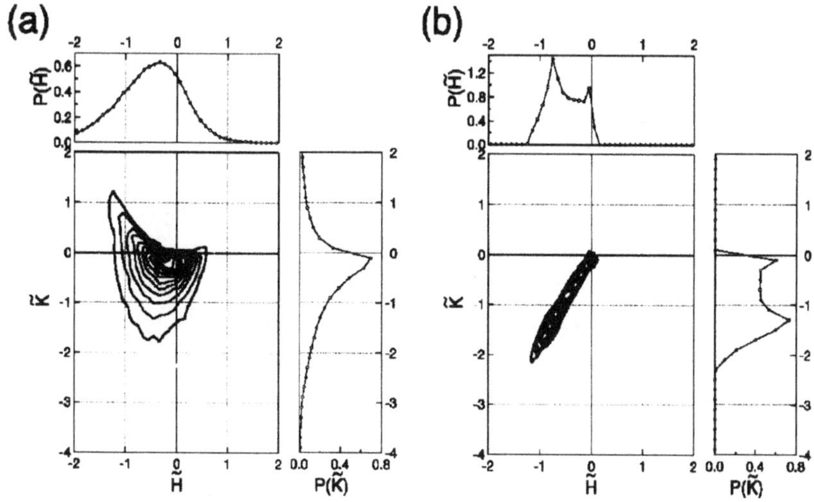

Fig. 29a,b Contour map of the scaled joint probability density, $\tilde{P}(\tilde{H},\tilde{K})$, and its marginal probability densities, $\tilde{P}_H(\tilde{H})$ and $\tilde{P}_K(\tilde{K})$, for the G morphology of the SIS triblock copolymer **(a)** and the constant-thickness model of the G surface (Eq. 34) **(b)**.

face. Secondly, $\tilde{P}(\tilde{H},\tilde{K})$ of the constant-thickness model exhibits two sharp maxima, in marked contrast to $\tilde{P}(\tilde{H},\tilde{K})$ from the G morphology in the SIS copolymer, which possesses a single broad maximum near $\tilde{K}=0$. In this sense, the so-called G morphology in the SIS copolymer differs markedly from the mathematical G surface.

5.3.1
Comparison Between Experiment and Theory in Terms of Interfacial Curvature Distribution

On the basis that interfacial tension constitutes the dominating factor for structure formation in microphase-separated block copolymers, Thomas et al. [87] have proposed that the complex nanostructures formed in block copolymers correspond to area-minimizing surfaces. From the extensive SCFT calculations, Matsen and Bates [94, 95] find that an equally important, but thus far disregarded, factor in block copolymer nanostructure stability is packing frustration [96]. For the minority blocks of an ordered copolymer to fill space uniformly, the interface self-adjusts so that no blocks are excessively stretched. This entropic consideration causes the interface to deviate from CMC (with $\sigma_H\approx 0$), in which case σ_H provides a measure of packing frustration and nanostructural stability. Although predicted $\langle H\rangle$ and σ_H are only available for diblock copolymers [95] (which differ from the present triblock copolymer in molecular architecture), it is worthwhile to compare the experimental interfacial curvature data obtained here with SCFT predictions.

The unperturbed statistical end-to-end distance of the SIS triblock copolymer (R_0) is discerned from $R_0 = \sqrt{2l_S^2 N_S + l_I^2 N_I}$, where l_i and N_i (i=S or I) denote the statistical segment length and block degree of polymerization, respectively. Since l_S=0.70 nm and l_I=0.65 nm, $R_0 \approx 22$ nm. The probability density measured for the G morphology in the SIS copolymer yields $\langle H \rangle$=0.034 nm^{-1} and σ_H=0.042 nm^{-1}, which can likewise be expressed as $\langle H \rangle = 0.74 R_0^{-1}$ and $\sigma_H = 0.91 R_0^{-1}$. Assuming that the unperturbed chain length R_0 remains constant, we find that $\langle H \rangle$ and σ_H from the constant-thickness model of the G morphology are $0.92 R_0^{-1}$ and $0.55 R_0^{-1}$, respectively. According to SCFT predictions [95] for an AB diblock copolymer with f_A=0.34, $\langle H \rangle = 0.70 R_0^{-1}$ and $\sigma_H = 0.12 R_0^{-1}$ at χN=20. Here, χN is a measure of the copolymer segregation power, wherein χ represents the Flory–Huggins interaction parameter and N=2N_S+N_I. Since a microphase-ordered triblock copolymer can be envisaged as a diblock copolymer of half chain length, we treat the SIS copolymer examined here as its SI analog of molecular weight 41,500, in which case χN is estimated [101] to be about 64. The value of $\langle H \rangle$ derived from the probability densities in Fig. 29a for the G morphology in the SIS triblock is in reasonably good agreement with that predicted by SCFT, whereas the value of σ_H obtained here is higher than what is predicted. This discrepancy may reflect noise inherent in the TEMT reconstruction or, alternatively, the large difference in χN (Matsen and Bates [95] predict that σ_H should increase with increasing χN). Further analysis of the factors influencing σ_H is needed for more accurate comparison between experiment and theory.

6 Geometrical Similarity/Difference Between Polymer Blend and Block Copolymer Structures

So far, we have shown that both the polymer mixture and block copolymer exhibit bicontinuous morphologies. The bicontinuous structure of the polymer mixture is a transient one of the order of microns, while that of the block copolymer is a nanometer-scale (equilibrium) structure. An introduction of the chemical covalent bond in order to connect dissimilar constituent sequences to form block copolymer makes such differences. At the same time, however, although these two morphologies are far different in size, they appear grossly similar. In the present section, let us compare the two polymeric morphologies in terms of the geometry to find out a possible effect of the chemical junction in polymer chains.

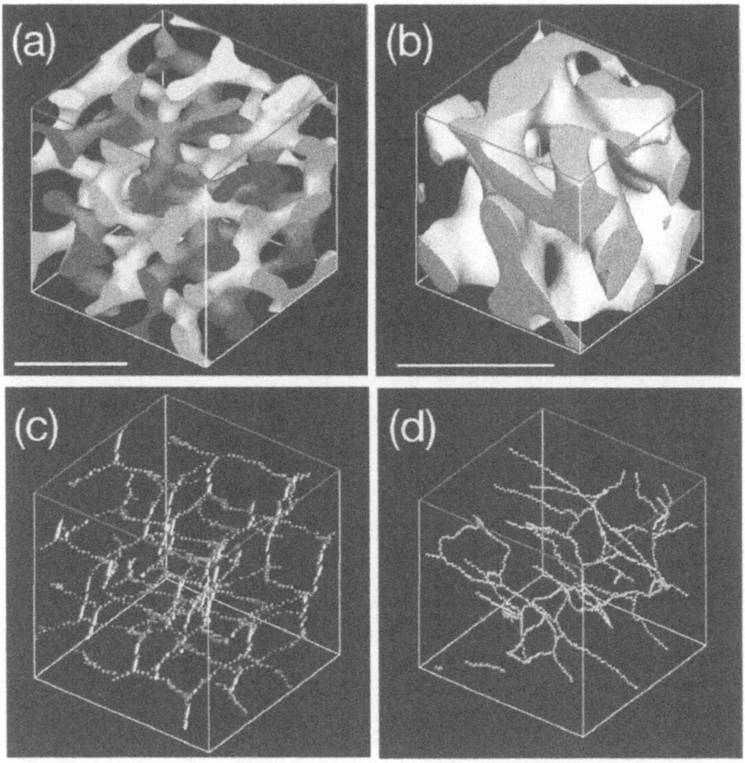

Fig. 30a–d Three-dimensional images of bicontinuous morphologies in two composition-matched polymer systems. **a** The G nanostructure in a microphase-ordered block copolymer (*bar*=74 nm), **b** The SD morphology in an off-critical polymer blend (*bar*=20.4 μm). In **a**, the nonintersecting light and dark channels correspond to the minority microphase, while the majority microphase is transparent. In **b**, the minority phase is shown, while the majority phase is transparent. Corresponding skeletal networks generated after channel thinning are displayed in **c** and **d**, respectively

Figure 30a shows the reconstructed 3D image of the SIS triblock copolymer nanostructure (same as Fig. 27a), which consists of two nonintersecting channel networks within a continuous matrix. For the sake of clarity, the volume-filling I microphase is transparent. Displayed in Fig. 30b is the reconstructed 3D image of the bicontinuous SD morphology of the DPB/PB blend. In this case, the channels correspond to the DPB-rich phase, the volume fraction of which is measured to be 0.37. As in Fig. 30a, the volume-filling matrix (PB) is rendered transparent. Note that the two polymer systems possess similar composition to facilitate comparison. Unlike the copolymer nanostructure, the DPB/PB blend morphology consists of a two-channel bicontinuous network in which Λ=20.4 μm.

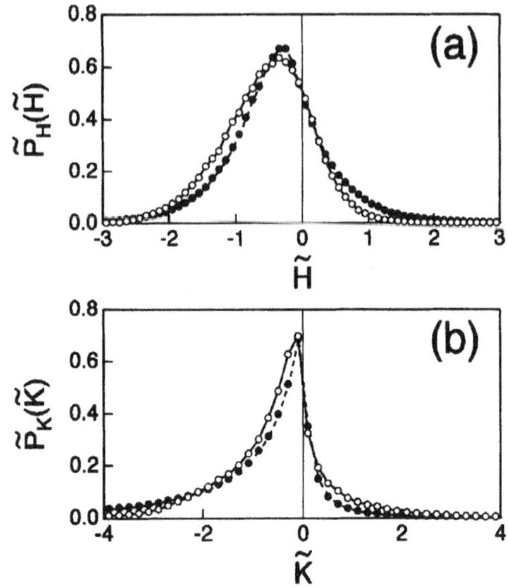

Fig. 31a,b Scaled probability densities **a** $\tilde{P}_H(\tilde{H})$ and **b** $\tilde{P}_K(\tilde{K})$ measured from the bicontinuous block copolymer G (*open circles*) and polymer blend SD (*filled circles*) morphologies

The local shape of the interface in each of the 3D images provided in Fig. 30 can be described by the probability densities of the mean and Gaussian curvatures—$P_H(H)$ and $P_K(K)$, respectively—and can be calculated from $P(H,K)$ [72]. The curvature is arbitrarily chosen to be positive if the center of the osculating circle resides within the I microphase of the copolymer or the PB phase of the polymer blend. To facilitate comparison, $P_H(H)$ and $P_K(K)$ have been scaled in the same manner as described in Sect. 4.4. Σ was equal to 0.070 nm^{-1} for the copolymer and 0.136 μm^{-1} for the blend. The $\tilde{P}_H(\tilde{H})$ and $\tilde{P}_K(\tilde{K})$ determined from the two bicontinuous morphologies shown in Fig. 30 are displayed in Fig. 31 and exhibit surprising similarity.

In Figure 31a, $\tilde{P}_H(\tilde{H})$ for each morphology exhibits a broad maximum at $\tilde{H} < 0$, in which case the area-averaged mean curvature is nonzero. Most of $\tilde{P}_K(\tilde{K})$ for each morphology in Fig. 31b resides at $\tilde{K} < 0$, indicating that each interface is mainly hyperbolic. Despite the minor differences evident in Fig. 31, it is intriguing that the interfacial curvature distributions of these two bicontinuous morphologies differing in characteristic size by a factor of over 300 are so strikingly similar.

A 3D thinning algorithm [47, 43] briefly described in Sect. 3.3 has been applied to the bicontinuous morphologies shown in Figs. 30a and 30b. Cursory examination of Fig. 30 suggests that the skeletal network of the G nanostructure in Fig. 30c is more regularly arranged than that of the SD

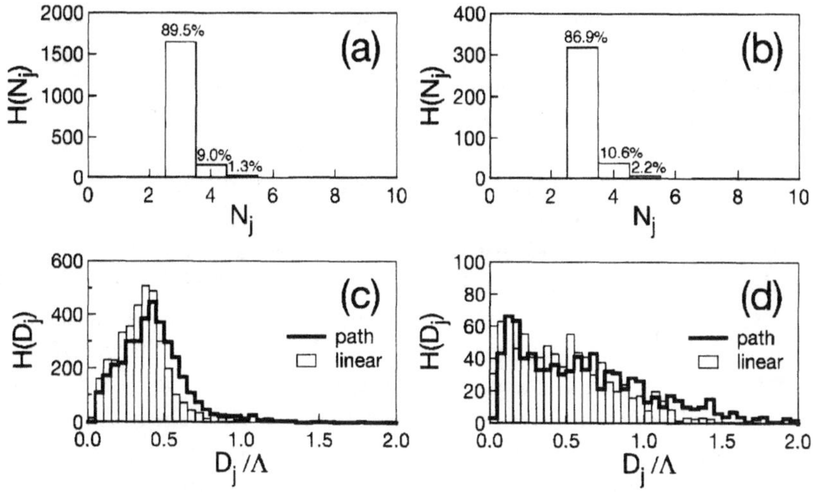

Fig. 32a–d Histograms of the coordination number (N_j) for **a** the G nanostructure and **b** the SD morphology. In both cases, the dominant N_j is 3. Corresponding histograms of the normalized interjunction distance (D_j/Λ) are provided in **c** and **d**, respectively. The *thick line* denotes D_j evaluated along the network path, whereas the *thin line* represents the linear distance between junctions

morphology in Fig. 30d. The average number of junctions per crystallographic unit cell, N_{cell}, for the nanostructure is 11, which is in fair agreement with N_{cell}=15 measured from a computer-generated constant-thickness model surface of the G morphology based on the Schoen G surface [30]. Note that the mathematically predicted value of N_{cell} for the constant-thickness surface is 16. The discrepancy in N_{cell} between the G morphology and composition-matched constant-thickness surface is attributed to the presence of defects in the grains or along the grain boundaries of the G morphology (the constant-thickness Schoen G surface is free from such defects). In contrast to the G copolymer nanostructure, N_{cell} discerned from the SD blend morphology is only 1.9, thereby confirming that the junctions are more densely packed in the nanostructure. Distributions of the coordination number, N_j, are displayed in Fig. 32a for the G nanostructure and in Fig. 32b for the SD morphology.

Both morphologies possess, for the most part, three branches at each junction. Higher coordination numbers account for less than ca. 13% of each N_j distribution. Since the requisite N_j for the G morphology is 3, the statistical result that N_j~3 provides further experimental evidence that the copolymer morphology is G, which is consistent with previous (but limited) observations [28].

Shown in Fig. 32 are distributions of scaled D_j for the copolymer (Fig. 32c) and polymer blend (Fig. 32d). Two measurements of D_j are provided in each figure: one represents the shortest distance between junctions (*linear* distance) and the other corresponds to the distance along the skeletal network (*path* distance). Although common features between the G and SD morphologies are evident in the interfacial curvature distributions (Fig. 31) and the coordination number distributions (Figs. 32a and 32b), this is clearly not the case with respect to D_j. According to the data presented in Fig. 32b, the G morphology exhibits maxima in D_j at 0.40 Λ (linear) and 0.45 Λ (path). These values are in favorable agreement with those derived from the constant-thickness model surface (0.4 Λ). The SD morphology of the polymer blend, however, exhibits a shoulder-like broad maximum in D_j at 0.5–0.7 Λ (linear and path distances). The data displayed in Fig. 32d also reveal that, unlike the G nanostructure, the SD morphology consists of a large population of junctions separated by surprisingly short distances in the range of 0.1–0.2 Λ. Thus, introduction of the chemical junction in a polymer chain to form block copolymers makes the resulting phase-separated structures not only smaller but also more regular than that of the polymer mixture.

The Euler–Poincaré characteristic, χ, can be estimated from the total number of junctions, N, and the total number of branches, B, through $\chi=2N-B$ [49]. Here, $B=\sum_{i=0}^{N} N_{j,i}$, where $N_{j,i}$ denotes the coordination number at the i-th junction. Values of χ per unit cell are −12.1 for the G nanostructure and −3.3 for the SD morphology. Corresponding values of the genus are 7.1 and 2.6, respectively. To put this result in perspective, a surface with genus g is topologically equivalent to a sphere with g handles. For comparison, the constant-thickness model surface yields $\chi=-14.7$ (−16.0) and $g=8.4$ (9.0) (values in parentheses are mathematically predicted quantities). These values agree well with the corresponding values discerned for the G morphology in Fig. 30a. Differences in χ and g between the G nanostructure and the composition-matched constant-thickness Schoen G surface are again ascribed to defects and grain boundaries. According to the Gauss–Bonnet theorem of differential geometry, χ is related to the Gaussian curvature by $2\pi\chi=\int K da=\langle K\rangle S$, where da denotes the area element of the surface and S is the interfacial area, if the surface is closed. From this theorem, we estimate χ to be −11.0 for the G nanostructure and −4.3 for the SD morphology (and −15.0 for the Schoen G surface). Since these values are reasonably close to those derived from the skeletonization analysis, the Gauss–Bonnet theorem appears to be applicable to these two bicontinuous morphologies, which can therefore be considered as closed surfaces.

7
Concluding Remarks

This review article summarizes recent advances in methodology to study polymer structures. A particular emphasis was placed on three-dimensional (3D) microscopy. With the new microscopy, it is now possible to image complicated structures, e.g., bicontinuous structures, in 3D. At the moment, laser scanning confocal microscopy (LSCM), transmission electron microtomography (TEMT), and high-resolution X-ray computerized tomography (X-ray CT) are available. TEMT uses tomography on the transmission electron microscope (TEM) and hence it can be used to obtain 3D images of the phase-separated structures of the order of nanometers, while LSCM and X-ray CT are useful to image micron-scale morphologies. We note that the X-ray CT appears to be quite useful for opaque samples. It may be even possible to perform real-time and in situ 3D imaging of the phase-separating morphologies, which would truly unveil interface dynamics during SD in the near future. We also hope that advances of such new methodologies will develop new academic fields. For example, the bicontinuous morphologies, ranging from nano- to macroscale in both material and biological fields, have been studied for design possibilities and thermodynamic origins of the structure [102].

From the 3D volume data array obtainable from the 3D microscopes, some basic structural parameters, e.g., volume fraction of one of the constituents (ϕ) and interfacial volume per unit volume (Σ), can be directly and relatively easily measured. These structural parameters are basic, but they are so far inferred from the two-dimensional (2D) images and thus possibly inaccurately. Characteristic lengths, being readily measured by use of scattering techniques, are also estimated by taking Fourier transforms of the 3D images. Crystallographic analysis from the 3D images can be done with at least similar precision as the scattering, sometimes better than the scattering when a single grain can be obtained. (Note that multiple grains in laser, X-ray, and/or neutron beam smear the scattering pattern and accordingly ruin the spot-like pattern).

Although it would be useful enough for experimentalists to measure the above rather basic structural parameters from the 3D volume data, those are just a small advantage of the 3D microscopy. The 3D digital data array contains rich structural information that can be extracted with the aid of "cutting-edge" quantitative image analysis. It is possible to go further to evaluate a new set of parameters. Interfacial curvatures measured by the PSM and SFM described in Sects. 3.2.2 and 3.2.3 are such examples. Connectivity of network domains of the morphologies is likewise important. We would like to emphasize that these structural parameters have never been evaluated by any other (conventional) experimental techniques besides 3D microscopy. For an off-critical polymer mixture in which droplet domains of minor components exist inside a matrix of the majority phase, the diameter of the droplets and their distribution are important structural parameters characterizing such systems. Thus the 3D microscopy together with 3D digital analysis seem to have great potential for analyzing polymer structures and we hope the methodology will unveil hidden physics in polymer science and open up new research areas in the future.

Acknowledgments H.J. wishes to thank Prof. R.S. Spontak at North Carolina State University and Prof. T. Hashimoto at Kyoto University for their valuable discussions and collaboration. The authors are grateful to NEDO for support of this research through a Japanese National Project "Nano Structure Polymer Project" by the Ministry of Economy, Trade and Industry. Support given by the Grant-in-Aid for Scientific Research on Priority Areas (A), " Dynamic Control of Strongly Correlated Soft Materials" (No. 413/13031057 & 14045246) from the Ministry of Education, Science, Sports, Culture, and Technology is also gratefully acknowledged.

References

1. Gunton JD, San Miguel M, Sahni PS (1983) In: Domb C, Lebowitz JL (eds) Phase transition and critical phenomena, vol 8, p 269. Academic, New York
2. Hashimoto T (1993) Structure of polymer blends. In: Cahn RW, Haasen P, Kramer EJ (eds) Materials science and technology, vol 12. Structure and properties of polymers, p 251. VCH, Weinheim
3. Bates FS, Fredrickson GH (1999) Phys Today 52:32
4. Ito M, Nakamura T, Matsumoto T, Tsurusaki T, Hayashi K (1998) Bone 23:163
5. Wilson T (1990). In: Pawley JB (ed) Handbook of biological confocal microscopy, p 167. Plenum, New York
6. Wilson T (1990) In: Wilson T (ed) Confocal microscopy. Academic, London
7. Verhoogt H, van Dam J, Posthuma de Boer A, Draaijer A, Houpt PM (1993) Polymer 34:1325
8. Ribbe AE, Hashimoto T, Jinnai H (1996) J Mater Sci 31:5837
9. White WR, Wiltzius P (1995) Phys Rev Lett 75:3012
10. Jinnai H, Nishikawa Y, Koga T, Hashimoto T (1995) Macromolecules 28:4782
11. Jinnai H, Koga T, Nishikawa Y, Hashimoto T, Hyde ST (1997) Phys Rev Lett 78:2248
12. Ribbe AE, Hashimoto T (1997) Macromolecules 30:3999
13. Vorobyova O, Winnik MA (2001) J Polym Sci B 39:2317
14. Hopkinson I, Myatt M (2002) Macromolecules 35:5153
15. Li L, Sosnowski S, Chaffey CE, Balke ST, Winnik MA (1994) Langmuir 10:2495
16. Kumacheva E, Li L, Winnik MA, Shinozaki DM, Cheng PC (1997) Langmuir 13:2483
17. Terada Y, Hirano S, Ikehara T, Nishi T (2001) Polymer J 33:371
18. Loren N, Altskaer A, Hermansson A-M (2001) Macromolecules 34:8117
19. Magoshi T, Ziani-Cherif H, Ohya S, Nakayama Y, Matsuda T (2002) Langmuir 18:4862
20. Thomas JL, Olzog M, Drake C, Shih C-H, Gryte CC (2002) Polymer 43:4153
21. Raviv U, Frey J, Sak R, Laurat P, Tadmor R, Klein J (2002) Langmuir 18:7482
22. Frank J (1992) Electron tomography: three-dimensional imaging with the transmission electron microscope. Plenum, New York

23. Spontak RJ, Williams MC, Agard DA (1988) Polymer 29:387
24. Koster AJ, Ziese U, Verkleij AJ, Janssen AH, De Jong KP (2000) J Phys Chem B 104:9368
25. Midgley PA, Weyland M (2003) Ultramicroscopy 96:413
26. Spontak RJ, Fung JC, Braunfeild MB, Sedat JW, Agard DA, Kane L, Smith SD, Satkowski MM, Ashraf A, Hajduk DA, Gruner SM (1996) Macromolecules 29:4494
27. Laurer JH, Hajduk DA, Fung JC, Sedat JW, Smith SD, Gruner SM, Agard DA, Spontak RJ (1997a) Macromolecules 30:3938
28. Laurer JH, Ashraf A, Smith SD, Spontak RJ (1997b) Langmuir 13:2250
29. Radzilowski LH, Carragher BO, Stupp SI (1997) Macromolecules 30:2110
30. Jinnai H, Nishikawa Y, Spontak RJ, Smith SD, Agard DA, Hashimoto T (2000a) Phys Rev Lett 84:518
31. Cramer H, Wold H (1936) J London Math Soc 11:290
32. Smith PR, Peters TM, Bates RHT (1973) J Phys A Math Gen 6:361
33. Radermacher M (1992) Weighted back-projection methods. In: Frank J (ed) Electron tomography. Plenum, New York
34. Bates RHT, McDonnell MJ (1986) Image restoration and reconstruction. Oxford University Press, New York
35. Crowther RA, DeRosier DJ, Klug A (1970) Proc R Soc London A 317:319
36. Koster AJ, Grimm R, Typke D, Hegrel R, Stoschek A, Walz J, Baumeister W (1997) J Struct Biol 120:276
37. Lorensen WE, Cline HE (1987) Computer graphics SIGGRAPH '87 21:163
38. Hyde S, Andersson S, Larsson K, Blum Z, Landh T, Lidin S, Niham BW (1997) The language of shape. Elsevier, Amsterdam
39. Nishikawa Y, Jinnai H, Koga T, Hashimoto T, Hyde ST (1998) Langmuir 14:1242
40. Omitted
41. Nishikawa Y, Koga T, Hashimoto T, Jinnai H (2001) Langmuir 17:3254
42. Kabel J, Odgaard A, Rietbergen B van, Huiskes R (1999). Bone 24:115
43. Jinnai H, Watashiba H, Kajihara T, Takahashi M (2003) J Chem Phys 14:7554
44. Aksimentiev A, Fiakowski M, Hoyst R (2002) Adv Chem Phys 121:141
45. Saito T, Toriwaki J (1994) IEICE Trans INF & SYST, Japan, E77-D:1005
46. Toriwaki J, Mori K (2001) Distance transformation and skeletonization of 3D pictures and their applications to medical images. In: Bertrand G, Imiya A, Klette R (eds) Digital and image geometry, vol 2243 of LNCS, p 412. Springer, Berlin Heidelberg New York
47. Nishikawa Y, Jinnai H, Hasegawa H (2000) Kobunshi Ronbunshu 58:13
48. Saito T, Toriwaki J (1994) Pattern Recogn 27:1551
49. Hyde ST, Ramsden R (2000) Chemical frameworks and hyperbolic tilings. In: Hansen P, Fowler P, Zheng M (eds) Discrete mathematical chemistry: DIMACS workshop discrete mathematical chemistry, vol 51 of DIMACS series in discrete mathematics and theoretical computer science, p 203. American Mathematical Society, Providence
50. Furukawa H (1979) Phys Rev Lett 43:136
51. Nishi T, Wang TT, Kwei TK (1975) Macromolecules 18:227
52. Reich S (1986) Phys Lett 114A:90
53. Chakrabarti A, Toral A, Gunton JD, Muthukumar M (1989) Phys Rev Lett 63: 2072
54. Oono Y (1991) IEICE Transactions E74:1379
55. Shinozaki A, Oono Y (1993) Phys Rev E 46:2622
56. Koga T, Kawasaki K (1993) Physica A 196:389
57. Snyder HL, Meakin P (1983) J Chem Phys 79:5588

58. Izumitani T, Hashimoto T (1985) J Chem Phys 83:3694
59. Sato T, Han CC (1988) J Chem Phys 88:2057
60. Wiltzius P, Bates FS, Heffner WR (1988) Phys Rev Lett 60:1538
61. Hashimoto T, Takenaka T, Jinnai H (1991) J Appl Crystallogr 24:457
62. Schwahn D, Hahn K, Streib J, Springer T (1990) J Chem Phys 93:8383
63. Jinnai H, Hasegawa H, Hashimoto T, Han CC (1991) Macromolecules 24:282
64. Jinnai H, Hasegawa H, Hashimoto T, Briber RM, Han CC (1993a) Macromolecules 26:182
65. Jinnai H, Hasegawa H, Hashimoto T, Han CC (1993) J Chem Phys 99:4845
66. Jinnai H, Hasegawa H, Hashimoto T, Han CC (1993) J Chem Phys 99:8154
67. Cahn JW (1962) Acta Metall 10:179
68. Cook HE (1970) Acta Metall 18:297
69. Binder K, Stauffer D (1974) Phys Rev Lett 33:1006
70. Bates FS, Wignal GD, Koehler WC (1985) Phys Rev Lett 55:2425
71. Nose T (1989) In: Tanaka F, Doi M, Ohta T (eds) Space-time organization in macromolecular fluids. Springer, Berlin Heidelberg New York
72. Jinnai H, Nishikawa Y, Morimoto H, Koga T, Hashimoto T (2000) Langmuir 16:4380
73. Hashimoto T, Izumitani T, Takenaka M (1989) Macromolecules 22:2293
74. Jahn W, Strey R (1988) J Phys Chem 92:2294
75. Hashimoto T, Kumaki J, Kawai H (1983) Macromolecules 16:641
76. Kawasaki K, Ohta T (1983) Physica A 118:175
77. Koga T, Jinnai H, Hashimoto T (1999) Physica A 263:369
78. Takenaka M, Izumitani T, Hashimoto T (1990) J Chem Phys 92:4566
79. Läuger J, Lay R, Maas S, Gronski W (1995) Macromolecules 28:7010
80. Tomita H (1984) Prog Theor Phys 72:656
81. Porod G (1982) In: Glatter O, Kratky O (eds) Small angle X-ray scattering. Academic, New York
82. Hashimoto T, Jinnai H, Hasegawa H, Han CC (1994) Phys Rev A 204:261
83. Jinnai H, Kitagishi H, Hamano K, Nishikawa Y, Takahashi M (2003) Phys Rev E 67:021801
84. Jones RAL, Norton LJ, Kramer EJ, Bates FS, Wiltzius P (1991) Phys Rev Lett 66:1326
85. Ball RC, Essery RLH (1990) J Phys Condens Mat 10:303
86. Thomas EL, Alward DB, Kinning DJ, Martin DC, Handlin J, Fetters DL, Fetters LJ (1986) Macromolecules 19:2197
87. Thomas EL, Anderson DM, Henkee CS, Hoffman D (1988) Nature 334:598
88. Schulz ML, Bates FS, Almdal K, Mortensen K (1994) Phys Rev Lett 73:86–89
89. Hajduk DA, Harper PE, Gruner SM, Honeker CC, Kim G, Thomas EL, Fetters LJ (1994) Macromolecules 27:4063
90. Hajduk DA, Ho R-M, Hillmyer MA, Bates FS, Almdal K (1998) J Phys Chem B 102:1356
91. Schick M (1998) Physica A 251:1
92. Hajduk DA, Harper PE, Gruner SM, Honeker CC, Thomas EL, Fetters LJ (1995) Macromolecules 28:2570
93. Hyde ST, Andersson S, Larsson K, Blum Z, Landh T, Lidin S, Ninham BW (1997b) The language of shape. Elsevier, Amsterdam
94. Matsen MW, Bates FS (1996) Macromolecules 29:7641
95. Matsen MW, Bates FS (1997) J Chem Phys 106:2436
96. Gruner SM (1989) J Phys Chem 93:7562
97. Anderson DM, Bellare J, Hoffman JT, Hoffman D, Gunther J, Thomas EL (1992) J Colloid Interface Sci 148:398

98. Jinnai H, Nishikawa Y, Hashimoto T (1999) Phys Rev E 59:R2554
99. Fung JC, Liu W, DeRuijter WJ, Chen H, Abbey CK, Sedat JW, Agard DA (1996) J Struct Biol 116:181
100. Harris JL (1977) Appl Optics 16:1268
101. Förster S, Khandpur AK, Zhao J, Bates FS, Hamley IW, Ryan AJ, Bras W (1994) Macromolecules 27:6922
102. Jinnai H, Nishikawa Y, Ito M, Smith SD, Agard DA, Spontak RJ (2002) Adv Mater 14:1615

Editor: Akihiro Abe

Received: November 2003

APS (2004) 170:169–273
DOI 10.1007/b94405

Two-Photon Photopolymerization and 3D Lithographic Microfabrication

Hong-Bo Sun[1, 2] (✉) · Satoshi Kawata[1, 3] (✉)

[1] Department of Applied Physics, Osaka University, Suita, 565–0871 Osaka, Japan
hbsun@ieee.org
kawata@ap.eng.osaka-u.ac.jp
[2] PRESTO, Japan Science and Technology Corporation (JST), Japan
[3] RIKEN (The Institute of Physical and Chemical Research), Hirosawa, Wako, 351–0198 Saitama, Japan
kawata@ap.eng.osaka-u.ac.jp

Abstract This chapter attempts to give an overview of the historical development and current progress of femtosecond laser micro-nanofabrication based on multiphoton absorption, and particular emphasis is placed on two-photon photopolymerization. Femtosecond laser interaction with matter differs essentially from those with longer pulses or CW lasers in its significant nonlinearity, ultrafast characteristics and the possibility of highly localization of reaction volume. These features enable three-dimensional (3D) micro-nanofabrication in solid and liquid media. In two-photon photopolymerization, when a near-infrared femtosecond laser is tightly focused into a photopolymerizable resin, 3D polymer micro-nanostructures are produced by pinpoint photopolymerization of liquid precursory resins. Using this direct laser writing scheme, various photonic, micro-optical components and micromechanical devices have been readily produced. The two-photon photopolymerization technology is expected to play a similar role to that played by lithography for planar semiconductor device processing, but for micro-nanofabrication of 3D polymer-based optoelectronic devices as well for microelectromechanical systems.

Keywords 3D lithography · Two-photon photopolymerization · Femtosecond laser · Micro-nanodevice · Micro-nanofabrication

Abbreviations and Symbols

2D	Two-dimensional
3D	Three-dimensional
AFM	Atomic force microscope
B1536	1,2-Dicyano-1,2-bis(2,4,5-trimethyl-3-thienyl)ethane
BCC	Body-centered cubic
BSA	Bovin serum albumin
CAM	Computer-aided manufacturing
CAD	Computer-aided design
CCD	Charge coupled device
CW	Continuous wave
DBR	Distributed Bragg reflection
DFB	Distributed feedback
DMF	Dimethyl formamide
FCC	Face-centered cubic
FWHM	Full width at half maximum
HCP	Hexagonal close packing
IR	Infrared
LD	Laser diode
LED	Light emitting diode
μCP	Microcontact printing
MEMS	Microelectromechanical system
MMA	Methyl methacrylate
MW	Molecular weight
NA	Mumerical aperture
NIR	Near-infrared
NSOM	Near-field scanning optical microscope
PAG	Photoacid generator
PBG	Photonic bandgap
PDMS	Poly(dimethyl siloxane)
PhC	Photonic crystal
PMMA	Poly(methyl methacrylate)
PSF	Point spread function
PVK	Poly(vinyl carbazole)
PZT	Lead zirconate titanate

R	Radical
RB	Rose Bengal
S	Photosensitizer
SC	Simple cubic
SDL	Sub-diffraction-limited
SEM	Scanning electron microscope
SLI	Square of light intensity
SLM	Spatial light modulator
STM	Scanning tunneling microscope
TE	Transverse electric
THPMA	Tetrahydropyranyl methacrylate
TM	Transverse magnetic
TPA	Two-photon absorption
TPE	Two-photon excitation
UV	Ultraviolet
Voxel	Volume element
XUV	Extreme UV
n	Refractive index
δ	Two-photon absorption cross-section
ϵ	Electrical permittivity
T_g	Glass transition temperature
E	Electric field strength; Young's modulus
I	Light intensity
λ	Wavelength
ν	Lightwave frequency
l_c,	Coherence length
Q	Quality factor
ω_0,	Beam waist of Gaussian beam
Z_R	Rayleigh depth
ϕ_{H+}	Quantum efficiency of proton generation
G_s	Shear modulus

1 Introduction

The last decade has witnessed rapid progress in high-performance ultraviolet (UV)-curable systems [1–5], which have resulted in a growing number of industrial applications, including paints, optical adhesives, medicine, coatings, graphic arts, microelectronics, optics, manufacturing, and so forth. It is estimated the world consumption of UV curable products in the year 2000 is around 200,000 tons, which corresponds to a two billion dollar market. Requirements of monomers and oligomers of novel function, active photoinitiators, and more efficient photosensitizers are increasing, which will prompt market growth.

Among the above applications, *computer-aided manufacturing (CAM)* using UV curable resin, generally called *laser rapid prototyping*, is a new and expanding technology [6–9]. It converts three-dimensional (3D) objects of complex shape, designed via *computer-aided design (CAD)*, from designs into real products. The resin used for fabrication is photocured at the spot exposed to UV laser, a *single-photon photopolymerization* process. By scanning the laser beam, one slice of the 3D structure is first hardened according to the design patterns; and then a thin-layer liquid resin is added and a new patterned slice is polymerized. The entire structure is sequentially created the same way. This technology is suitable for manufacturing devices that are difficult or costly to prepare by conventional mechanical methods.

Commercial laser rapid prototyping machines have a fabrication precision greater than 10 μm. This accuracy cannot fully satisfy the modern requirements for device multifunctionalization and miniaturization that demand sub-micron feature size. The emergence of a new technology in 1997, *two-photon photopolymerization* [10], has brought the light curable resin into the realm of nanofabrication. As indicated by the name, the resins are polymerized not by absorbing one UV photon, but by simultaneously absorbing two photons at longer wavelength, usually in the red-infrared (IR) spectral region. The two-photon process [11–13] has at least two advantages compared to single-photon absorption used in conventional rapid prototyping. First, common polymers have negligible linear absorption in the red-near-infrared (NIR) region, so the laser penetrates deeply into materials and directly induces polymerization from inside without contaminating outside of the focal volume; secondly, the quadratic dependence of polymerization rate on the light intensity enables 3D spatial resolution, and the accuracy is better than that achieved in single photon process. Actually a near 100-nm lateral spatial resolution has been reported [13, 14]. Two-photon polymerization, as currently the only microprocessing approach that has intrinsic 3D fabrication capability, has been successfully applied to production of a variety of photonic and micromechanical devices [15–20]. It accomplishes manufacturing that is otherwise not accessible and brings new scientific possibilities to nano-research.

The current research effort in two-photon photopolymerization is largely devoted to the synthesis of high-efficiency photoinitiators and sensitizers [21–25], about which good reviews have been published in this series. Nevertheless, as a new technology, there is a lot of work that has been done to establish it as a nanoprocessing tool, which is the major content of this review. So, in the next section, we will introduce the general stereolithography concept, which we define as the technology that is utilized to produce stereostructures using lasers, although here we are more concerned with processes that address submicron features using multiphoton processes.

Following that, we will discuss the principles and materials of two-photon photopolymerization, as well as the systems used, and ground-work performed in this area. Then we turn our focus in the next section to the advanced technologies that have been developed, and in the final section we look at some applications of the technique. You should note, however, that our content selections for the advanced technologies section are quite subjective, since it is difficult to know which approaches will withstand the test of time.

2
General Stereolithography Using Femtosecond Lasers

Stereolithography historically refers to the technology of creating 3D objects from CAD patterns by adding and exposing photopolymerizable resin layer by layer. Nowadays, materials that are useful for 3D laser modeling have extended to gas phase [26, 27], chemical solutions [28], metal powders [29, 30] and transparent glassy or crystalline solids [31, 32]; the lasers used as an irradiation source range from extreme UV (XUV) to NIR wavelengths, operating at continuous wave (CW) to pulsed mode at nanosecond (ns), picosecond (ps) and femtosecond (fs) widths [33–35]; 3D patterns are created using either multi-beam interference [36–39] or direct laser writing, and the writing does not necessarily start from the surface layers but is accomplishable from inside materials via multiphoton pinpoint addressing [13–20]. Correspondingly, the concept of stereolithography has been broadened. Here we focus our topics only on the latest progress that involves the use of femtosecond lasers. These technologies share common features such as the dominant role of nonlinear effects in laser material interactions and the similarity in experimental techniques. An overview of the entire family should facilitate an understanding of the origin, the current status, and future direction of the two-photon photopolymerization technology.

2.1
Material Processing with Femtosecond Lasers

The basis of laser fabrication is laser material processing [40, 41], which started soon after the first demonstration of the ruby laser in 1960. Due to the poor beam quality and reproducibility of lasers, initial works were largely qualitative, mainly devoted to simple research like material evaporation. With the emergence of new-type lasers such as Nd (Nd-doped YAG or glass), CO_2, Ar ion, and excimer lasers, and the improvement of laser performance, the field was expanded to laser annealing, crystallization of amorphous layers, compound synthesis, plasma formation, and laser cutting, hole-drilling, welding, jointing, and so forth. The commercialization of tunable solid ultrashort pulse lasers (for example the Ti:Sapphire laser) in the 1990 s pushed forward these applications and opened new domains such as controlling, manipulation and processing of biological and nanoscale species. Today's commercial laser systems have already been able to provide output power up to 10^{20} W in pulse duration with a good beam quality, and up to several kilowatts in CW mode, although usually with worse beam quality. Laser pulse duration has reached less than 5 fs, wavelengths cover a region from a few nm in the XUV to the far IR with several tens of μm, and pulse energies reach up to 10^4 J, while frequency stability and resolution better than 10^{-13} is already available. With the excellent laser beam quality, coherence, power and frequency stability, short pulse duration, and high transient power, almost any kinds of material can be processed by lasers in a well-controlled manner.

Nonlinear processes [33, 42, 43], for example, *multiphoton absorption* including *two-photon absorption (TPA)* [11, 12, 25], has come to play a dominant role in nanofabrication. In order to produce a lasting effect on a material, photons must first be absorbed. The energy and momentum are exchanged between the optical fields and molecules through absorption and emission. In such a process, the imaginary part of nonlinear susceptibility represents the energy transfer from the light field to a medium. The light-matter energy change per unit time and unit volume is:

$$\frac{dW}{dt} = \left\langle \vec{E} \cdot \dot{\vec{P}} \right\rangle \tag{1}$$

where $\vec{E}$ is the electric field vector and the brackets denote time average. The value of material polarization $\vec{P}$ is:

$$P = \chi^{(1)}E + \chi^{(2)}E^2 + \chi^{(3)}E^3 + \cdots \tag{2}$$

where the quantities of $\chi^{(1)}$, $\chi^{(2)}$, $\chi^{(3)}$ are second-, third-, and fourth-rank tensors, representing linear, second-order and third-order optical susceptibilities. In resonant processes, there is no contribution from the even-order susceptibilities like $\chi^{(2)}$ and $\chi^{(4)}$. Therefore, the nonlinear absorption is described by the imaginary parts of $\chi^{(3)}$, $\chi^{(5)}$, of which typical effects are two-photon and three-photon absorptions, respectively. Particularly, for degenerate TPA, that is, the process of photons of identical energy are simultaneously absorbed, the energy absorption rate is:

$$\frac{dW}{dt} = \frac{8\pi^2\omega}{c^2n^2} I^2 \,\mathrm{Im}\left[\chi^{(3)}\right] \tag{3}$$

It is seen that the TPA rate quadratically depends on the light intensity, which is an important mechanism to improve the spatial resolution in two-photon fabrication. A high capability of materials to absorb photons via TPA is desired, which is described by *TPA cross-section*, δ, defined by

$$\frac{dn_{\mathrm{P}}}{dt} = \delta N F^2 \tag{4}$$

where N and n_p are the number density of absorbing molecules and number of absorbed photons, respectively, and $F=I/h\nu$ denotes photon flux. According to Eq. 4), the TPA cross-section is:

$$\delta = \frac{8\pi^2 h\nu^2}{c^2 n^2 N} I^2 \mathrm{Im}\left[\chi^{(3)}\right] \tag{5}$$

The design of molecules that have a large TPA cross-section is an important task of stereolithography using two-photon photopolymerization. We will discuss work related to this in Sect. 3.1.3.

In order to take advantage of nonlinear effects, the use of ultrashort laser excitation is essential. In the early 1990 s, it was recognized that laser-matter interactions for femtosecond pulses were fundamentally different from interactions resulting from longer pulses or CW lasers [44–46]. First, a femtosecond laser carries much greater peak power. With conventional light sources the strength of the light field is in the range of 1 V/cm and the resulting elongation of dipole is smaller than 10^{-16} m, much smaller than atomic or molecular diameters (10^{-10}~10^{-7} m). With femtosecond laser irradiation, the field strength could be as intense as 10^8 V/cm, sufficient to induce direct bond breaking. Various nonlinear effects could be easily launched, among which (the most important for laser fabrication) is multiphoton absorption. Multiphoton absorption has an extremely small cross-section; it is confined to occur only in a small 3D volume around the close vicinity of the laser focus, less than the cubic wavelength (λ^3). Hence, a quite high 3D spatial resolution can be achieved in the pinpoint exposure. Secondly, when materials are irradiated with a femtosecond laser pulse, the photon energy is deposited much faster than electrons could transfer it to the lattice, or molecule/atom oscillations through phonon emission, meaning that the excitation is a heat insulation process [47–49]. This provides an ideal optical excitation means for many photochemical or photophysical reactions where thermal effects, a process difficult to localize, are not desired. In addition, for many dielectric materials, there is a transparent window in the red-NIR spectral region, which is covered neither by electronic band-band absorption nor by atomic/molecular oscillation absorption. It happens that the fundamental wavelength of general femtosecond lasers is located in this regime, for example, 680–1000 nm for Ti:Sapphire laser. Therefore a femtosecond laser can penetrate into and tailor desired structures from inside transparent materials, giving an intrinsic 3D processing capability.

2.2 Femtosecond Laser 3D Micro-Nanofabrication

Although multiphoton absorption was predicted as early as 1931 [50] and experimentally observed immediately after the invention of lasers [51], the technology found limited application, for example, solely as a spectroscopic tool [52], due to the extremely small absorption cross-section of most materials. With the advent of convenient femtosecond lasers the utility of multiphoton excitation processes has undergone a rebirth in the past decade. Fluorescent imaging of biological molecules is one of the most important uses [24, 25, 53–55]. Additional usage includes fluorescence up-conversion [56], power limiting [57], photodynamic therapy [58], two-photon lasing [59], cell surgery [60], ultrasonic generation [61], and certainly micro-nanofabrication. As a stereolithographic approach, the femtosecond laser has found considerable use in optical memory, micro optical components, photonic crystals and complicated 3D patterning.

2.2.1 3D Optical Memory

Conventional optical recording media like compact disks (CDs) and magneto-optical (MO) disks register binary bits, spots with optically modified phase status or refractive index, in an active layer of the disk, by which information of about 10^8/cm^2 is recordable using visible light at the diffraction limit. The information capacity in a disk volume can be expanded by means of multi-layer recording [11, 62, 63]. A simple scheme is to focus the laser at a series of different depths in an optically thick active media, and at each depth, a bit plane is recorded. Estimated from the diffraction-limit-defined volume, a storage density as high as 10^{12} bits/cm^3 is possible. However, in linear recording-reading (single-photon absorption for both processes), the same amount of photon energy is absorbed in each plane transverse to the optical axis since nearly the same amount of photon flux crosses them. This strongly contaminates the planes above and below the particular focal plane to be addressed, causing the issue of crosstalk. Therefore it is quite difficult to realize the multilayer recording strategy by a single photon process.

This problem was solved by using multiphoton absorption. First the intrinsic nature of deep penetration allows a laser to address a certain depth inside materials without power dissipation; and secondly, the excitation depends roughly on n-order of the intensity so that net excitation per distant plane falls off sharply, where "n" means n-photon absorption. This enables recordings in well-separated layers and minimizes the crosstalk. Compared to other approaches to 3D optical data storage, like holographic recording on photorefractive media [64], hole burning [65], photon echo [66], and encoded thin disc stacking, the multiphoton method has the merits of highest storage capacity (up to terabits/cm^3), random access, fast write-read times, possibility of erasing and rewriting, and low cost.

Multi-photon-induced modification of optical properties of materials including refractive index, absorbance, polarization and fluorescence appearance or wavelength shifting could be utilized for 3D optical memory in various media, including polymers, inorganic crystals, or glasses.

2.2.1.1
Isomerization of Photochromic Materials

Photochromic molecules [11, 67] exist in two chemically stable isomer forms, and the two isomers are inter-switchable by photochemical reactions after absorbing light of different wavelengths. This implies that they can be used for rewritable optical memory. The two isomers differ in their absorption, refractive index, fluorescence wavelength, and even molecular orientation-induced polarizations, permitting recorded binary bits of one isomer status embedded in a matrix of the other isomer status. To be a good 3D storage medium, the material should have high sensitivity and fast response to excitation, stable isomers at both two states, and high resistance to fatigue during cyclic writing and erasing. Three classes of molecules, spirobenzopyran [11], diarylethene [68] and azobenzene [63, 69] and their derivatives are found to be promising for this purpose. Particularly, diarylethene derivatives with heterocyclic rings exhibit no thermochromicity up to 200 °C; their colored close-ring forms are stable for more than three months at 80 °C; and no significant fatigue has been observed even after 10^4 cyclization/ring open

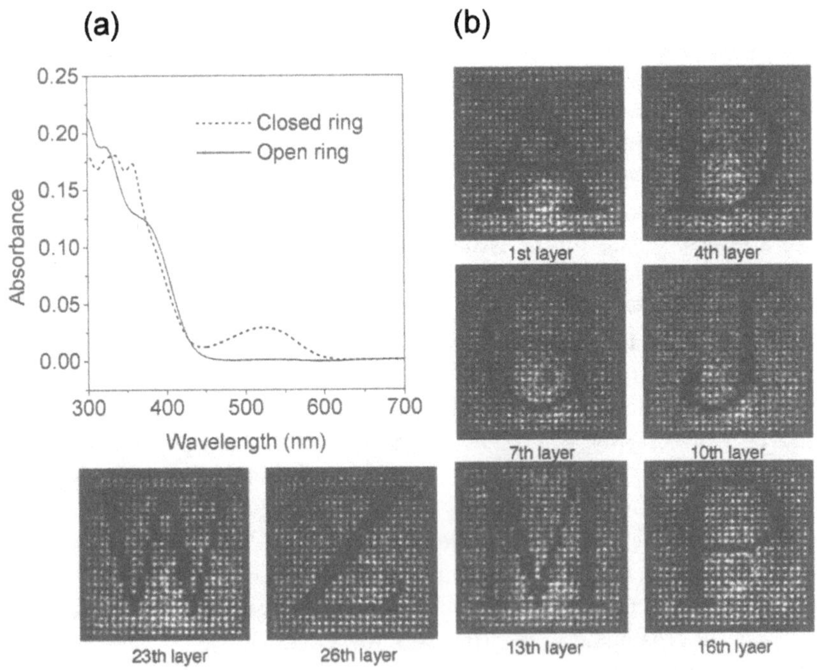

Fig. 1 Photochromic materials for 3D optical memory. **a** Absorption spectra of open-ring and closed-ring diarylethene derivative B1536. **b** Bit patterns written by femtosecond two-photon absorption and readout using a reflection confocal microscope. Refer to Fig. 60b for ring close-open reactions

reaction cycles [67]. As an example of multilayer writing and reading, Fig. 1 shows absorption spectra of two-form 1,2-dicyano-1,2-bis(2,4,5-trimethyl-3-thienyl)ethane (B1536) (Fig. 1a) and several two-photon recorded bit patterns out of 26 sequential bit planes (Fig. 1b) [63, 70]. Bits were recorded by exciting the 380-nm absorption of open-ring isomers (red color) using a 760-nm femtosecond laser. The bits consist mostly in the form of close-ring isomers (yellow). The recording layer and bit intervals are 5 μm and 2 μm, respectively. The refractive index change around 10^{-4} was distinguished and readout by a reflection confocal microscope.

The current major issue in photochromatic memory is that, although isomers are stable in respective single-phase bulk form, when the bits are embedded in the matrix with opposite isomer status, they are liable to relax, losing contrast. Further work on material stability is apparently needed.

2.2.1.2
Photorefraction, Photopolymerization, Photobleaching and Photoreduction Effects

The photorefraction effect [64, 34] has been utilized for holographic memory for many years. The mechanism is that, when a photorefractive material such as $BaTiO_3$, $Bi_{12}SiO_{20}$, $LiNbO_3$, $KNbO_3$, GaAs, SBN or nonlinear functional polymers is exposed to light, free charge carriers (electrons or holes) are generated by excitation, in most cases from impurity energy levels, and these diffuse, leaving behind fixed charges of opposite sign. This charge distribution creates an internal electric field that causes local refractive index change by virtue of Pockel's effect. Since photoinduced spatial charges (and therefore the refractive index) depend on only on the irradiation history of a local site, it is possible to induce refractive index change either by frame (holographic) or on a bit basis. Then single bit and single frame addressing that is not possible in holographic memory becomes natural in the bit recording scheme. Bit-orientated two-photon memory in inorganic $LiNbO_3$ crystal [71], poly(vinyl carbazole (PVK) [72] and poly(methyl methacrylate) (PMMA)-based [73] polymeric photorefractive materials has been realized using this technique. The attribute of charge distribution implies, on one hand, the possibility of erasing and rewriting, on the other hand, the bit information is easy to lose by uniform illumination or by heating.

The refractive index change in photorefractive materials is generally small, of the order of $\Delta n/n<10^{-3}$. Large change of n can be achieved in a different scheme, photopolymerization [62]. The mass density increases after polymerization gives rise to $\Delta n/n\sim 1\%$. In this case, two-photon solidified small volume elements (voxels) that are suspended in unpolymerized matrix are treated as binary bits. However, the gel status and light sensitivity of the recording media make the technology impractical.

Different from the need to sensitively detect the refractive index change, bits recorded as points of emergence or disappearance of a certain fluorescence band make for easier reading. Fluorescent dye can be photobleached by strong excitation, generally by long pulses or CW laser [74]. The bleached volumes do not produce fluorescence. As a consequence, the places that have been bleached show up as darker areas than their surroundings when the recorded information is read out in a fluorescence microscope.

A direct use of the fluorescence approach is the photoreduction of noble [75] (for example Au^{3+} and Ag^{+}) and or rare-earth (for example Sm^{3+} and Eu^{3+}) [76, 77] metals. For instance, Sm^{3+} in glass can be space-selectively photoreduced with an IR femtosecond laser. It is found that after photoreduction from Sm^{3+} to Sm^{2+}, a pronounced fluorescence band appears at 650~775 nm, which has been used as readout signal of 3D memory [77]. Another outstanding characteristic is that photoreduced Sm^{2+}, stable at room temperature, can be converted back to Sm^{3+} by photo-oxidation with a CW laser, such as an argon-ion laser or a semiconductor laser. This enables rewritable optical memory with two stable statuses.

2.2.1.3
Photodensification and Cavitation

Other than the particular photochemical reactions discussed above, there is a universal mechanism that is useful for 3D optical memory. Due to their extremely large transient power, tightly-focused femtosecond laser pulses interact with almost any kinds of materials and pinpoint mark bits inside them, provided that the media are transparent to the laser wavelength. These bits are generally visible under an optical microscope, implying a large variation of the refractive index. Although a detailed mechanism of the photoinduced changes is still an open problem, it is already well-accepted that highly-excited electron-ion plasma is produced at the focal volume during excitation [48, 49]. At relatively lower transient power, the plasma condensation induces local mass densification of materials, in case of amorphous silica up to ~3%, and at higher transient power, the plasma releases energy in an explosive way, leaving a void surrounded by a densified crust [31, 32, 48, 78]. Both the densified and cavititated spots have been utilized as the 3D memory bits in polymers [79], vitreous and active glasses [80–82], and other optical materials like diamond and sapphire. Readout is basically from the refractive index change; for example, in the case of complete cavitation of silica to vacuum, Δn~0.45 [31, 32]. For silica glass, a broad fluorescence band appears at 400~700 nm, which may arise from laser-induced defects (oxygen vacancy, E' center, peroxial bonds [78, 83]). The fluorescence signal was successfully employed for detecting bits [80–82]. The above mechanism is useful only for recordable (not erasable and rewritable) memory.

2.2.2 Micro Optical Components

Optical memory is the simplest application of 3D lithography. One just needs to focus the laser inside a recording media, irradiate it by a single pulse or by multipulses for a short duration, and then a bit is naturally produced at the focal site. More complicated optical elements are created by scanning the laser focus in three dimensions. The merit of using laser fabrication is the potential to integrate components of different dimensions and different functions into one chip, simply by one-step laser writing. For instance, on some occasions, integrating devices of different functions in conventional optoelectronic systems is quite difficult, since they are produced by different technologies and materials, (for example laser diode, LD: GaAs, ~100 μm, by epitaxy; modulator: $LiNbO_3$ ~10 mm, by "diffusion+lithography+deposition"). In contrast, these devices could be written by a laser in a single matrix chip, implying the ease of integration. Although there is a lot of work to do to realize a workable integrated optical system, separated devices have been intensively studied.

2.2.2.1 Waveguides and Couplers

Optical fiber is one of the most useful light guiding devices. However, it is not easy to provide optical connection if the light emitting, detecting, and controlling devices are prepared in-chip. The refractive index difference between the core and cladding layers of a single-mode fiber is generally of the order of $\Delta n/n$~0.1%. It is noteworthy that the ratio can be one order larger in the laser-irradiated region compared to the background in typical polymers or glasses [84]. This large contrast is sufficient to confine light inside a laser irradiated fiber line by total internal reflection, as employed in conventional fibers. Glass is a widely-used material for optical components. To write a fiber or any other structure inside silica, of which the bandgap energy can be as large as >8 eV, it is necessary to use regenerative amplification ($\times 10^4$) of pulse energy to launch multiphoton absorption. By this means, waveguides have been recorded inside various glasses such as fused and synthetic silica, Ge-doped silica, borosilicate, borate, phosphate, fluorophosphate, fluoride, and chalcogenide glasses. It was experimentally observed [84] that 15-mm-long fluoride glass waveguides of diameters 8 μm, 17 μm and 25 μm support 800-nm LP_{01}, LP_{11} and LP_{22} modes, respectively. Sikorski et al. observed a 3dB/cm gain at 1062 nm from an active waveguide inside a neodymium-doped glass substrate, from which it would be possible to produce active laser waveguides like fiber lasers or fiber amplifiers.

With the development of single-mode waveguides, it is possible to fabricate more complicated micro-optical devices such as X [85], Y [86] couplers

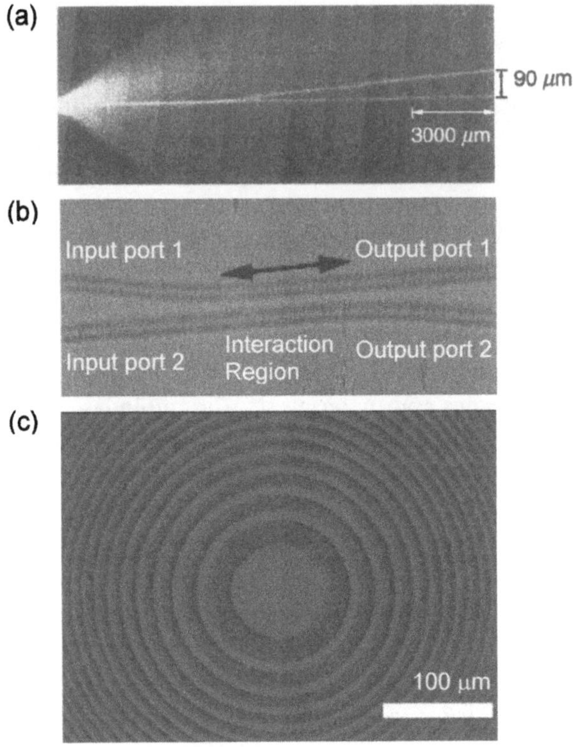

Fig. 2 Laser written micro-optical components. **a** Optical microscopic image of a Y coupler drawn in pure fused silica, which guides 514.5-nm light from an argon-ion laser. The scattered radiation from the coupled argon-ion light is observed in the photograph. The vertical direction is magnified with respect to the horizontal direction for clarity. **b** Phase contrast microscopic image of one of the two X-couplers that make up a Mach-Zehnder interferometer. **c** Optical microscopic image of a Fresnel zone plate

(Fig. 2a). An X coupler was written in glass from two identical waveguides that were crossed with each other by a small angle, α. Experimentally it was found that the individual branches are single mode and at the output the power was split into two output waveguides with ratio of 1:1 for α=1 °and 16:1 for α=4°[4], both at 800-nm wavelength [85].

If the two single-mode waveguides are instead placed parallel with and adjacent to each other at a certain length, they exchange power by mode coupling in this interaction length. This structure is a directional coupler. A directional coupler with splitting ratio of 1.9 dB at 633 nm was written in glass using a 400-nm, 25-femtosecond laser oscillator [87]. An even more complicated photonic device is a Mach-Zehnder interferometer filter. It consists of two X-couplers (Fig. 2b) placed back-to-back. Light coupled into the interferometer is split into two arms at the first X coupler, travel different path lengths, and will either constructively or destructively interfere at the second X coupler [87]. The unbalanced Mach-Zehnder interferometer could act as a spectral filter.

2.2.2.2
Gratings and Zone Plate

Gratings are useful optical structures that could be written by two-beam interference. For femtosecond laser pulses, the coherence length is short, which needs sensitive adjustment of pulse overlapping at both spatial and temporal domains. Various gratings [88, 89] were produced in a variety of materials either by surface ablation or surface relief mechanisms. Both form phase gratings, but the former arises from mass removal and the latter is caused by mass migration. The grating periods, Λ, are determined by $\Lambda=\lambda/[2\sin(\theta/2)]$, where λ is the laser wavelength and θ the two-beam angle. The smallest grating groove that has ever been reported is 15 nm [90]. Such a fine feature was producible only in amorphous materials, implying the creation mechanism of densification. Due to the deep penetration effect, it is feasible to encode 3D grating structures inside materials [88, 89]. These works are quite similar to those we will introduce in Sect. 4.6. Another interesting micro optical component that was produced by femtosecond laser is the Fresnel zone plate (Fig. 2c). The recording femtosecond laser was focused 300 μm beneath the surface of 3-mm thick glass sample, where the silica was bombarded by regeneratively amplified laser pulses [91]. The refractive index change in the irradiated ring zones provides the phase modulation that is necessary for the plate function. In one fabricated structure, the primary focal spot size has been measured at 6.1 μm, agreeing with design, and diffraction efficiency was 2.0%.

One major problem with femtosecond laser processed micro-optical devices is the large power dissipation due to Rayleigh scatterings from the particle-like fine structures, which is difficult to control in the fabrication process because of an intense laser-matter interaction at the femtosecond time scale. Suitable post-irradiation treatments should be found to improve performance of devices [31].

2.2.3
Photonic Crystals

The *photonic crystal (PhC)* [92. 93], the optical analogue of electronic crystals, consists of two or more kinds of materials. Blocks of materials of different refractive indexes appear alternately, producing a periodic structure. The emergence of photonic bandgap (PBG) is the most important phenomenon that is associated with PhCs. Light with wavelength matching the structural periodicity is rejected in a certain direction incident upon the periodic structure. In the case where the rejection occurs at all directions, the PhC has a *full bandgap.* To achieve the full bandgap, structures should be optimized in (i) lattice type: in other words how the primitive units repeated in three dimensions, for example, face-centered cubic (FCC), body-centered cubic (BCC), and so on; (ii) filling ratio: what percentage of the volume is occupied, for example, by the high-index material; (iii) refractive index contrast of the two components; and (iv) material connection: in other words either the high or low index material blocks, the repetitive unit, are spatially isolated [94]. A number of technologies have been proposed for PhC fabrication, but the four requirements are very difficult to satisfy simultaneously due to the limitation of each technology. Multiphoton laser processing is very promising because the lattice type, filling ratio and material connection can be arbitrarily designed from computer programs. Sufficient refractive index, at least 2.0 for diamond lattice [95] – the best lattice to achieve the full bandgap – may be accomplished by high refractive index material doping in photopolymers by using a novel mechanism like metallization.

PhCs written in currently-available transparent solids doesn't produce a sufficient contrast of refractive indexes to open a full bandgap. However, the weak bandgap effect is still interesting for PhC physics, like various nonlinear optical phenomena, and for applications that don't need a full bandgap, such as filters or attenuators. In self-organization of colloidal PhCs [96–98], microbeads are arranged in three dimensions with limited lattice type – FCC or hexagonal close packing (HCP) – and the filing ratio of beads is restricted to 74%. Note that in the optical memory work, bits that are recorded as the binary information unit were arranged layer by layer. If voxels were organized in three dimensions the same way as atoms exist in real-world crystals, the voxels would function as photonic atoms and a PhC would be formed. Photonic atoms can be configured to various lattice points, in size scaled-up from 7 systems, 32 classes, and 230 space groups of generic crystals.

The above idea has been realized inside glass with irradiation of 800-nm, 150-fs IR wavelength [31]. It was found that if the deposited energy of laser and focusing conditions were properly chosen, the voxels take the form of a well-defined near-spherical shape. A 3D FCC lattice with a lattice constant of approximately 1.0 μm has been written, from which a pronounced transmission dip that shows a bandgap effect was measured at 3490 cm^{-1}. Figure 3a illustrates the FCC lattice, (b) one fabricated layer of the FCC lattice, (111) plane, and (c) a defected PhC, a waveguide structure.

An alternative way to create microstructures is by continuously scanning the laser focal spot along a line so that a cylinder is produced [32]. If high repetition rate output is selected, the nominal spot spacing can be very small, for example, 10 nm for 1 kHz repetition rate and 10 μm/s scanning speed. Therefore interwalls between bits neighbored in the same line are partly or completely crushed and ejected by the ensuing pulse shocks. Hence a hollow cylinder is produced. The cylinders are arranged to pack

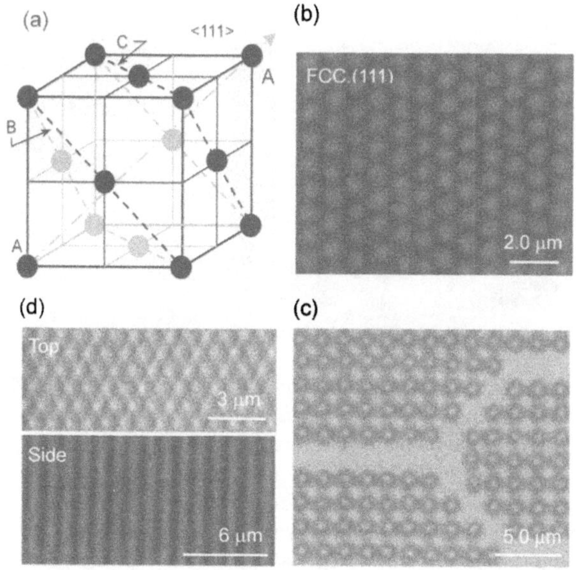

Fig. 3 Femtosecond laser written photonic crystals. **a** a schematic FCC lattice with femtosecond laser-modified material voxel as photonic atoms **b** one (111) photonic atom plane in a FCC lattice written in silica glass **c** a waveguide structure embedded in a simple cubic (SC) photonic lattice, and **d** a 2D triangular lattice photonic crystal consisting of rods continuously scanned inside silica glass

into three dimensions so that logpile-like and triangular lattices were produced. Since the cavititated dot or cylinders were producible in various material matrices, the technology can be widely utilized.

It is noteworthy that no PBG effect is visible from the as-irradiated samples in both dot- and rod-PhCs. After post-irradiation annealing, transmission dips show up, attributed to the smoothing of internal structures which reduced scattering

2.2.4
Use of Complicated Material Recipes

Light, as an energy source, can trigger photochemical reactions, which is the basis of photochemistry. Focused femtosecond laser irradiation confines these reactions into a sub-wavelength tiny volume. Localized photochemical reactions provide a diverse mechanism for fabricating microstructures besides those discussed previously.

It is already known that porous SiO_2 prepared by a sol-gel method has interconnected 3D network structure. At the wet-gel stage of the sample preparation, the internal solution may be exchanged for solutions of various compositions. Actually, the use of solvent exchange to alter the chemistry of

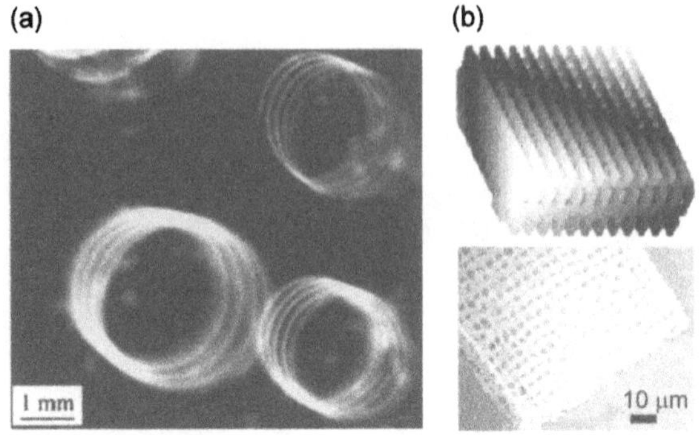

Fig. 4 Femtosecond laser written structures with a complicated material recipe. **a** A 3D spiral structure within silver-doped sol-gel materials. The latent image was produced using multiple exposures with an 800-nm, 120 fs laser. **b** Image of the actual 3D silver structure fabricated in a nanoparticle-seeded polymer nanocomposite by two-photon laser exposure; the upper image was reconstructed from a series of two-photon fluorescence microscopy images obtained at various depths in the sample and the lower SEM image of the free-standing 3D silver structure is the same as the upper one after removal of unexposed material using dichloromethane

the pore liquid is a well-developed method for using sol-gel glass as chemical sensors. Wu et al. [99] incorporated an aqueous solution of silver salt (like $AgNO_3$) into the pores of a SiO_2 matrix of approximately 20-nm diameter, and then wrote 3D structures inside the glass using a 800 nm 120 fs laser beam. Following the trace of focal spot scanning, the silver ion was photoreduced to elemental status by a multiphoton absorption process. After a final developing process that was used to enhance the structure, a 3D metallic structure was produced. Figure 4 shows the optical microscopic image of a spiral coil produced like this with a 3 mm diameter.

Experiments to date using metal ions photoreduced in a matrix have resulted only in patterns of isolated metal particles. Continuous metal structures are desired for better self-support of produced structures, and in order to provide electrical conductivity needed for microcircuits. This target was hampered because crystal nuclei of metals appear and grow into isolated particles in random sites of the irradiated volume. These particles are less easily combined into a bulk due to the limited doping concentration and due to steric resistance to particle movement. Noticing the fact that growth rates are generally much larger than nucleation rate and the former depends on the number of nucleation centers, Stellacci et al. [100] solve this issue by introducing nanoparticle seeds into the composite to be irradiated. The metal seeds are equivalent to high concentrations of ions. Metal ions would be consumed mostly by growth of the existing nanoparticles instead of producing new dispersed nuclei. The growth can be used to fill inter-seed intervals to provide continuous metal structures. Experimentally they used an organic solvent-soluble silver salt ($AgBF_4$) as a precursor to metal atoms, and to achieve homogeneous dispersion of the nanoparticles and high optical quality film, they used ligand-coated particles to avoid aggregation. The role of the particles was proved by the fabricated results; 3D shelf models of copper and gold were produced the same way as silver (Fig. 4b).

3
Fundamentals of Stereolithography using Two-Photon Photopolymerization

Photopolymerization is one of the most important types of photochemical reactions that have been used for laser fabrication [1–5]. This is because the material resins undergo a significant phase transition after laser irradiation, from liquid to solid, and non-polymerized liquid is easily removed by a developing process so that solidified 3D structures stand out [10, 101–103]. Photopolymerized structures have real physical shape, contrasting with those image-like structures recorded in solid matrixes. Hence, not only optical components, but also micromechanical devices as well as microelectromechanical systems (MEMS) could be produced. By using two-photon-induced photopolymerization, it is possible to polymerize structures with sub-micron features, implying a more diverse use of the technology.

3.1
Two-Photon Photopolymerization

3.1.1
Photoinitiation and Photopolymerization

Photopolymerization refers to the process of using light as an energy source to induce the conversion of small unsaturated molecules in the liquid state to solid macromolecules thorough polymerization reactions. Although other radiations, including x-ray, γ-ray, microwave, and even electron and ion beams can induce similar curing reactions [4], photopolymerization deals with those that are induced by light in the UV, visible to IR spectral region. The basic components of the starting liquid material are monomers and oligomers (or prepolymer). Upon light excitation, the monomers or oligomers may be solidified by two means: polymerization and crosslinking [1–5]. An important feature of polymerization is the chain reaction by which macromolecules are created; while cross-linking is concerned more with the formation of crosslinks with chemical bonds (different from the entangling of polymer chains [19]). An important difference of these two kinds of reaction lies in their quantum yield, which is defined as the ratio of number of polymerized monomer units to the number of photons that are needed to cause this polymerization. In the case of photocrosslinking, addition of each monomer unit requires absorption of a photon, leading to a quantum yield less than 1. In contrast, photopolymerization is realized via chain reactions as shown in the following equation (Eq. 6), so the quantum yield can reach several thousands [5].

$$M \xrightarrow{M} M_2 \xrightarrow{M} M_3 \cdots M_{n-1} \xrightarrow{M} M_n \tag{6}$$

Here M is the monomer or oligomer unit, and M_n, the macromolecule containing n monomer units.

For practical photopolymer systems, more components are included, most importantly photoinitiators and photosensitizers [1–5]. The quantum yield of general monomers and oligomers is low. In order to increase the initiating efficiency, one or several low-weight molecules that are more sensitive to light irradiation are added. They form initiating species of radicals or cations by absorbing photons. Such small molecules are called *photoinitiators*. The production of active species that attack monomers or oligomers is called *photoinitiation*, the most important step in photopolymerization. Take the radical case for example, with the following initiation step:

$$I \xrightarrow{\hbar\nu} I^* \rightarrow R\cdot \tag{7}$$

where symbols denote photoinitiator (I), radical ($R\cdot$) and I^*, an intermediate state of the photoinitiator after absorbing a photon. Therefore the polymerization process is more precisely described by the following equation:

$$R\cdot + M \rightarrow RM\cdot \xrightarrow{M} RMM\cdot \cdots \rightarrow RM_n\cdot \tag{8}$$

The photoproduced radicals react with monomers or oligomers, producing monomer radicals, which combine with new monomers, and so on; so the monomer radicals expand in a chain reaction, until two radicals meet with each other. This chain propagation stops in either of the following channels:

$$RM_n\cdot + RM_m\cdot \rightarrow RM_{m+n}R \tag{9}$$

$$RM_n\cdot + RM_m\cdot \rightarrow RM_n + RM_m \tag{10}$$

Therefore the polymerization process consists of several steps: (i) photoinitiation (Eq. 7), (ii) chain propagation, (Eq. 8), and (iii) termination, (Eqs. 9, 10). We can see from the above description that a good photoinitiator should be (i) easily reduced to an initiating species upon light irradiation, and (ii) provide photoproduced radicals or cations active enough to react with monomers or oligomers.

In many cases, the energy collection (i) and triggering chain polymerization (ii) are cooperatively accomplished by multi-type molecules. A *photo-*

sensitizer is a molecule that absorbs light and then transfers the energy to a photoinitiator. With such a scheme, the photoinitiation process is expressed as:

$$S \xrightarrow{h\nu} S^* \cdots \xrightarrow{I} I^* \rightarrow R\cdot \tag{11}$$

where S is the photosensitizer. A *coinitiator* itself doesn't absorb light, but it is involved in the production of radical species.

The above descriptions of polymerization are based on radical initiators. Actually, photopolymerization reactions are basically classified into two categories: radical polymerization and ionic polymerization. Among these two types of photopolymerization, reactions that are typically used for laser fabrication are [1–5]:

(i) double-bond addition of acrylates (radical-type)

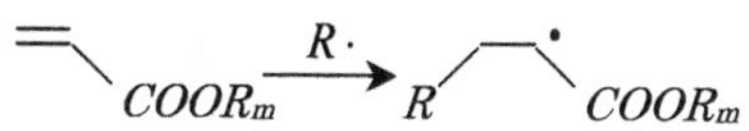

(ii) ring-opening of expoxides (cationic-type)

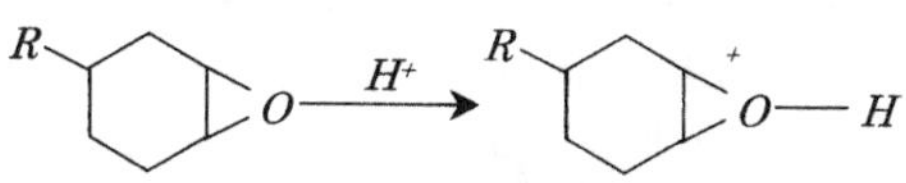

These two types of reactions require triggering by different initiators. For a radical type initiator, benzoyl is the most widely used chromophore since it exhibits good absorption in the UV region.

Although radicals may be produced by various photochemical conversion processes like photoscission, abstraction of intramolecular hydrogen, and electron and proton transfer, the most efficient radical initiators developed so far work via bond cleavage [5], for example:

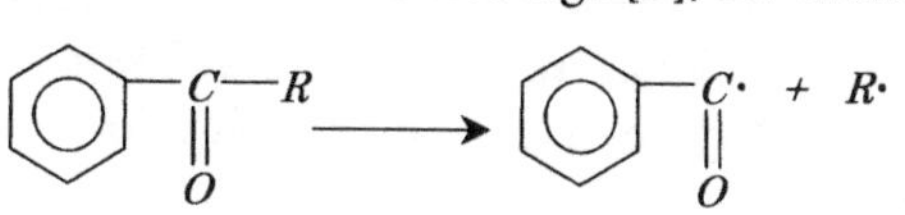

Good reviews of the synthesis, performance, and general research into various radical type photoinitiators can be found in [3, 4, 104].

Cationic photopolymerization is much less used than the radical type. The photoinitiation is generally based on the ring opening of the oxirane group [1–5]:

$$I \xrightarrow{h\nu} C^+ \tag{15}$$

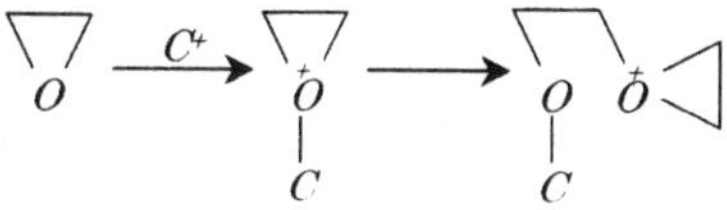

Three classes of molecules are found to be valuable for practical use here: diazonium salts, onium salts and organometallic complexes, about which detailed discussion have been published [4, 105]. Compared to radical type reactions, cationic polymerizations feature (i) low curing speed, (ii) lower viscosity, (iii) small shrinkage after polymerization, and (iv) severe post-irradiation dark polymerization. Sometimes extra thermal processing is needed to increase the conversion of monomers [106]. The above general information is instructive for choosing a suitable material system for laser fabrication.

After polymerization, the oligomer constitutes the backbone of the polymer network. The physical, chemical and mechanical properties of the solidified resin strictly depend on the nature and structure of the oligomer. Oligomers generally contain at least two reactive groups, from which both cross-linking and polymerization could occur. For example, oligomers possessing two acrylate groups may have many different backbones due to different components of R: polyester, polyurethane, polyether, epoxy, and so forth.

Monomers have a much smaller molecular weight and consist of one or several reactive groups [104–106]. They polymerize similarly to oligomers and are an important factor in determining the efficiency of polymerization. In addition, monomers are also useful for diluting resins so that the polymer is easier to handle for a particular use. For 3D micro-nanolithography, a suitable viscosity is of particular importance due to the opposite requirements in different steps of processing: a high viscosity is needed for keeping early produced volumes where they are created; while a low viscosity facilitates removal of unsolidified resin from intervals. Strictly speaking, resin is the oligomers that have a molecular weight ranging from 500 to 3000, and exhibit a viscosity of 5 to 25 Pa·s. However, to keep the convention of word usage, we call the mixture of the starting liquid as resin. In addition to the viscosity, among many, the following bahaviors are preferred for a successful fabrication: (i) high polymerization efficiency upon light irradiation, (ii) lower shrinkage after polymerization, (iii) fast reaction time and low dark polymerization.

3.1.2
Photopolymerization Induced by Two-Photon Absorption

Selection rules for single-photon and two-photon excitation (TPE) are different [42, 107, 108]; however, most resins that polymerize under UV (λ) exposure can undergo similar reactions when two photons (2λ) are absorbed simultaneously (two-photon photopolymerization), provided that the light intensity is large enough.

Electron excitations that need absorbance of two-photon energy can occur stepwise or simultaneously [33] as shown in Fig. 5a,b. The former relies

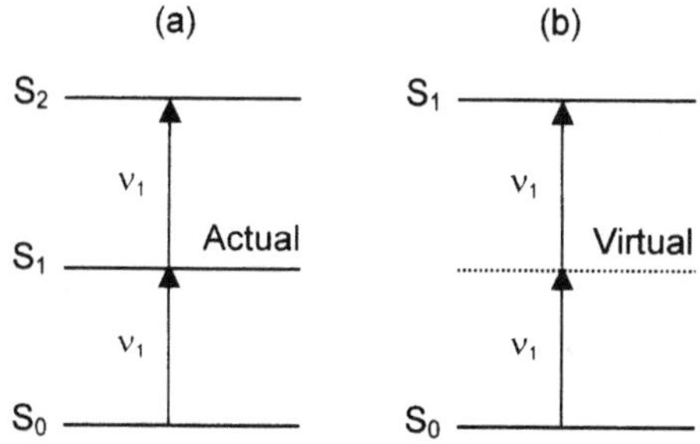

Fig. 5 Illustration of two-photon absorption schemes. **a** stepwise TPA with an actual intermediate energy level, and **b** simultaneous TPA with a virtual energy level. The former could be treated as two sequential single-photon absorption processes. For femtosecond laser micro-nanofabrication, simultaneous TPA is more relevant. All works discussed in this review are based on simultaneous TPA

on the existence of a real intermediate state, from which an excited population is further pumped to a higher energy level by absorbing photons of the same energy as the ground state (excited state absorption). Such a process, although on some occasions also called TPA, is better termed as stepwise absorption, or stepwise TPA to avoid confusion. Compared to simultaneous TPA, stepwise TPA doesn't require coherence of the incident light, and may be considered as two sequential single photon absorptions. The excitation efficiency can be high enough to provide intense up-conversion fluorescence, which is one of the important mechanisms of IR sensor cards.

Simultaneous TPA, most generally referred to TPA, is a quantum mechanical three-body process, where an electron absorbs two photons simultaneously to transcend the energy gap in one excitation event [42, 43]. An intuitive physical scenario is, as light passes through a molecule, a virtual state is formed when the first photon is absorbed. It persists for a very short duration (of the order of several femtoseconds as prescribed by Heisenberg's Uncertainty Principle), which contrasts with the long lifetime of the actual intermediate energy level in stepwise absorption. TPA can result if the second photon arrives before the decay of this virtual state. If the energy of the two photons are identical, the process is referred as *degenerate TPA*, otherwise, the process is a non-degenerate one. TPA was theoretically predicted as early as 1931 [50] and was experimentally observed for the first time in 1961 [51]. For a long time, a practical use for it was not known except as a spectroscopic tool for determining the positions of energy levels that are not connected to the ground state by single photon absorption [108, 109].

TPA can be utilized for inducing photopolymerization. The difference between one-photon [101–103] and two-photon induced photopolymerization lies in how the energy for activating initiators is provided. In the case of TPA photopolymerization, initiators are excited to triplet states by absorbing combined two-photon energy,so correspondingly Eqs. 7 and 11 should be re-written as:

$$I \xrightarrow{2\hbar\nu'} I^* \rightarrow R\cdot \tag{17}$$

$$S \xrightarrow{2\hbar\nu'} S^* \cdots \xrightarrow{I} I^* \rightarrow R\cdot \tag{18}$$

where $\nu' \sim \nu/2$ denotes photon frequency in the two-photon excitation beam. From our experience, resins that have been developed to polymerize at UV or visible wavelengths, ca. λ, were generally polymerizable under 2λ irradiation only if the photon flux density is sufficiently large. However, the usability, TPA efficiency, thresholds of polymerization and laser-induced breakdown, tolerance to exposure dose variation, and so on depend on the natures of the specific materials.

Two-photon photopolymerization was experimentally reported for the first time in 1965 by Pao and Rentzepis [110] as the first example of multiphoton excitation-induced photochemical reactions. They focused 694-nm laser from a pulsed Ruby laser into a sample of styrene that was cooled to 77 K. After developing in methanol, solid precipitate was extracted and confirmed to be polystyrene through IR absorption. In the particular experiment, no photosensitizer was used and the author tried to increase the two-photon absorbance by using monomers with added functional groups, such as para-isopropylstyrene and chlorine-substituted derivatives of styrene. As a result, much enhanced two-photon polymerization was observed. After this work, although there were some ensuing researches that were scattered among the literature [111–113] , no particular efforts were devoted to two-photon photopolymerization until this technology found value as a microfabrication tool.

3.1.3 High Efficiency Two-Photon Materials

Molecules of large TPA cross-section [21–25, 114–116] are very important for the broad application of two-photon photopolymerization technology. They would enable the use of inexpensive CW laser or nanosecond lasers, or multi-beam simultaneous polymerization for batch production by beam splitting. Also, the large TPA cross-section would open up a large dynamic power range for tailoring microstructure dimensions by power control. Polar molecules were found to have a large change of dipole moment ($\Delta\mu$>10D) upon excitation from ground state to excited state [24, 117]. Since both the ground and excited state can participate in the formation of the virtual energy level, the transition probability is proportional to $(\Delta\mu)^4$ [24]. Correspondingly, the TPA cross-section can be larger than $\delta \sim 100\times10^{-50}$cm ^{4}s photon^{-1}.

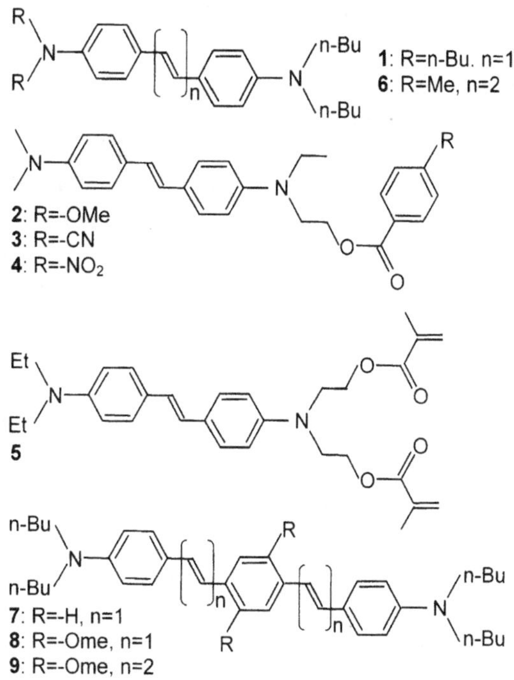

Fig. 6 Molecular structures of high-efficiency two-photon photochromophores

One important effort in the molecular design of large δ is searching for a molecular structure that potentially has larger $\Delta\mu$. It was found that π-conjugated systems such as those with phenylethenyl [21], fluorenyl [117], or polyenyl constructs [118] were good candidates. In these molecules, electron-donating (D) and/or electron-withdrawing (A) moieties were separated by a conjugated π-electron system (A-π-A, D-π-D, D-π–A–π-D and A-π-D-π-A). These chromophores function on the basis of symmetric charge transfer, from the ends of a conjugated system to the middle or vice versa. It is theoretically predicated and experimentally found [22] that δ can be enhanced by increasing the conjugation length and the donor and acceptor strengths. Efficient electron transfer from excited 4,4'-bis(*N,N*-di-n-butylamino)-E-stilbene (**1**, Fig. 6) was confirmed by a large biomolecular quenching rate and by reduction of fluorescence lifetime in covalently-linked chromophore-acceptor systems wherein one or more of the amine-bound alkyl groups of **1** was replaced with an electron acceptor (**2–5**, Fig. 6). The radicals produced by electron transfer from photoexcited **1** to an acrylate or subsequent radical products were found to initiate the polymerization of the acrylate with a much higher efficiency; for example, photopolymerization of a commercial resin SR9008 initiated by **1** and **6** has a threshold 30% that of most conventional sensitive UV initiators. In a further experiment, Marder et al. demonstrated for the above systems that δ was increased by extending the conjugation length (**7–9**, Fig. 6). δ as high as ~1250×10^{-50}cm^{4}s photon^{-1} was obtained from **9**. These initiators were successfully utilized for 3D microfabrication [22].

Another concern in two-photon molecule design is wavelength sensitivity [117, 119]. This arises from the fact that the most suitable femtosecond laser is solid wavelength-tunable Ti:Sapphire laser, of which the wavelength ranges from 680 ~ 840 nm (extendable to around 1 μm but subject to a low output power). Belfield et al. [24, 117, 119, 120] synthesized a series of fluorene derivatives with varying electronic characteristics. The UV-visible absorption peaks at the ideal wavelength of near 400 nm, and some of them possess δ~1300×10^{-50}cm^{4}s photon^{-1}.

In addition to increasing TPA cross-section of chromophores, there is another route to enhancing TPA: by increasing the chromophore number density without causing aggregation. Dendrimers functionalized with TPA chromophores at their periphery have been explored [121, 122]. It is found [123] that by attaching a soluble two-photon absorbing chromophore with a functional group to a dendritic backbone, the molecular TPA cross-sections could be doubled from one dendrimer generation to next. The dendrimers were chosen as highly soluble in common organic solvents. The linear correlation between the end chromophore number and TPA cross-section implies that neither cooperative nor deleterious effects due to the high local chromophore concentration occur.

3.2 Microfabrication Systems

3.2.1 Design Consideration of Optical Systems

The current laser micro-nanofabrication systems stem from the conventional laser prototyping method [6]. Figure 7a illustrates its concept, and for reference, three other layered manufacturing technologies are also shown (Fig. 7b–d) [6].

Laser rapid prototyping [6–9] involves the formation of a 3D object additively, in a layer-by-layer way. It couples the power of computer designs, through laser-initiated photopolymerization, to the formation of a unique, real plastic form. Figure 8 illustrates a practical prototyping system. In fabrication, 3D CAD patterns were first extracted into a series of two-dimensional (2D) slices corresponding to profiles at different height levels (Fig. 7a). Starting from the first layer, the lateral dimension is controlled by scanning a laser beam over the surface of the polymer film at a rate sufficiently low to form a gelled layer of polymer with desired and fixed depth. After the formation of the first layer that was affixed to the substrate, a new

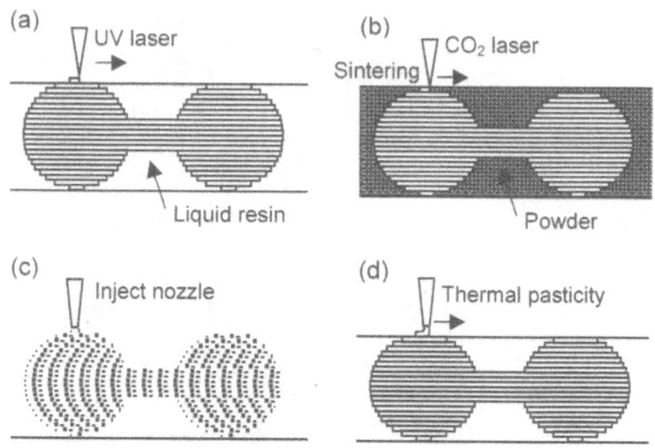

Fig. 7 Illustration of several types of layered manufacturing technologies. **a** Laser rapid prototyping using photopolymerizable resin, **b** powder sintering, **c** ink injecting and writing and **d** resin squeezing and writing

thin layer of polymer was added by casting (Fig. 8a), by immersing the platform (Fig. 8b), or by releasing the liquid surface (Fig. 8c) and then solidifying. This process is continued in an additive modeling fashion until the desired object is formed [6].

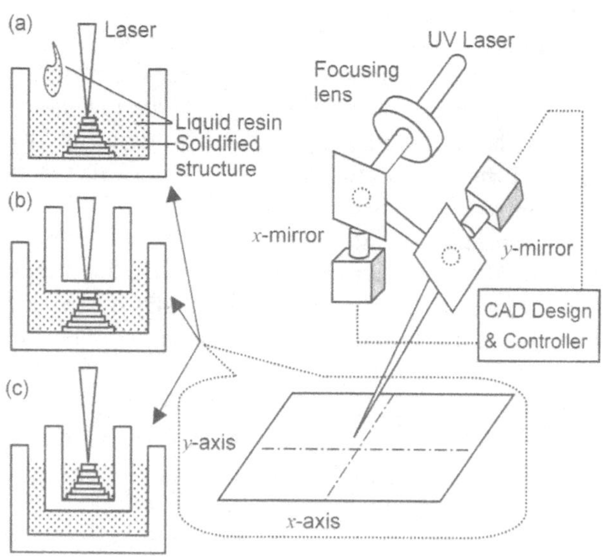

Fig. 8 A laser rapid prototyping system. The writing could be accomplished either by mirror angle scanning or by moving sample stages. The former mechanism is shown here. Three major approaches were utilized to control the longitudinal resolution, **a** film casting, **b** surface regulation, and **c** open surface methods

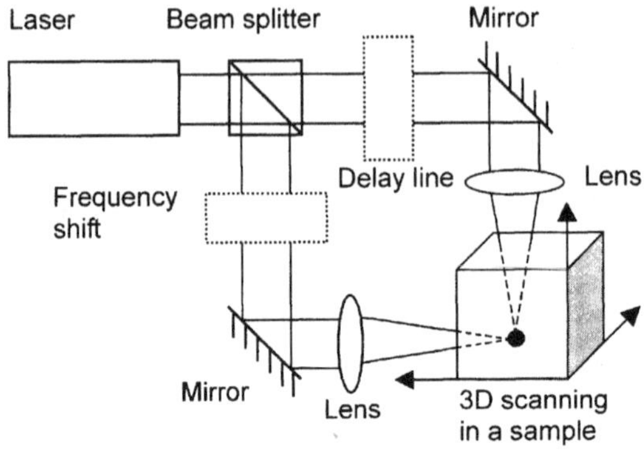

Fig. 9 A crossbeam two-photon two-color scanning laser microscopic system. Pulses from two beams split from an identical laser output should overlap in both time and temporal domains so that a TPA process could be launched by simultaneously absorbing two photons. Removal of the frequency shifter gives rise to a degenerative two-photon fabrication system

In the above system, patterning in the two horizontal dimensions is realized from a series of CAD slices, and the longitudinal spatial resolution is controlled by the thickness of newly-added resin film. There are two disadvantages that hinder applying the mechanism for micro-nanosize fabrication. First, it would be difficult to form film of thickness 1 μm or less due to the viscosity and surface tension of the resin, and second the layer-by-layer scanning method sometimes restricts the achievable geometry.

To solve this problem, Maruo et al. [10] proposed a two-photon laser rapid prototyping technology, now known as *two-photon photopolymerization.* In this scheme the laser was directly focused inside a liquid resin droplet and it polymerized the focal point volume by TPA. This technology firstly eliminates the requirement of thin additive liquid film and controls the longitudinal spatial resolution by focal spot size itself, and secondly, it provides the capability of writing arbitrary 3D patterns within the droplet volume, as can be done in 3D laser writing in solid matrix only if the resin viscosity is reasonably high. From this sense, the laser focus functions as a real 3D laser pen.

For two-photon excitation, crossbeam geometry was originally considered [11, 124] as shown in Fig. 9. Actually in the first proposal on 3D optical memory by Parthenopoulos and Rentzepis, the crossbeam two-color two-photon excitation was employed [11]. The system is difficult to handle in arranging optical components and in synchronizing laser pulses in the time domain, although it may be possible to reach a relatively better longitudinal spatial resolution. Afterwards, Denk et al. [12] applied two-photon excitation to laser scanning microscopes. They tightly focused a laser beam with a single lens to start degenerate TPA, by which high quality living cell images

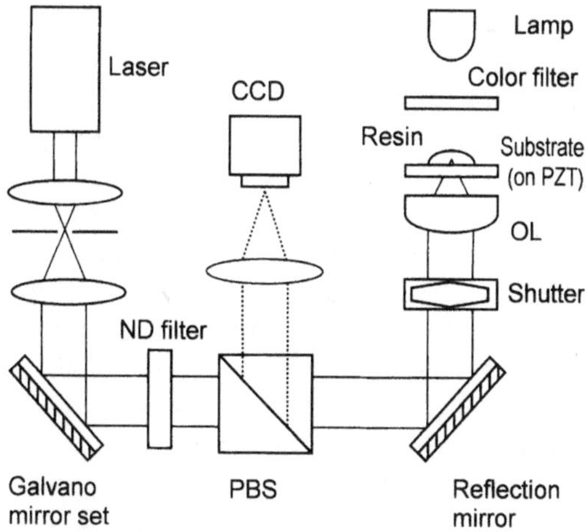

Fig. 10 Two-photon direct writing laser microfabrication system. The Galvano mirror set is used for scanning the laser beam in the two horizontal dimensions, and along the longitudinal direction a PZT stage is used. The laser power was continuously adjusted by a neutral density (ND) filter. The polarization beam splitter (PBS) lets the laser beam pass but reflects the illumination light to the CCD monitor for in-situ monitoring of the fabrication process. OL: objective lens

were obtained. This configuration was widespread due to its simplicity and has been taken as the standard for two-photon fluorescence microscopy. Now most two-photon photopolymerization systems take the form of a laser scanning microscope with enhanced 3D scanning capability.

Figure 10 shows a typical setup of a two-photon photopolymerization CAD-CAM system. It consists of three parts, fulfilling functions of (i) CAD and scanning control, (ii) laser output and beam control, (iii) two-photon exposure and in-situ monitoring.

Laser beam focusing, realized by a microscopic objective lens, is of primary importance to the entire fabrication system. Large magnification is advantageous for in-situ monitoring of the fabrication process. Comparatively speaking, a high resolution is more meaningful to fabrication. The resolution of a given objective is governed by the laser wavelength and its numerical aperture, NA [125, 126]. Shorter wavelength will proportionally reduce the diffraction-limit focal spot size. In addition, TPA efficiency is generally large at short wavelengths. For example, it has been experimentally found [127] that the two-photon photopolymerization threshold of a commercially-available resin was decreased from 3.2 TW/cm^2 (800 nm) to around 0.6 TW/cm^2 (660 nm). However, the use of short wavelength is limited by the availability of laser source and attention should be paid to avoid resin degrading due to monomer or oligomer bond cleavage. *NA* is an expression related to the maximum angle that light rays are collected from the object plane (for imaging) or the maximum angle that a laser is converged onto the focal spot (for excitation). For example, the 1.4-NA oil-immersion (n~1.512) objective lens has a beam convergence angle of 67.2 °. In 3D fabrication, the size of excited volume, *voxel*, depends on $1/(\mathrm{NA})^4$. Therefore choosing large NA optics is essential for realizing a high fabrication resolution.

In imaging optics, the lens aberration is an important issue to address [125, 126]. Among basic aberrations, coma, astigmatism, curvature of field and distortion influence off-axis points. Therefore they don't take effect in the sample stage scanning system, but restrict the scanning range in the mirror-scanning configuration. The other two kinds of aberration, spherical and chromatic, affect the whole field. In a bandwidth-limited laser system, the shorter the laser pulse, the wider the spectrum. The *chromatic effect* (focal length changes versus wavelength) is well corrected with modern apochromatics to a negligible level. *Spherical aberration* (light rays passing through objective lens from different radial zone is focused at different depths) is a serious problem in high-NA objective lenses. This problem was partly solved: (i) many dry lenses were already corrected by considering beam focusing through a coverslip of a certain thickness; (ii) oil immersion is generally adopted in high NA lens design to improve the optical homogeneity between objective front and the coverslip. However, in both cases, if a laser is focused deep (for example, >20 μm) into samples having refractive indexes significantly different from the cover glass (for example, Δn>0.1), measures should be taken to reduce spherical aberration. Recently Sun et al. (results to be published) solved this problem by using a deformable mirror to pre-compensate the aberration.

After the laser focal spot is focused into resin, it should be scanned relative to the sample. This is accomplished using two possible means: (i) along horizontal dimensions by beam scanning by Galvano mirror, plus vertical sample or objective lens movement by lead zirconate titanate (PZT) piezo-electronic stage [13, 14, 18–20] and (ii) sample movement in all three dimensions by PZT [15–17, 101]. The former features high scanning speed because there is no need to reserve time for stage stabilization, but the scanning range is limited by spherical aberration, (to e.g. ~ 20-μm lateral dimensions for general 1.4-NA objectives). In the 3D PZT scheme, the object dimensions are determined by the scanning range of stage to hundreds of microns (range decreases if higher accuracy is needed). In both cases, the vertical dimension is restricted by the working distance of the objective lens, for example to 100~200 μm in a general 1.4-NA oil-immersion objective lens. All-PZT configuration has the shortcoming that a residing time has to be reserved between individual voxel's exposure for stage stabilizing, which can occupy a large percentage of working time, for instance 50% in the case of 1 ms exposure and 1 ms residing.

3.2.2
A Comparison with Other Microfabrication Technologies

It is projected that miniaturization [128], particularly MEMS [129], will bring a societal revolution, so development of new manufacturing technologies has attracted intense research efforts.

3.2.2.1
Photolithography

In addition to its indispensable role in microelectronics and optoelectronics, photolithography has been widely utilized for versatile fields including MEMS, micro sensors and actuators, micro-chemical reactions and analytical systems, and micro-optical systems [128, 129]. Currently UV lithography has feature sizes down to 250 nm, and it is expected to be reduced to ~100 nm in the near future by use of a combination of deep UV light (for example 193 nm ArF excimer laser or 157 nm F_2 excimer laser) and improved photoresists.

To push the resolution of conventional lithography into the sub-100 nm regime, new irradiation sources with short wavelengths have been utilized, including: XUV lithography, soft X-ray lithography, e-beam writing, atom beams, focused ion beam (FIB) writing, and proximal-probe lithography [128-131]. Better resolution, up to several nanometers has been achieved by some of these technologies. However, substantial efforts are needed to improve the development of reflective optics and/or new types of masks, and arrays of beams. In addition, all of these systems need to operate in vacuum, which causes high costs, difficult maintenance, and low fabrication efficiency.

3.2.2.2
Soft Lithography

Soft lithography [132, 133], including microcontact printing (μCP) [134], embossing [135], microtransfer molding [136], cast molding [137], injection molding [138], and replica molding [139], have a common feature. They all use elastomer, typically PDMS or thiol, patterned from a master plate created by lithography, as the stamp, mold, or mask to generate micropatterns and microstructures. This technology circumvents the limit of optical diffraction and achieves resolution up to 20 nm [132, 133]. Different from lithography, that is intrinsically suitable for use on planar structures, soft lithography can be feasibly used to produce surface structures of different heights (2.5 dimensions).

3.2.2.3
Two-Photon Photopolymerization

Compared with the above micro-nanofabrication schemes, two-photon photopolymerization has unique merits:

- First of all, it has intrinsic ability to produce 3D structures. In addition, the long wavelength chosen for TPA has less absorption and less scattering, which gives rise to the deep penetration of light; use of ultrashort pulses can start intense nonlinear processes at relatively low average power, without thermally damaging the samples.
- The two-photon photopolymerization system resembles a laser scanning microscope, which doesn't need vacuum condition for operation. The system is easy to operate and maintain.
- No mask, mold, or stamp is needed for fabrication. It directly converts computer-designed patterns into matter structure. The rapid turnaround time for fabrication allows one to quickly iterate and modify design.

3.3
Early Works in Microfabrication

3.3.1
Initial Proposals

The first reported real 3D microstructure [10] is a 7 μm-diameter and 50 μm-long spiral coil with a line cross-section of 1.3 μm×2.2 μm (Fig. 11). Due to the use of a relatively low NA (0.85) objective, the feature size is larger than the potential limit provided by the technology. However, it confirmed the feasibility of two-photon photopolymerization in three dimen-

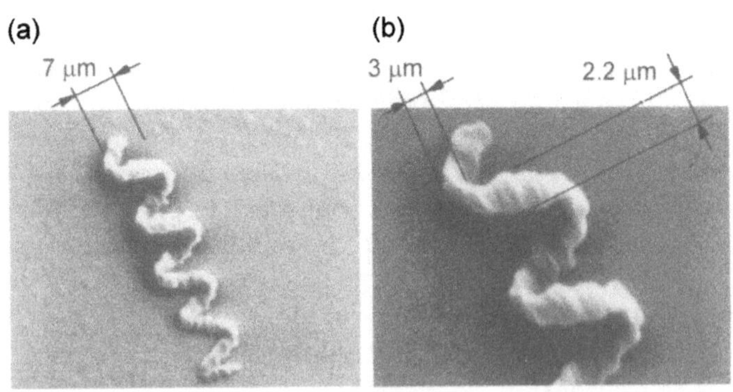

Fig. 11 Scanning electron microscopic (SEM) image of a spiral coil structure made by two-photon photopolymerization after removal of unsolidified resin. **a** A view of the entire structure and **b** the manified top end

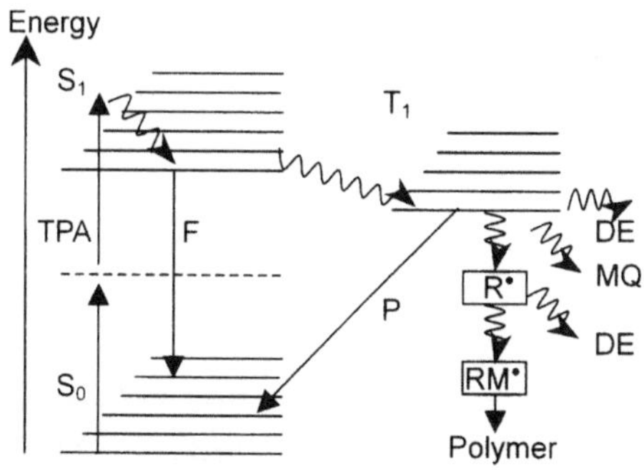

Fig. 12 Energy diagram for two-photon excitation and processes occurring in the excited states. Valence electrons of an initiator are excited from the ground (S_0) to the first excited (S_1) singlet state by simultaneously absorbing two photons. The excited electrons then relax by transition to the triplet state (T_1) via intersystem crossing, where the initiator is liable to undergo bond cleavages, producing radicals for photopolymerization. The excited states can also be relaxed by radiative processes: fluorescence emission from singlet states (F) or phosphorescence emission from triplet states (P). Both the triplet state and photoproduced radicals can be deactivated (DE), for example by monomer quenching (MQ) for the former, and by radical quenching (RQ) for the latter. Efficient photopolymerization generally needs these competing processes (F, P, DE) minimized but there are a few cases, for instance RQ, which find use for it in reducing voxel sizes. The dashed line denotes a virtual energy level for TPA

sions. In this experiment, a 790 nm, 200 fs laser was focused into SCR 500 [Japan Synthesis Rubber Company, JSR], a commercial urethane acrylate resin that is often used in the group.

Figure 12 shows the scheme of TPA and processes occurring in the excited states. Due to a long excited-state lifetime, the triplet state (where the sum of electronic spin in a molecule is $2s+1=3$, where $s=\pm 1/2$ is the quantum number of the electronic spin) is the major transient state that is responsible for the generation of radicals.

The carbonyl group contained in the initiators that were used, a mixture of benzoyl cyclohexanol and morpholino phenyl amino ketones (inset of Fig. 13a), exhibits two kinds of triplet states: $\pi\pi^*$ and $n\pi^*$. However, according to selection rules of electronic transition, $n\rightarrow\pi^*$ transitions are symmetry forbidden and $\pi\rightarrow\pi^*$ transition are symmetry allowed. Therefore the molar extinction coefficient of the $n\rightarrow\pi^*$ transition is much smaller than that of the $\pi\rightarrow\pi^*$ transition. Figure 13a shows the absorption spectrum of the resin, which consists of urethane acrylate oligomers with different molecular weights (480 and 1200), urethane acrylate monomer as a dilutor, and the initiators mentioned above.

Actually, in earlier research, Wu et al. [140] already proposed exposing thick photoresist film using TPA by substituting UV light with lasers of longer wavelength, and scanning the laser focal spot in three dimensions in-

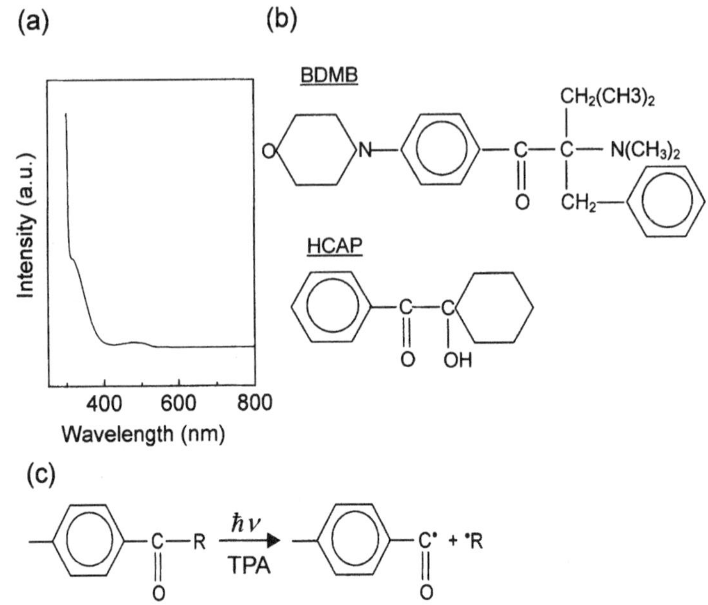

Fig. 13 **a** Absorption spectrum of SCR 500 resin, which was major precursor material for two-photon photopolymerization in the work of Sun et al. [13, 14, 18–20] **b** Molecular structures of two kinds of benzoyl initiators: BDMM and HCPA. **c** Typical reaction of radical production via an α-cleavage process. Radicals are produced in HCAP entirely by, and in BDMB mostly by, this mechanism. Note negligible single photon absorption at the working wavelength of 780 nm

stead of illuminating a photomask with parallel beam. Experimentally the photoresist was spin-coated on a cover glass to form a 3~4 μm film and then baked at about 80 °C. After 3D scanning with 100 fs, 620 nm laser light, a trench structure with an aspect ratio of eight and trenches with undercut profile (for example 0.3 μm at the top and 0.8 μm at the bottom) were obtained. Although this research was different from laser prototyping due to the use of conventional solid film, it is already quite similar to the current two-photon polymerization technology.

3.3.2 Evidence of the Two-Photon Process

The absorption of UV curable resin due to valence electron transition is situated in the UV or the blue side of the visible. Generally it is safe to attribute polymerization reactions, when exposed to IR, to a two-photon process from the transmission or absorption spectra of the starting materials. However, it is still helpful to obtain direct evidence of the occurrence of TPA. A simple test could be conducted using a general mode-locked femtosecond laser. Linear exposure needs a certain photon flux, D_{th} (photon/cm^2), expressed by the product of irradiation duration (Δt) and photon flux intensity (photon/s cm^2) that is proportional to light intensity (W/cm^2). D_{th} is reachable either by intense irradiation for short time or weak irradiation for long time. However, experimentally there is no polymerization observed when lasers work in CW mode, where the power is uniformly delivered over time [141]. On the other hand, polymerization occurs with the same irradiation dose under pulsed mode. When switched to pulsed mode, a 82 MHz, 100 fs and 1 mW mean power at diffraction limit focal spot has a photon flux density of 10^{29} photons/s cm^2 in the pulse duration, 1.2×10^5 times larger than that in CW mode. Since single-photon polymerization relies only on photon flux, while both photon flux and photon flux density are critical for two-photon photopolymerization, the above simple test provides solid proof of the TPA process.

Wu et al. [142] studied two-photon exposure using photographic films based on silver halides. An 800 nm, 120 fs laser was slightly focused with the waist crossing the film. By extrapolating the polymerization thresholds at 350 nm (2.8×10^{-6} J/cm^2) and at 470 nm (3.8×10^{-4} J/cm^2) to 800 nm, it is predicted that the single photon exposure at 800 nm needs 10^9 J/cm^2. This value is larger than the actually used value (~ 0.4 J/cm^2) by many orders, excluding the possibility of single-photon process at 800 nm wavelength. In further experiments, it was observed that the photon flux that was needed to induce a diffraction-limit spot using 120 fs pulses was four times smaller that needed by 240 fs pulse. This result, consistent with the CW/mode-lock polymerization test, supports the theory that the excitation is by a two-photon process.

3.3.3 Pulse Energy Issues, Laser Oscillator and Regenerative Amplification

The immergence of two-photon photopolymerization is a direct outcome of the progress of laser technology. For example, pulse regeneration technology can increase energy of single pulses by more than four orders of magnitude, by which polymerization thresholds of almost any kinds of resin become accessible. Take Spectra Physics Ti:Sapphire laser amplifier (Spitfire) as an example. The single pulse energy of a 80 MHz, 1 W average input power laser beam is 12.5 nJ, while after regenerating amplification, the repetition rate become 1 KHz with the average power unchanged, giving rising to a single pulse energy of 1 mJ. Increase of single photon energy multiplies photon flux density by the same order ($>10^4$), and TPA probability is squared (10^8). With such huge single pulse energy, almost any UV curable resins can be polymerized by TPA. That is a major advantage of the pulse regenerative amplification system.

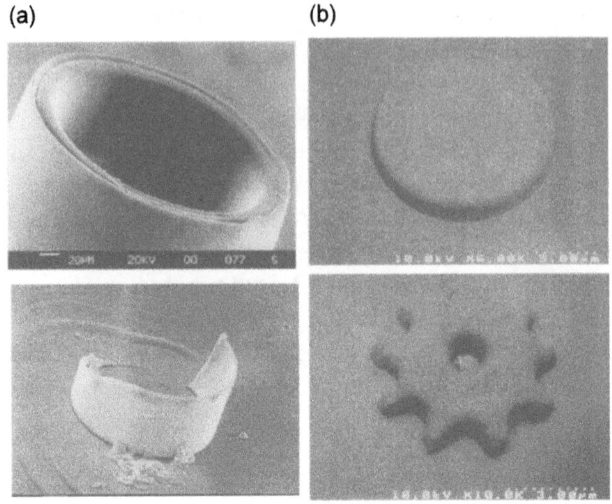

Fig. 14 Microstructures produced using regeneratively magnified laser pulses, where voxels are created by single-shot pulse irradiation. **a** a tube (top) and a spiral tube (bottom), showing the large longitudinal fabrication resolution in the case of low-NA focusing (25-mm focal length), and **b** a micro-disk (top) and a micro-gearwheel (bottom)

The difference in laser pulse energy slightly influences the fabrication processes. First, the average power needed to induce photopolymerization is much smaller for a pulse-amplified beam than for an oscillator beam. Second, large pulse energy enables formation of voxels by single pulses. Witzgall et al. [127] used SU-8, developed for high aspect ratio MEMS fabrication. 4-μm diameter and 80 μm height voxels were created by focusing with a lens of 25-mm focal length (very low NA). Consecutive single-shot laser irradiation produced a spiral structure by fast axial translation and simultaneous sample rotation (Fig. 14a). Also by single shot laser irradiation enabled by pulse regeneration, Sun et al. [15–17] fabricated various micromechanical and PhC structures; as an example, a micro gearwheel is shown in Fig. 14b.

If the photopolymerization threshold could be reached, a voxel could be produced either by exposing an 82 MHz pulse for 1 ms, that is 82,000 shots, or by 1 kHz single-shot (1 shot/ms) irradiation. Therefore, the fabrication speed is not much increased by use of currently available regeneration systems. Another point that should be mentioned is that in this review we don't compare laser power or laser pulse energies that were reported from different groups and different works because of a lack of a standard method to precisely measure the focal spot power, particularly when the laser is tightly focused. The difficulty arises from the fact that the average power at a focal spot is generally small (hundreds of nanowatts) and quite divergent (67.2 ° for 1.4 NA). Calculation from lens transmission is less accurate because the power loss in objective lenses arises largely from the rejection at the entrance pinhole and the pinhole used to tune NA, and therefore the net output power is related to the incident beam shape and size.

3.3.4 Dynamic Power Range

The polymerized voxel size increases with the increase of the irradiation duration (Δt) and the square of light intensity, SLI, (I^2), implying the possibility of tuning voxel dimensions by controlling Δt and I. However, the tuning range or *dynamic power range* is defined by the window between the two-photon polymerization threshold and the laser-induced breakdown threshold [49]. This range is quite small for general resins, several times the polymerization threshold [16, 127], leaving little room for power tuning.

The photopolymerization threshold is determined by the production efficiency of initiating species from excited triplet states, which is characterized by the quantum yield of polymerization. The reactions that produce radicals should compete with monomer quenching, oxygen quenching and other pathways of deactivation of the excited states like phosphorescence emission. The threshold is also determined by the reactivity of radicals and monomers.

When laser irradiation is greater than a particular value, intense damage is induced in materials. This phenomenon is called *laser-induced breakdown.* The breakdown is dominated by a thermal process when pulse width is long (e.g. >10 ps for most transparent materials [49, 143]). This was evidenced by observations that the breakdown threshold scales approximately with pulsewidth by $\tau^{1/2}$ for τ >10 ps [143, 144]. For shorter wavelength (<1 ps), it is believed that the breakdown occurs in various materials via plasma generation [145]. Plasma can be produced via an avalanche process whereby free electrons are accelerated by the incident light field, causing an explosive cascade growth in electron density. The generated plasma can, in turn, absorb and defocus the remaining incident light field. The breakdown causes the ablation process at sample surface and micro-explosion inside bulk, both accompanied by vaporization and atomization of the sample constituents. Laser induced breakdown spectroscopy is a useful tool for semi-quantitative elemental analysis of surface components in completely unknown samples. However, it is detrimental to photopolymerization because breakdown-induced resin bubbling damages existing structures and prevents further reactions.

The laser-induced breakdown is not necessarily related to the multiphoton absorption process [145] while photopolymerization depends on it strictly [24, 25]. Therefore, measures could be taken to increase the quantum yield of polymerization, and therefore the dynamic power range. A simple and effective method is choosing laser wavelength so that radicals are produced more efficiently. It was experimentally observed [127] that the two-photon exposure threshold at 660 nm is roughly half of that at 700 nm, and approximately five times lower than that at 800 nm, while the breakdown threshold didn't vary significantly. The most efficient method is use of high TPA cross-section initiators. By using photoinitiator 8 in Fig. 6, the dynamic power range of radical initiated polymerization of acrylate resin reached 50 [22]; and by using an isopropylthioxanthone/diarylidonium salt initiating system, the polymerization of epoxides was increased to more than 100 [146]. With such a large dynamic power range, the voxel size could be tailored by almost two orders.

3.3.5 Viscosity of Resins

In two-photon polymerization scanning should start from the substrate surface so that polymerized structures can be fixed. If the structure/substrate contact area is small, the adhesion may not be strong enough against post-fabrication washing. This issue was largely alleviated by pre-casting a thin layer of resin to be used, and then fully polymerizing it by single photon or thermal processes, since the polymer-polymer adhesion is better than that from polymer-glass connection. During the course of fabrication of real 3D structures, earlier-produced structures may move from where they were created, or even float away due to the flow of resin itself caused by resin spreading on the substrate surface or by ambient vibration. This is an intrinsic shortcoming of using liquid droplets.

This problem was solved by choosing resins of relatively high viscosity [1–5], which offer a clamp function to the isolated components that were temporarily not yet connected to the main body. The high viscosity is attainable by reducing the concentration of monomers in the resin. For some commercial resins with ready ingredients, it was found effective to use a pre-polymerization method [16]. That is, partly exposing the resin via single-photon absorption before TPA fabrication to induce short-chain photopolymerization, which equivalently increased the viscosity of resins. Since photons can't penetrate the surface to a significant depth, a strenuous stirring of resist was utilized to afford better uniformity.

To test this idea, Sun et al. [16] pre-exposed NOA-800 (Norland Products Inc.) resin under a 150 W xenon lamp for 60 seconds with a quick stirring. With this partly exposed resin, they fabricated a micro-gearwheel affixed to a shaft (Fig. 15). First the shaft on the hemispherical base was constructed, and then the gearwheel was co-axially polymerized at a level of the top end of the shaft. Due to the high viscosity of resin, the gearwheel didn't move even if it was not connected to any part of the shaft that was adhered to the substrate surface. The gearwheel gradually fell in an in-situ developing, and was finally trapped by the axle.

Fig. 15 SEM image of a real 3D device, a microgear wheel affixed to a shaft, which was produced using the pre-photopolymerized resin to increase its viscosity. The gearwheel was observed to be rotatable during the developing process

3.3.6
Two-Photon Fluorescence-Induced Photopolymerization

Novel polymer systems with much enhanced TPA initiators have been synthesized. However, it is still attractive to use well-known systems that were developed for conventional UV polymerization due to their excellent chemical, physical, and mechanical properties. Initiators in these resins generally have small TPA cross-sections, which require very high laser pulse energy for two-photon use. A strategy for circumventing this situation is to induce photopolymerization not by direct TPA of initiators, but by TPA-induced up-conversion fluorescence. An efficient fluorophore that emits at blue or shorter wavelengths was doped into resin. By TPA, the excited fluorescence polymerizes the resin by a single photon process. Since the fluorescence was three-dimensionally confined at the focal spot, the polymerization was also restrained to occur at this volume.

This technology uses well-developed UV polymers and takes advantage of TPA. Joshi et al. [147] applied this method to NOA 72 resin, a well-known urethane acrylate oligomer that was cured by absorption light of 315~450 nm. The resin was doped with AF183 dye by 2 wt%. The fluorophore AF183 has a very strong TPA at near IR, and emits blue up-converted fluorescence peaking at ~465 nm, with the blue side covering the absorption of NOA 72. By exciting an 800 nm, 100 fs laser, they wrote a splitter-type channel waveguide (Fig. 16), and succeeded in coupling a He-Ne laser into the waveguide. Similar experiments [148] were conducted using AF 380 dye doped NOA 72, producing two-beam interfered gratings.

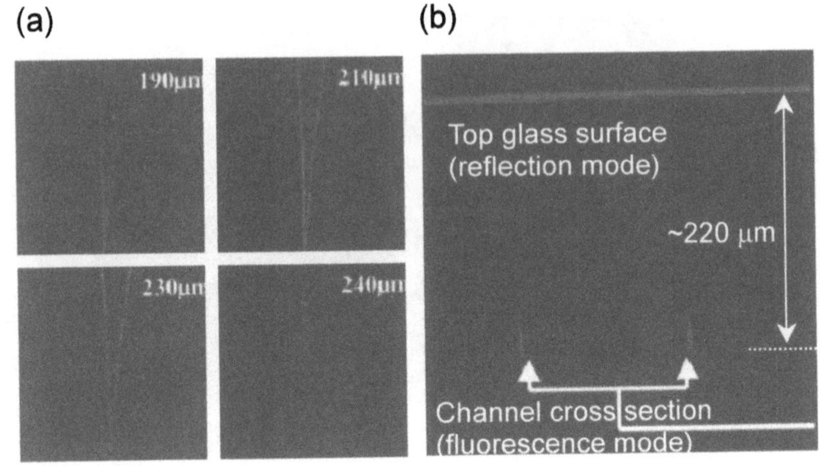

Fig. 16 **a** Different confocal images of a splitter-type channel waveguide polymerized inside a polymer gel by two-photon-induced up-conversion fluorescence. **b** Vertical cross-section of the channel waveguide imaged using both the fluorescence and reflection modes in a confocal microscope

3.3.7
Cationic Photoinitiated Polymerization

Although less used, cationic photoinitiated polymerization of epoxides have some merits in microfabrication [1–5, 104–106]: for example, less shrinkage after polymerization, which may facilitate high fidelity 3D writing; low viscosity, which is important for developing complicated internal structures; and high stiffness. These resins may be useful in fabricating 3D PhCs which have sufficient self-supporting due to the 3D periodicity but impose high requirements in washing away unsolidified liquid monomers and oligomers.

Commercially-available diaryliodonium (CD-1012, Sartomer) and triarylsulfonium (CD-1010, Sartomer) salts were found to initiate polymerization of multifunctional epoxide and vinyl ether monomers [24]. Figure 17a shows a line structure achieved by polymerization of a mixture of poly(bisphenol A-co-epichlorohydrin) glycidyl end-capped and 3,4-epoxycyclohexylmethyl 3,4-epoxylcyclohexanecarboxylate (K126, Sartomer) initiated by CD-1012. To check the potential spatial resolution, we focused the laser using high-NA optics to polymerize commercial resin SCR701 (JSR) and a lateral spatial resolution of 100 nm was obtained (Fig. 17b).

The refractive index increases with the increase of degree of photopolymerization. Therefore a monitoring light beam passing through the center of a voxel would undergo a phase shift relative to that through the periphery of the voxel due to the high central light intensity in a Gaussian field. Interference patterns formed due to different portions of the monitoring beam

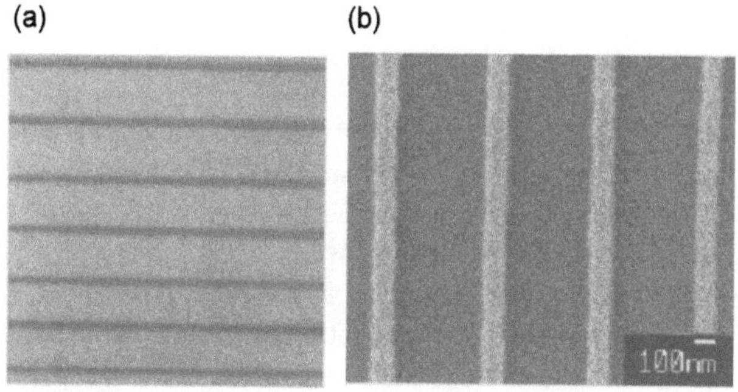

Fig. 17 Line structures produced by cationic photoinitiated polymerization. **a** by polymerization of a mixture of poly(bisphenol A-co-epichlorohydrin) glycidyl end-capped and 3,4-epoxycyclohexylmethyl 3,4-epoxylcyclohexanecarboxylate (K126, Sartomer) initiated by CD-1012, and **b** by commercial resin SCR 701. Note the 100-nm lateral spatial resolution in (b)

would vary with the progress of polymerization [146]. This effect was implemented for studying the polymerization threshold (I_{th}), the polymerization rate (R) and its dependence on various parameters of the optical system and resin components. By using this technology. Boiko et al. [146] found that R of an isopropylthioxanthone/diarylidonium salt initiated epoxide polymerization is proportional to $m = 1.7$ power of the intensity:

$$R = [C(I - I_{th})]^m \tag{19}$$

where C is a constant and I the light intensity. This is different from radical polymerization, where the dependence is much more complicated. The difference is possibly related to the different termination mechanism in the two types of polymerizations.

4 Advanced Techniques in Two-Photon Micro-Nanofabrication

The progress of femtosecond laser technology and exploration of high efficiency photoinitiators and photosensitizers have fueled the progress of two-photon polymerization lithography. Its basic principle and potential applications have been demonstrated. However, to establish this technology as a nano-processing tool, a lot of work on optics, materials and electronic controlling has been done to reduce the writing laser power, evaluate and improve the fabrication accuracy and efficiency, and launch new applications.

4.1 Circumventing the Diffraction Limit

Compared with projection lithography (photo, x-ray and electron beam) and soft lithography technologies, two-photon photopolymerization is unique in its 3D processing capability. However, the use of relatively long wavelength worsens the spatial resolution, for which the bottleneck is set by the optical diffraction limit [125, 126]. This is a limiting factor for the wide application of this technology. In this section, we will introduce how to circumvent the restraint and achieve SDL accuracy in 3D fabrication.

4.1.1 The Diffraction Limit

Optical diffraction plays an essential role in the resolving power of a focusing light microscope, which was revealed by Abbe's work on image formation [126]. His insight initiated imaging approaches such as electron and scanning tip microscopes. Spatial resolutions in these techniques were improved either by decreasing probe beam wavelengths using atom, electron or ion waves, or by using a sharp tip-end to reduce the probe-specimen interaction area as done in scanning tunneling microscopes (STMs) [149], atomic force microscopes (AFMs) [150] and near-field scanning optical microscopes (NSOMs) [151]. The resolution reported for these systems range from hundreds of nanometers down to the remarkable value of less than one nanometer. However, almost all of these methods abandon the use of light (STM, AFM) or the propagating modes of light (NSOM) and are therefore applicable only to surface imaging and fabrication.

In two-photon 3D lithography, the nominal Gaussian output of the laser underwent beam expansion and focusing, and was spatially filtered, producing a relatively flat wave front. Light distribution at the focal plane arises essentially from Fraunhoffer diffraction on the aperture of an objective lens [126], of which the diffraction pattern is shown by the inset of Fig. 18.

For TPA, the SLI, I^2, is directly related to the photopolymerization rate. A near-focus SLI distribution is shown in Fig. 18. The light intensity in the focal plane was assumed to follow a formula due to Airy [126]:

$$I(x) = (2J_1(x)/x)^2 I_0 \tag{20}$$

where $x=2\pi a\omega/\lambda$, w is the coordinate in the diffraction pattern, and $I_0=ED/\lambda^2R^2$ is the intensity at the center of the pattern, E is the total energy incident upon the aperture and $D=\pi a^2$ is area of the aperture with an effective radius a. For imaging, the resolving power was limited by Rayleigh's criterion, $\delta r=k_1\lambda/NA$, due to the signal overlapping from neighboring object points, where $k_1=0.4\sim0.6$ is a constant depending on the laser linewidth and projection geometry. The spatial resolution could be improved by utilizing either

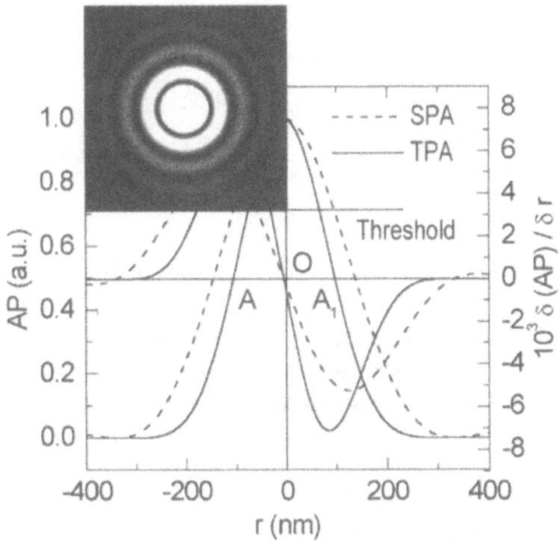

Fig. 18 Light intensity analysis for understanding the achievement of sub-diffraction-limit spatial resolution. Focal plane light intensity (dashed line) and the square of light intensity (solid line) distribution are associated with single-photon and two-photon excitation, respectively. Their derivative distribution is also shown. The inset is the diffraction pattern at the focal plane

shorter wavelengths or larger NA focusing, but the diffraction limit cannot be circumvented. For direct laser writing, a single focused beam is employed, producing an Airy pattern. The issue of light intensity overlapping is eliminated and the light intensity at the focal point was continuously adjustable. Therefore, even if the focal spot size is fixed for a given optical system (wavelength and objective lens), the diffraction limit can be circumvented, provided that the photochemical processes responsible for the formation of voxels have a threshold response to light excitation. Here the *threshold* was a level of light intensity (Fig. 18), above which the photochemical reactions become irreversible (for example, photopolymerized). In this case, the diffraction limit becomes just a measure of the focal spot size; it does not put any actual restraint on voxel sizes.

4.1.2
A Thresholding Mechanism: Radical Quenching Effects

Thresholding performance depends on the individual mechanism of photochemical reactions. In radical type photopolymerization, the oxygen molecules play an important role in the reaction process [5, 152] as described by Fig. 19. Oxygen quenches polymerization via two possible routes. The first is triplet state quenching (T-quenching), where the triplet state of the initiator molecules can be directly consumed by reacting with oxygen molecules without generating any radicals. This is a relatively weak effect. For example,

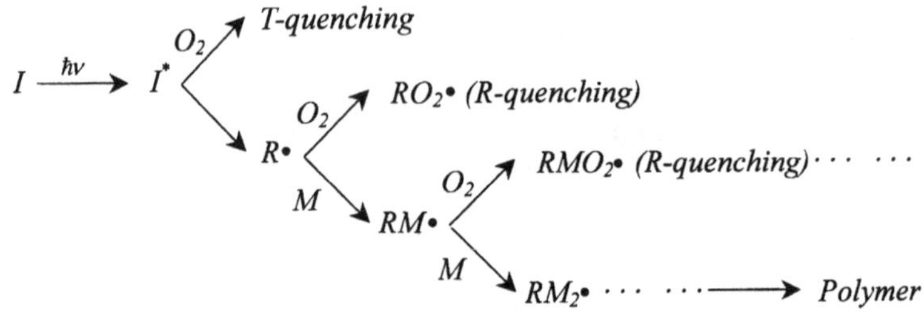

Fig. 19 Photopolymerization quenching by oxygen molecules under two mechanisms: triplet state quenching (T-quenching) and radical quenching (R-quenching). The latter is the major factor that works in the sub-diffraction-limit fabrication

given an oxygen concentration $[O_2]\sim 10^{-3}$ M^{-1} and a typical rate constant of bimolecular quenching in fluid solution of 10^9 $M^{-1}s^{-1}$, the reaction rate is of the order of 10^6 s^{-1}. A triplet state shorter than 100 ns would not be significantly influenced by the presence of oxygen. In resins, particularly acrylate resins, the high viscosity makes the rate constants even smaller. Therefore triplet quenching doesn't play a significant role. The second effect is radical quenching (R-quenching), where radicals combine with oxygen molecules, producing much less active peroxial radicals. This phenomenon, common in radical type polymerization, is the origin of the reaction induction period and it reduces polymerization efficiency [153]. The two-photon photopolymerization threshold is closely related to the radical quenching effect.

Due to the quenching effect, the existence of oxygen is generally considered as detrimental factor in polymerization. Measures were taken to prevent it, including N_2 gas blowing, addition of O_2 barriers like paraffin waxes [154], addition of oxygen scavengers [155]: thiols, phosphines [156], and acrylate amines [4]. However, as a competing process of polymerization, the quenching effect may be utilized as the thresholding effect to confine the polymerized voxel size [13, 14, 102]. By tailoring the light intensity at the focal volume, it is possible to reach a state where TPA-induced radicals survive and initiate polymerization only at the region where exposure energy is larger than the threshold. The intensity of high-order diffraction features (see the subsidiary maxima in the inset of Fig. 18) was low, and therefore easily controlled under the TPA threshold. It is interesting to study the region A_1A in Fig. 18. Judging from the derivative distribution of absorption probability, the voxel size more sensitively depends on variation of the light intensity for TPA than for single-photon absorption, implying that the former possesses a more pronounced threshold effect, although both benefit from it. For the same reason, TPA provides an important mechanism for optical power limiting and power stabilization.

4.1.3
Realization of Sub-Diffraction-Limit Features

The above physical scenario was experimentally evidenced [13, 14]. Figure 20a shows the SEM image of voxels formed under different exposure durations and laser pulse energies. Voxel sizes are quite reproducible, fluctuating within less than 8%. This was firstly due to the high stability of the output laser pulse energy, and secondly aided by the utilization of a diaphragm that served as a spatial filter, allowing a relatively flat field. A lateral spatial resolution down to 120 nm (the inset of Fig. 20a) has been achieved, which is much better than that attained by conventional TPA photopolymerization and that by laser rapid prototyping. A logarithmic dependence of voxel size on exposure time was obtained (Fig. 20b), which is a natural result of the exponential decay of the oligomer/monomer concentration versus exposure [152]. By linearly fitting the experimental data using the least-squares method and extrapolating the curve to zero diameter, threshold exposure times

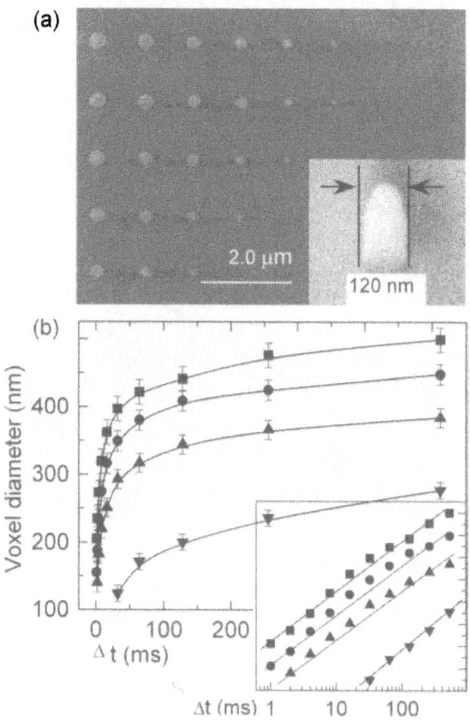

Fig. 20 Achievement of sub-diffraction-limit voxels. **a** SEM image of voxels formed under different exposure durations, and **b** an exposure time-dependent lateral spatial resolution. The right-lower inset represents the dependence in half-logarithm coordinates. For different curves, the laser pulse energies are 163 pJ (filled squares), 137 pJ (filled circles), 111 pJ (filled triangles) and 70 pJ (up-side-down filled triangles), respectively

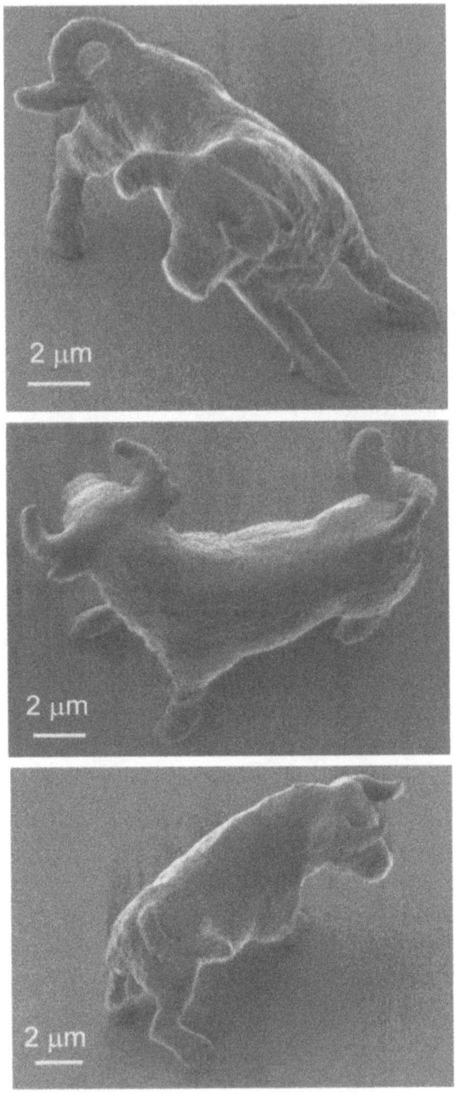

Fig. 21 Different view-angle SEM images of a micro-bull sculpture that was two-photon photopolymerized with sub-diffraction-limit accuracy. The 10-μm long and 7-μm high bull is believed the smallest animal sculpture that was ever fabricated

were determined to 0.1, 0.23, 0.6 and 18 ms for laser pulse energy of 163, 137, 111 and 70 pJ, respectively.

With the sub-diffraction-limit fabrication accuracy, it is possible to fabricate devices of nanoscale size or with nano-features. Figure 21 is the SEM image of a micro-bull sculpture, which consists of smooth and rough surfaces, curvatures and sharp horns. It is good proof of the feasibility of creat-

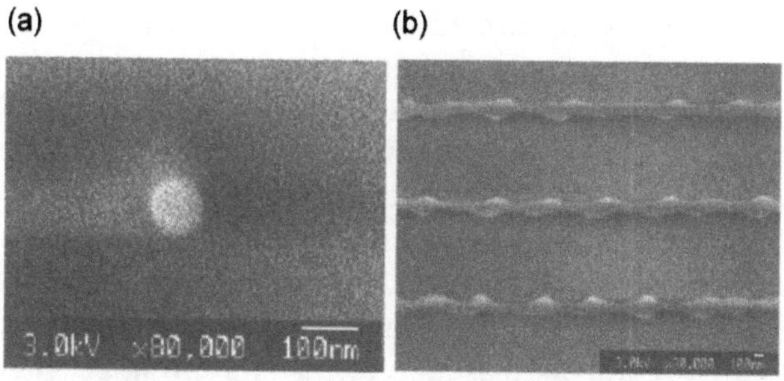

Fig. 22 Achievement of 100-nm lateral spatial resolution achieved after adding quenchers into SCR 500 resin. SEM images of a single voxel **a** and a line-dot structure **b**

ing sub-diffraction-limit features by two-photon photopolymerization. The 10 μm long and 7 μm high bulls are the smallest animal sculptures ever made artificially, and are as small as a red blood cell. Their volume allows us to send them, or actually micromachines of this size, to any location inside the human body through blood micro-vessels to make clinical treatments.

Any chemical species that tends to prohibit photopolymerization reactions, even in small concentrations, is called quencher. Oxygen is just one of many choices of quencher [154–156]. By attentively adding prescribed quenchers into resin solution, it is much easier to control polymerization than use of dissolved oxygen. Figure 22 shows line structures of 100 nm width that were two-photon photopolymerized with the same resin, SCR 500, except for an additional quencher. It is believed that by properly choosing quencher species and optimizing their concentration, a further decrease of polymerized voxel size is possible.

4.1.4 Point Spread Function Engineering

Besides the utilization of specific material properties like the thresholding effect, another promising route to tailoring voxels is by designing the point spread function (PSF) at laser focus using adaptive optics. Dry optics is inevitable if long work distance lenses are needed, and high refractive index resins, either intrinsic or doped, are important for realizing full-PBG PhC. Each case suffers from aberrations, which are induced because of the refractive index mismatch between the objective immersion medium (air or oil) and a cover glass, or between a cover glass and the resins [157]. The induced aberrations, consisting primarily of spherical aberration, increase linearly with focusing depth and cause a lateral broadening of the focal spot and, more importantly, a lengthening in the axial direction [158]. The larger size of the focal spot means a worse spatial resolution.

Aberration can be compensated by pre-shaping the light beam with an equal but opposite aberration using adaptive optics, ensuring an aberration-free focal spot [159]. An adaptive optics system consists primarily of a wavefront sensor for measuring aberrations, a wavefront correction element and a control system to interface the wavefront sensing and correction. A number of existing wavefront sensing technologies can be used for this purpose. Typical wavefront correction elements include pixellated spatial light modulators (SLMs), pixellated and deformable mirrors. A ferro-electric liquid-crystal SLM can be configured as an arbitrary wavefront generator, however, its low throughput (~1% at 800 nm) makes it impractical to provide sufficient energy for voxel polymerization. We utilized deformable membrane mirrors, which are micromachined aluminized silicon nitride membranes suspended above an array of electrodes. Voltage applied between the electrodes and the membrane allows control of the mirror shape. These mirrors are particularly convenient for the correction of low order aberrations such as those predominate in two-photon polymerization fabrication.

4.2 Characterization of 3D Focal Spots

A focal point was originally defined in ray optics as a geometrical point at which parallel light rays, incident on a lens (or mirror), are focused after refracting or reflecting. Further treatment involved diffractions at an aperture, on which primary imaging theories were established [126]. 3D lithography including two-photon photopolymerization is pursuing sub-diffraction-limit accuracy, and a spatial resolution 20% of the diffraction limit has been achieved (Figs. 20 and 22). Apparently, it is not appropriate to consider the focal spot as a geometric point any more, but we should be concerned with its internal structure, most importantly, the shape and size of a focal spot that is related to two-multiphoton absorption.

4.2.1 Two-Photon Excitation Related Focal Spots

The linear light intensity distribution near focus had been theoretically investigated by Lommel and Struve [126]. Their method can be approximately applied to two-multiphoton processes. For simplicity, we discuss hereafter only two-photon excitation without loss of generality. To study two-photon excitation-related focal spots, one is interested in the SLI at the focal volume. To experimentally investigate a two-photon focal spot in 3D, one needs (i) a medium that responds proportionally to the SLI; and (ii) a technique that is capable of revealing 3D features of focal spots. Direct observation of two-photon fluorescent spots with a CCD camera tends to give less accurate and distorted images due to insufficient resolving power and sometimes due to

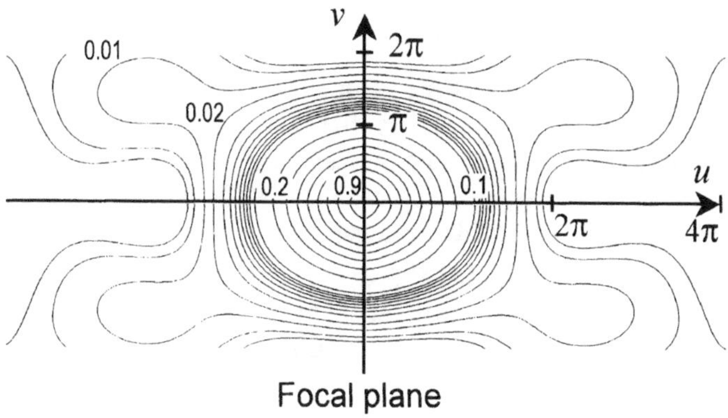

Fig. 23 The SLI isophotes of a focused laser beam. Unitless parameters, u and v are normalized axial (z) and lateral (r) coordinates. For details, see [126]

CCD pixel size limitation and possible fluorescence saturation. A femtosecond time-resolved optical polarigraphy method for visualizing laser pulse propagation was recently demonstrated by Fujimoto et al. [160], from which, however, pure information on the focal spots is difficult to extract.

Starting from the work by Lommel [126], it is easy to calculate the near-focus SLI contour. A complicated SLI distribution was found near the focal region, but a tubular structure of high light intensity (Fig. 23) exists in the central portion of the diffraction pattern. Generally only this regularly formed volume is useful for microfabrication and imaging. This two-photon excited tubular volume can be several times larger than the diffraction limit, and theoretically could be as small as several tens of nanometers, limited by the polymer molecular size. Since diffraction occurs at circular apertures imposed by lenses, in the neighborhood of the focus the intensity distribution, and therefore SLI, is rotationally symmetrical about the optical axis (u axis in Fig. 23), and mirror-symmetrical about the geometrical focal plane. The inflence of beam polarization will be discussed below in Sect. 4.3.2.

To experimentally investigate TPA focal spots, one needs a direct two-photon responsive medium. When IR ultrashort laser pulses were tightly focused in a resin, initiators were decomposed into radicals by simultaneously absorbing two photons. The number of photons absorbed per molecule per unit time by means of TPA is proportional to the TPA cross-section, δ and to SLI. The total number of photons absorbed per unit time is also a function of initiator concentration, C, and the excited sample volume, V. In the absence of saturation and photobleaching, the number of radicals generated per unit time, N, is given by:

$$N = C\delta\phi I_0^2 \int_V S^2(r,\theta,z)dV \tag{21}$$

where I_0 is the light intensity at the geometrical focal point, $S(r,\theta,z)$ is a unitless function used to describe the spatial distribution of the incident light, and φ is the quantum efficiency of radical yield. The shape of the TPE focal spot is determined by $S^2(r,\theta,z)$. The radical distribution would represent the SLI distribution. Mathematically, radicals play a role of "$k\sqrt{\ }$"*operator*, converting the squared light intensity to linear concentration, where k is a constant that can be deduced from Eq. 21. Physically, the oligomer and monomers act as a *3D "film"* to fix the distribution of the radical where its concentration is higher than the threshold of a solid voxel.

4.2.2
Ascending Scan Method

To practically realize the idea formulated above, a Ti: Sapphire mode-locked laser system that was operated at 76 MHz and delivered 780 nm, 150 fs laser pulses was employed. The laser beam was focused by a high NA (~1.4) objective lens. The SCR 500 resin was dropped on a microscopic cover-glass substrate. The sample stage was moved up and down along the optical axis using a piezo stage (Fig. 20a). An important issue for obtaining isolated, complete 3D voxels is substrate truncating, which caused a lot of observa-

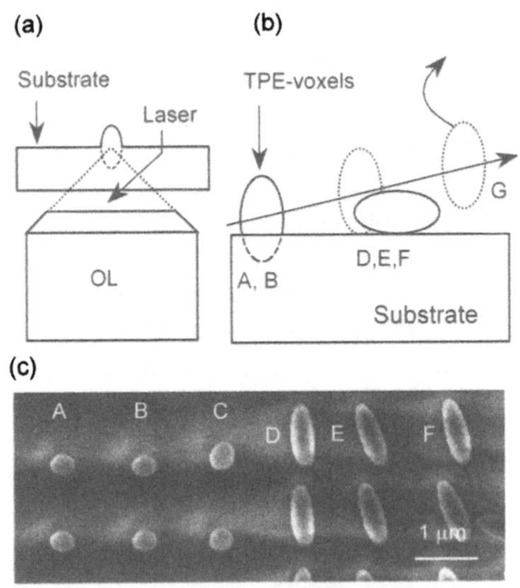

Fig. 24 Schematic ascending scan method for achieving isolated and complete 3D voxels. **a** laser beam focusing that illustrates the substrate truncation effect, **b** voxels formed at different focusing level, and **c** SEM image of voxels produced by scanning the laser focal spot from inside to above the substrate

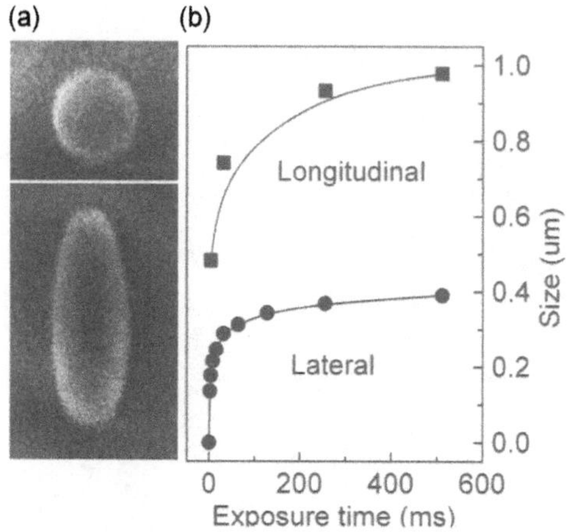

Fig. 25 **a** An isolated and complete voxel, the 3D image of two-photon excitation related focal spot, and **b** exposure time-dependent lateral and longitudinal voxel size

tion errors as discussed later. Truncation happens when the laser is focused too near the substrate surface (Fig. 24b); on the other hand, floating voxels would be formed which would be flushed away during developing if the laser was focused too far above the substrate.

To solve this problem, an ascending scan method was proposed by Sun et al. [20]. The laser focus was scanned along a slant line (the arrowed straight line in Fig. 24b). The arrow denotes the scanning direction from below to above the substrate surface, residing at a series of positions. At each position, the shutter in the light path was switched on and kept for a short term, the exposure time, to get voxels at an identical exposure condition. Figure 24c shows SEM images of the produced voxels. The left voxels (**a**, **b**, **c**) were truncated voxels. They stuck to and erected on the substrate, revealing only their lateral size information. The rightmost voxels were floated away (not shown). A transition state always existed between these two regions: the edge of voxels bordered at and weakly adhered to the substrate surface, and they were overturned during developing. That is the case for **d**, **e**, **f**, from which both lateral, and most importantly, longitudinal information could be attained.

Figure 25a shows top- and side-view SEM images of a voxel. It resembles a spinning ellipsoid with axis length of 2a=3.4 μm and 2b=1.4 μm. Therefore an axial ratio can be defined as $\mu=\frac{a}{b}=2.4$. The theoretical value for a voxel of the same lateral size is, however, μ=2.8 according to Fig. 1. The deviation may be due to the over-simplification of the model. However, a near-circular cross-section ($\mu=1$) of rods could be achieved, for example, by adaptive optics [157–159] or by multiple line-scannings [15, 17], where the scan positions were laterally offset.

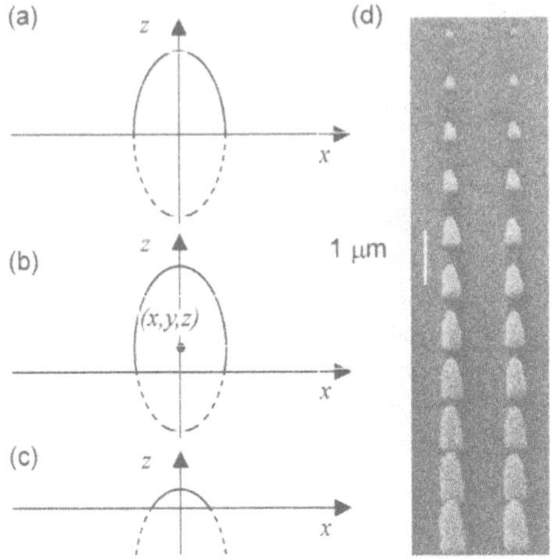

Fig. 26 Focusing depth dependent vertical and longitudinal voxel size, which is the origin of most observation errors. The laser was focused at different positions to produce **a** exactly half, **b** over half, and **c** less than half voxels. **d** SEM image of voxels corresponding to the situation of (a), (b) and (c)

Here the focal spot was defined as an energy-concentrated 3D volume demarcated by a series of intensity isophotes, while the voxel is the primary fabricating unit. Focal spots are sometimes not truly reflected by voxels: for example, when the laser pulse energy is so high as to induce breakdown [143]. In particular, in the case of micro-explosion [49], the shape of voxels was significantly different from that of the focal spot. In the current 3D photographing method, for a fixed optical system at a certain laser pulse energy, one can use the ascending scan method to map the focal spot layer-by-layer from the core center ($a,b \to 0$) to any large size that is limited by breakdown by changing exposure time (Δt). Figure 25b shows an example of Δt-dependent voxel sizes. Since only the appearance of voxels contributes to its dimensions, even if photobleaching or excitation saturation occurs at the central portion, where the intensity maximum exists, the imaging accuracy is not affected. It was observed at different laser pulse energy levels until breakdown that the longitudinal size of voxels was always proportionally related to its lateral size, but the axis ratio increases with the increase of exposure.

Now we can analyze how the conventional observation error of spatial resolutions originated. The situation of lateral size is slightly complicated. A critical case is that the geometrical center of the focal spot (x, y, z) falls exactly on the substrate surface (z=0, Fig. 26a), where the potential voxel is half cut. Focusing at this and positions higher than this level ($0<z<a$, Fig. 26b), the top observation would give rise to an identical lateral voxel size, $2b$. This is the case of voxels **a, b, c** in Fig. 24c. If less hemisphere is ex-

posed ($-a<z<0$), any values ranging from zero to $2b$ would be possible depending on how large a percentage of the focal spot volume comes out of the substrate.

The above analysis was experimentally tested using the ascending scan technology. The exposure started when the entire focal spot was immersed in the glass substrate, then the focusing height level was increased step by step. Although the exposure time and laser pulse energy were the same for all voxels, many different lateral and longitudinal sizes were obtained (Fig. 26d). For longitudinal resolutions, the situation is simple; any measurement of erected voxels gave a value less than the real one. We believe that most of the reported resolution measurements were performed without properly considering the truncation effect. Therefore, measured values are more or less smaller than the actual size, $2a$. It is the substrate truncation that causes most measurement errors, sometimes leading a the confusing conclusion.

4.2.3
Suspending Bridge Method

It is technically challenging to observe voxels of lateral dimensions of 100 nm or less using the ascending scan method because, first, the voxels themselves have difficulty surviving developing, for which special care has to be taken; and second, when size reaches this scale, voxels adhered to the substrate surface position themselves in random orientations instead of overturning. DeVoe et al. [161] proposed a suspended bridge method. Instead of producing a single voxel, a line suspended between two bridges was polymerized. The line width and height denote lateral and longitudinal resolution, respectively. In the test structure shown in Fig. 27, polymer lines were held between solidified polymer supports and suspended 15~20 μm above the substrate. The lines were written at scan velocity from 1.25 mm/min to 113 mm/min, and increased by $\sqrt{2}$ for each line, under constant average power. The supports were formed at lower speed scanning so that they were solid enough. The line width was designed to be quite long, ~100 μm, to avoid artifacts arising from stage acceleration and deceleration at the ends of lines.

The method was found effective for different types of photopolymers, cationic (SU-8 2025, MicroChem Corp.) and a radical resin. In the case of the PMMA/acrylate resin system, they obtained a resolution of 140 nm×200 nm. Although the technology is less effective than the ascending scan method for revealing more information about the focal spot, it may play a promising role in characterization of sub-100 nm spatial resolution. It is worth mentioning that, in this work, strong two-photon absorbers were added to diaryliodonium salts to yield a highly sensitive system for both cationic and

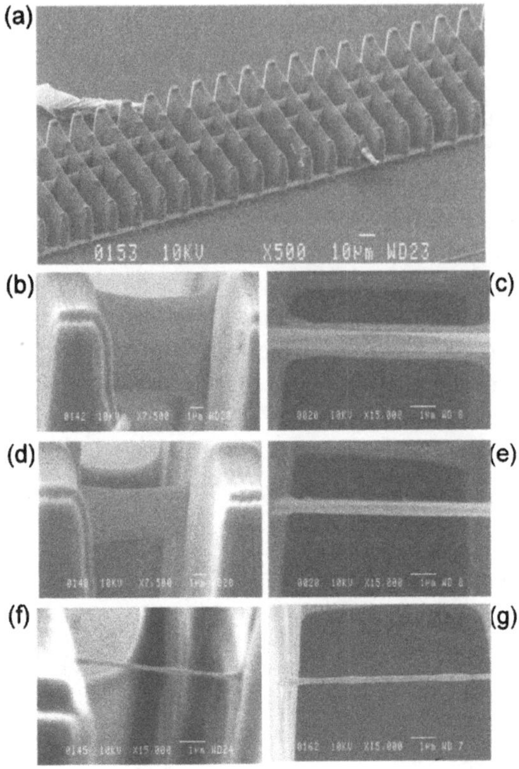

Fig. 27 SEM images of suspended crossbar structure in PMMA/acrylate resin for voxel shape determination. Note scale bars in each photo. **a** overview of entire structure. **b** and **c**, top and side views at 2.58 mW, 0. 12 mm/s; **d** and **e**, top and side views at 2.58 mW, 1.9 mm/s; **f** and **g** top and side views 1.07 mW, 1.9 mm/s

radical polymerization, which expanded the dynamic power range tby two orders and permitted voxel size tailoring of the same order.

Similar to the case of two-photon excitation, for multiphoton absorption the initiators would perform an operation of "$k\sqrt{[n]}$" with a much smaller coefficient k for n-photon absorption. The above two methods and knowledge acquired about the focal spot related to two or multiphoton excitation is essential for not only photopolymerization fabrication, but also important for understanding excitation behavior for various laser microfabrications.

4.3 Understanding the Role of Laser Parameters

Laser beams are described by their spatial, temporal, spectral and polarization distributions in addition to coherence properties [33–35]. At fundamental limits, the beams can be diffraction and bandwidth limited, linearly polarized and coherent. A Gaussian beam, the 3D solution of the wavefunction derived from the Maxwell function, represents the highest possible beam quality and satisfies the above requirements, and near-Gaussian output is already available from many commercial lasers. For example, a 780 nm wavelength laser with a 100 fs pulse width at a repetition rate of 80 MHz possesses a spectral full width at half maximum (FWHM) of approximately 10 nm, implying a time-bandwidth product of $\Delta t\ \Delta\omega=0.493$, near the transform limit of 0.441 of a Gaussian pulse shape. The beams can almost be utilized as they are, although sometimes beam expanding and spatial filtering is used to form a far-field pattern with a high symmetry of energy distribution. A short pulse width (<1 ps) is essential to provide a high transient power in order to launch nonlinear optical processes and exclude thermal effects, which are difficult to localize. The broad laser spectrum associated with the ultrashort pulse width brings about chromatic aberration, which is overcome by usage of apochromats.

4.3.1
Numerical Aperture

Compared with the roles of temporal, spectral and coherence performances discussed above, laser energy distribution and beam polarization directly determine the spatial resolution of fabrication. After spatial filtering and beam expanding, only the central portion of the laser beam is induced into an objective lens. It is already known that the focal spot size is estimated using the Rayleigh criteria, $Z_R=n\pi\omega_0/\lambda_0$ along the optical axis, where n is the refractive index of the medium into which the laser is focused, λ_0 is the vacuum wavelength, and $\omega_0=1.22\lambda/(n\mathrm{NA})$ is the lateral diffraction limit. It is clear that the objective lens strongly redistributes the beam energy, and the laser intensity distribution at the focal region sensitively depends on NA. For a better understanding of the role of NA in two-photon nanofabrication, a theoretical simulation was conducted, for which the vectorial Debye theory was utilized since the paraxial approximation does not hold any more when focusing using high-NA optics (NA>0.7). Figure 28 shows two-photon PSF of different NAs in both x and y directions (Fig. 28a), and along the optical axis, the z direction (Fig. 28b). The refractive indices of the cover glass and the resin were adopted as experimental values, 1.518 and 1.52, respectively, and the laser is assumed to polarize along the x direction. The incident power before the aperture, which has variable size, is assumed to be identical in each case.

The peak power increase with NA is natural, due to the increase of convergence angle. Spatial resolution is improved at high NAs. For example, using the FWHM as the criterion of resolution, NAs of 1.4, 1.2, 1.0 and 0.8 give rise to FWHMs of 252 nm (450 nm), 269 nm (525 nm), 304 nm (629 nm) and 343 nm (787 nm) in the x direction and 542 nm (1860 nm), 765 nm

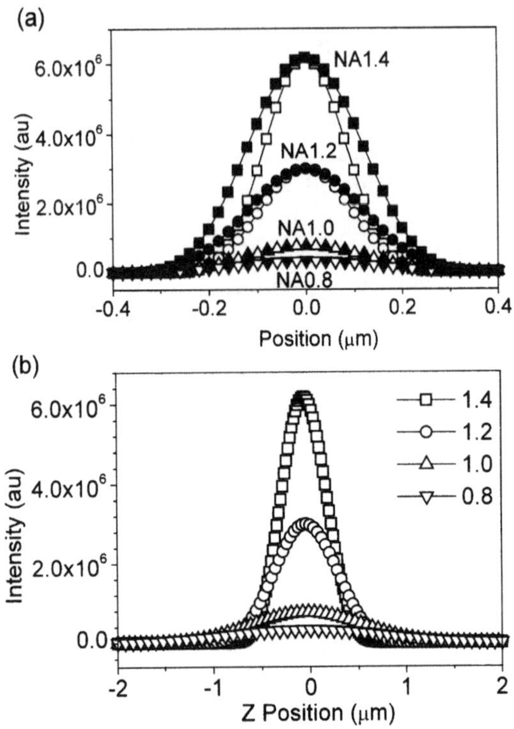

Fig. 28 Theoretical two-photon point spread function of different NA focusing calculated using vectorial Debye method. An identical incident power was assumed before the aperture. **a** Functions along x axis (solid symbols) and along y axis (open symbols). **b** The function in the longitudinal direction (z axis). Laser beam was assumed to be incident from the bottom and focused 10 μm above the glass substrate

(2525 nm), 1327 nm (3640 nm) and 1848 nm (5710 nm) in the *z* direction, respectively (the diffraction limits are listed in the parentheses, indicating the size of the central Airy pattern). It is seen from the figure that regardless of the absolute intensity level, low NA tends to give larger feature sizes in both lateral and longitudinal directions.

The above result is consistent with that predicted in imaging theories [162, 163]. It is interesting to know whether a similar rule holds true when a femtosecond laser interacts with a typical nonlinear material in a two-photon nanofabrication processes. The ascending scan method [20] was used to explore this, making it possible to intuitively discriminate how the focal spot size and shape are influenced by NA and other laser parameters. Shown in Fig. 29a are the side-view SEM images of voxels formed with objectives of NA 0.88, 1.12 and 1.4, respectively. In order to get comparable data, the exposure power measured at the objective output was kept identical and the

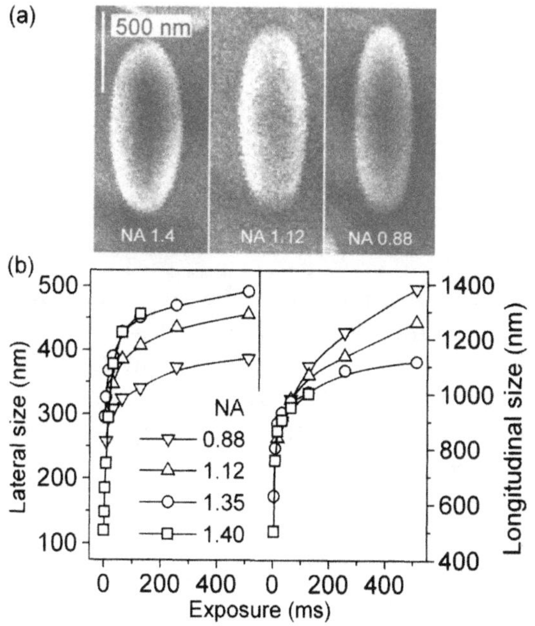

Fig. 29 NA-dependent voxels. **a** Side-view SEM images of voxels formed by focusing objectives of different NAs, 1.4, 1.12 and 0.88, where the power output from the objective lens was kept identical. **b** Exposure time-dependent voxel size in both lateral and longitudinal directions under different NAs

exposure time was chosen to be relatively long, 250 ms per voxel, so that the voxel size is reasonably large.

From Fig. 29a it is found that the axial ratio increases as NA decreases, as expected, from 2.3 (NA~1.4) to 2.7 (NA~1.12) and 3.3 (NA~0.88). This relationship is similar to that obtained from the theoretical FWHMs (Fig. 28): 2.2, 3.5 and 4.9 for 1.4, 1.12 and 0.88 NAs, respectively. In Fig. 28, it is implied that not only the longitudinal but also the lateral axis lengths of voxels obtained at low NAs are larger than those obtained from high NAs. However, it is seen from Fig. 29a that the low-NA voxels are smaller in lateral dimension. This contradiction, essential in choosing appropriate optics for laser nanofabrication, apparently needs clarification.

Various exposure conditions were tested and it was found that although at a medium irradiation level low-NA focusing gave rise to smaller lateral voxel size (350 nm at 0.88 NA versus 460 nm at 1.4 NA in Fig. 29a), the smallest visible voxels achieved with 0.88 and 1.4 NA optics were 260 nm and 120 nm, respectively. This phenomenon was interpreted by the threshold effect [13, 14]. In the case of low-NA focusing, the laser power is distributed to a larger volume, and then the solidified front demarcated by the threshold camber is vertically expanded and laterally shrunken. Therefore low-NA gives laterally smaller voxel size. The gradient of light intensity distribution in a high-NA focused light field is steeper and energy is more concentrated in the center of focal spots, which causes photopolymerization at smaller volume. Therefore, for pursuing high-accuracy nanofabrication, a high-NA objective is still preferable.

4.3.2 Polarization

In the above discussion and in reports published up to now, it is always considered that the appearance of a voxel resembles a spinning ellipsoid with axis lengths $a=b<c$. However, this is experimentally found to be not true if a general linearly-polarized laser beam is employed. Shown in Fig. 30a is the lateral SEM image of a voxel formed using 1.4 NA optics when the Ti: Sapphire laser output was utilized as it was. The voxel axis lengths are $2a$=325 nm (x direction) and $2b$ = 295 nm (y direction), giving rise to a lateral axis ratio $\mu_{\|}=a/b=1.1$.

This phenomenon can be understood by considering the depolarization effect predicted by electromagnetic focusing theory [163, 164]. It is already known that when the beam incident angle α is small, the focal field (E_x, E_y, E_z) is sufficiently described by a cylindrically symmetric function [$E(r)$, 0,0]

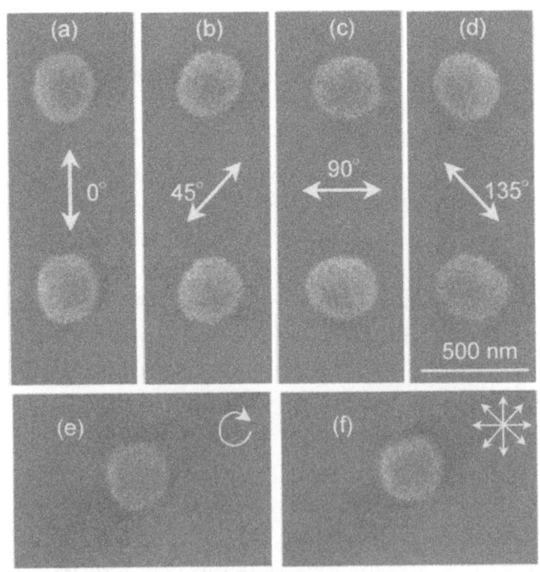

Fig. 30 Polarization effect. The laser was focused by oil-immersion 1.4-NA objectives. **a**, **b**, **c** and **d** are top-view SEM images of voxels formed with laser of different linear polarizations, from which a nearly 10% lateral axis ratio was observed. **e** Voxels produced at the same incident power but with circularly polarized laser and **f** nonpolarized laser

with $r=\sqrt{(x^2+y^2)}$, where the input beam is assumed to be along the x direction. Nevertheless, when α>40° (NA>0.7), the symmetry is broken and a field with significant E_y and E_z components appears. More precisely, the electric field can be expressed as:

$$E(r_2,\phi,z_2)=\frac{\pi i}{\lambda}\{[I_0+\cos(2\phi)I_2]\mathbf{i}+\sin(2\phi)I_2\mathbf{j}+2i\cos\phi I_1\mathbf{k}\} \tag{22}$$

where **i**, **j** and **k** are the unit vectors in the x, y and z directions, respectively, and variables r_2, z_2 and φ are cylindrical coordinates of an observation point. I_0, I_1 and I_2 are not constant-zero variables (see definitions in [163]). Hence, it is clear that the electric field at the focal region is depolarized. The calculation result is shown in Fig. 28, from which the lateral axis ratio is deduced to around 1.30, estimated using FWHMs.

The experimental lateral axis ratio is smaller than the theoretical expectation and depolarization can also be caused by lens imperfection or tension. To confirm the origin of a non-unit lateral axis ratio, the beam polarization direction was adjusted relative to the objective. As a result, the voxel orientation was found to be correspondingly rotated while the size and shape were kept unchanged (Fig. 30b,c,d). Similar experiments were conducted with low-NA optics and it is found that when NA is smaller than 1.0, the lateral eccentricity was not discernible.

With the depolarization effect experimentally confirmed, some technical issues could be clarified. For example, it had been previously observed that when scanning linearly-polarized laser in resin, the widths of photopolymerized lines along different directions are slightly different. This deviation can now be attributed to asymmetrical light field distribution due to the depolarization effect if the scanning speed of stage is equal in different directions.

A round-shape lateral cross-section of voxels is preferable in most fabrications. We tried to eliminate the non-unit lateral axis ratio by polarization modification. Circularly- and non-polarized beams were realized by inserting a 780 nm antireflection-coated c-axis-cut quartz $\lambda/4$ retardation plate and a visible-range double-plate-type quartz depolarizer into the light path, respectively. In both cases, there is no preferential direction of light intensity distribution at the focal spot. Experimentally a nearly perfect round shape was achieved for both cases, as shown in Fig. 30e,f.

4.4 Raster Scan versus Vector Scan

Laser scanning is the step needed to convert pre-designed CAD patterns into real structures. Two basic modes for direct laser scanning can be utilized, raster-scan mode and vector-scan mode, of which the concepts are illustrated in Figs. 31a, b.

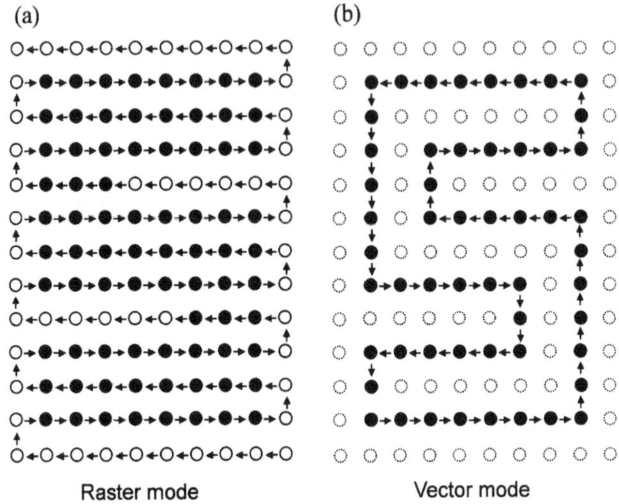

Fig. 31 Two basic scanning modes for two-photon photopolymerization microfabrication. Conceptive illustration of how the two scan modes could be utilized for writing a character "s". **a** raster scan and **b** vector scan. The solid and open circles denote exposed and unexposed dots, respectively, all scanned by the laser focal point. The dashed open circles in the right part means the dots that aren't scanned

In the raster mode, all voxels in a cubic volume that contains the microstructure are scanned by the actual/virtual laser focal spot, depending on having the shutter ON (actual)/OFF (virtual). In the vector mode, the laser focus directly traces the profile to be defined. Figure 31a and b respectively illustrate how a character "s" could be scanned with the two modes. Apparently the vector mode requires a smaller number of voxels. Depending on the structures, variations and combinations of these two basic modes could be used.

Experimentally the same object was fabricated using the two modes. The micro-bull in Fig. 21 was produced using a layer-by-layer raster-scanning scheme; all voxels that made up the bull were exposed point-by-point, line-by-line, and layer-by-layer by the two-photon process. As a result, it took three hours to complete the manufacturing. If we make a detailed analysis of the bull structure, it is found that the entire bull consists of 2×10^6 voxels. However, the bull profile can be well defined using only 5% of them. As a test, the bull was written once more using vector scanning as shown in Fig. 32. Astonishingly, we found it possible to depict the same structure within 13 minutes. In both cases the scanning step in three dimensions was 50 nm. However, the fabrication time in vector scan was reduced by more than 90%.

The TPA-produced bull crust was self-supported, standing on glass substrate either in liquid or air. To avoid possible distortion, we further solidi-

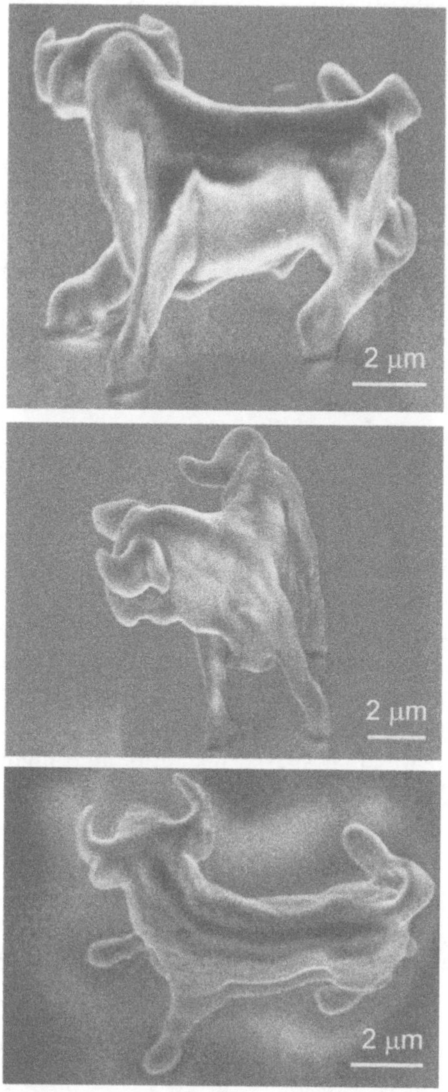

Fig. 32 A micro-bull scupture produced by vector scanning. Only the crust was defined by the two-photon process; the inside was solidified by illumination under an mercury lamp. In this particular structure, the two-photon scanning time was reduced by 90% due to the use of a vector scanning mode

fied the structure under a mercury lamp, which is a single-photon exposure process. This is an additional step but technically quite simple.

In the vector scanning method, voxels were actually classified into two categories: those on the surface layers are the least necessary points to define a structure and those inside the structure solely have the supporting function. Discriminating between these two kinds of points and rendering a separate exposure is critical for vector scanning to improve the fabrication efficiency. Vector scanning uses less exposure time, but the stage movement controlling is a little more complicated. In addition, this method is not suitable for resins that have a large volume change in the liquid-solid phase transition, since the volume variation-induced tension can't be released in the fabrication process as it can in a raster scanning. Comparatively speaking, raster scanning is more suitable for producing structures with complicated shape, at a higher accuracy, and with a greater percentage of the least necessary points.

4.5 Three-Dimensional Micro-Diagnosis

An important issue in two-photon photopolymerization micro-nanofabrication is how to evaluate the internal structure of devices. 3D functional micromechanical systems and micromachines need proper positioning, shaping and jointing. A pre-operation evaluation is critical for judging and optimizing designs and fabrications. Shown in Fig. 33 is a tube structure, which should be hollow according to design. However, it is not an easy task to confirm the internal status even if it is a quite simple device. A normal optical microscope doesn't have sufficient resolving power to distinguish details in three dimensions, particularly in the longitudinal directions [165]. Electronic microscopes have a higher spatial resolution and higher imaging quality, but they are only useful for observing the appearance of objects. A two-photon confocal microscope possesses 3D imaging capability with resolution better than conventional transmission or reflection optical microscopes [165]. To attain a high signal-to-noise ratio, fluorescence readout is preferable and so the structure is required to emit reasonably intense fluorescence, which is not available in general resins that are used for microfabrication.

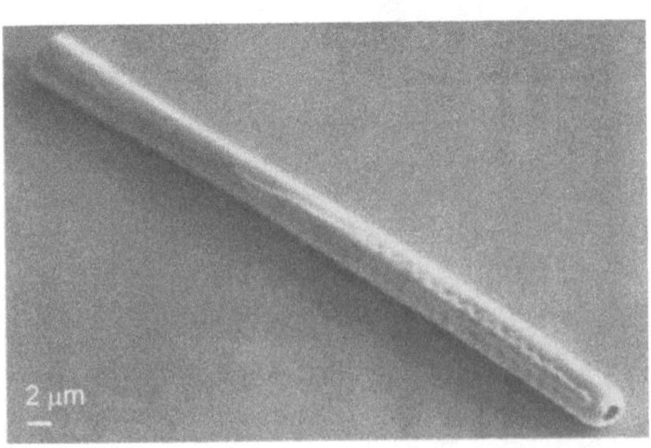

Fig. 33 A micro-tube structure, which is hollow according to design. However, it is difficult to confirm its internal status by conventional characterization technologies

4.5.1 Fluorescent Dye Doping

In order to use two-photon confocal 3D imaging, the fluorescence dye-labeling technology is found to be very useful. To induce fluorescence in photopolymerized structures, the laser dye of LD490 (Exciton Inc.), of which the major component is Rhodamine B, a fluorescent small molecule that is quite stable even under strong laser irradiation, was doped into resin before polymerization. The dye was introduced by dissolving SCR500 resin in an ethanol solution of LD490 until saturation. It absorbs and emits light with peaks at 396 nm and 474 nm, respectively (Fig. 34).

It is seen from Fig. 34 that the fluorescence wavelength of the dye was situated at the red end of the SPA absorbance curve, which may lead to fluorescence-induced polymerization. This problem can be solved by optimizing the fabrication-laser-pulse energy and the concentration of Rhodamine B so that local radical concentrations initiated by the TPA fluorescence are lower than the polymerization threshold, and then the radicals are quenched by dissolved oxygen.

A test structure was photopolymerized using the dye-doped resin following the normal procedure. For two-photon confocal readout, the system as used for fabrication was configured by adding a confocal pinhole and a CCD detector at the reading area. The laser pulse energy for reading was chosen to 1.5 pJ, two-orders smaller than that for fabrication, 130 pJ, in order to avoid optical damage. For minimizing imaging aberration arising from the mismatch of refractive indexes, structures were immersed in oil during the confocal scanning.

Figure 35 shows the two-photon confocal fluorescence images of a 5.4 μm×5.4 μm×5.4 μm cubic cage. From the design (the top illustration of Fig. 35), it is seen that different heights of the structure correspond to various cross-sections. This was clearly shown by the sliced fluorescent images

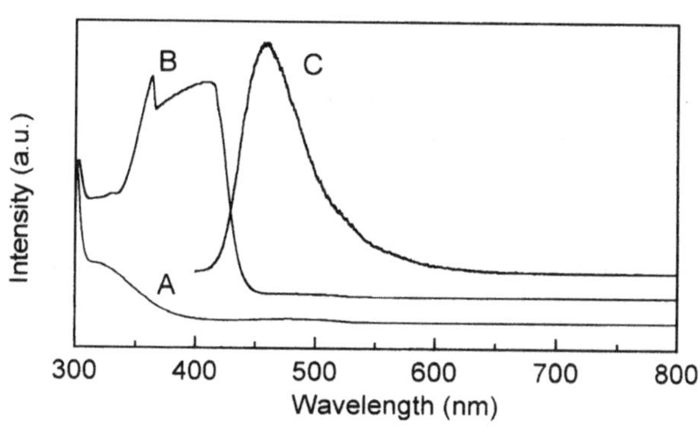

Fig. 34 Absorbance (**a**, **b**) and photoluminescence (**c**) spectra of undoped (**a**) and dye-doped (**b**, **c**) SCR500 resin

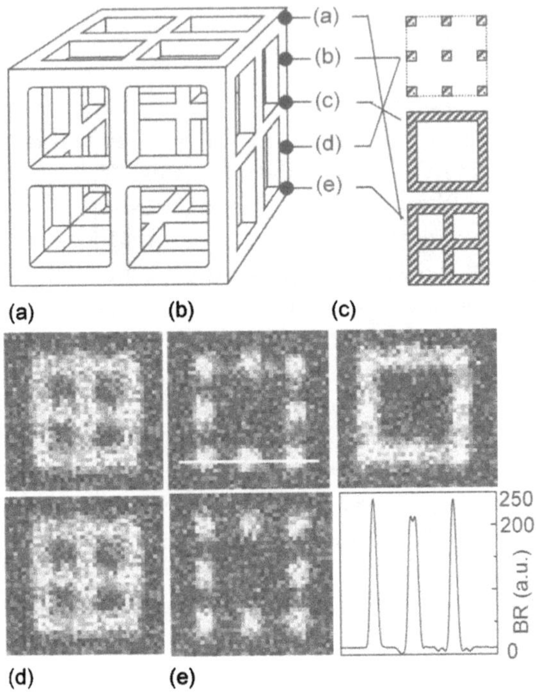

Fig. 35 Design (the top line drawing) and fluorescence images of a cubic cage fabricated by two-photon photopolymerization of dye-doped resin (**a**-**e**). The heights are **a**0.0, **b**1.35, **c**2.7, **d**4.05 and **e**5.4 μm, respectively. A fluorescence intensity distribution was extracted from **b** and given by **f**

(Fig. 35a–e). The fluorescence signal is more intense than that needed for reading. A signal/noise ratio of more than 20 (Fig. 35f, extracted from Fig. 35b) was obtained. In contrast, features of an identical cage solidified from unadulterated SCR resin can't be resolved due to a low signal-to-noise ratio.

Two conclusions can be reached from the above results. First, the fluorescence activity of the doped dye remained after a strong writing laser irradiation; and second, the fabrication spatial resolution was not much degraded after the dye doping. The seemingly poorer quality of the cage in Fig. 35 is due to the weaker resolving power of the optical microscope than SEM. Actually fabrication was not found to be degraded up to a Rhodamine B concentration of 1.0×10^{-4} M, and the cage was as smooth as those produced from non-doped resins under SEM.

4.5.2
Micro-Diagnosis in Three Dimensions

The above results indicate that the dye may act as a fluorescence label and dye-stained polymers will show themselves under laser irradiation, resulting in an effective fluorescence diagnosis technique with high spatial resolutions. In a confocal detecting scheme, the effective PSF, as a good approximation, is given by the square of the focal illumination intensity distribution in the objective lens so that, in a simplistic photon picture, only photons from the closest vicinity of the diffraction-limited spot contribute to the signal [165]. Aided by the intrinsic deep penetration capability of the TPA process, the effective focus acts as a 3D probe, and details of structures were shown by the distribution of fluorescence. Figure 36 shows confocal fluorescence images of axial and radial cross-sections of a tube similar to that shown in Fig. 33. The absence of internal fluorescence confirms its hollowness. The 1.6 μm internal diameter agrees well with the designed value.

The 3D confocal probe can be used not only for optical sectioning (Figs. 35 and 36) but also for reconstructing an entire 3D image of objects. Shown in Fig. 37 are reconstructed confocal fluorescence images of **a** a gear and **b** an icosahedron.

It is noteworthy that resin and dye were not optimized. If initiators with larger TPA cross-sections (10^3 larger or $1,250\times10^{-50}$ cm^4s/photon [22]) were utilized, the laser power necessary to launch photopolymerization would be decreased by the same order. This would effectively reduce the influence of

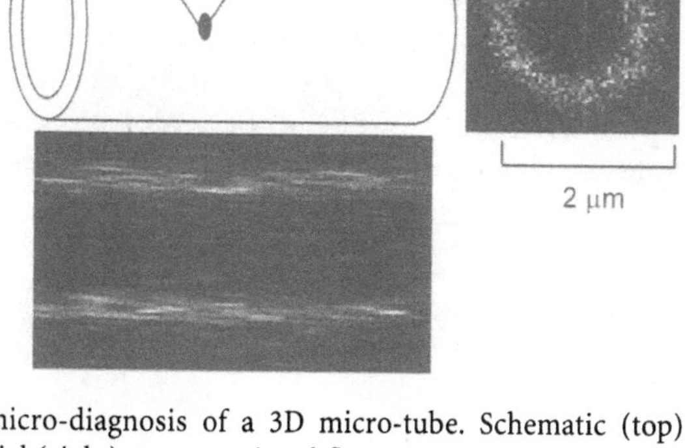

Fig. 36 Internal micro-diagnosis of a 3D micro-tube. Schematic (top) as well as axial (bottom) and radial (right) cross-sectional fluorescence images. The absence of the fluorescence inside the cross-sections indicates that there is no resin inside the tube

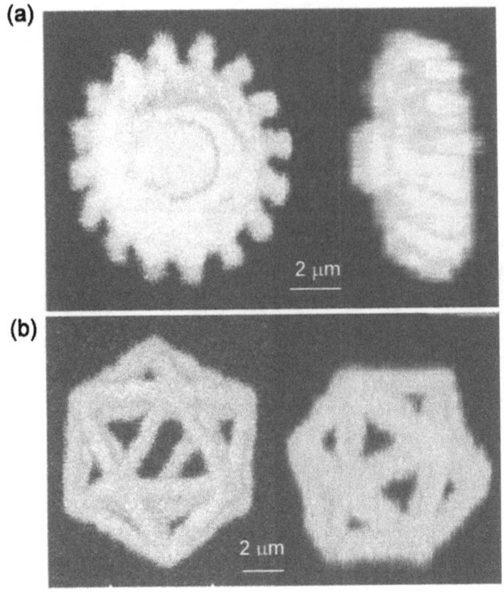

Fig. 37 Reconstructed 3D two-photon fluorescence images of **a** a gearwheel and **b** an icosahedron

two-photon fluorescence on polymerization, or equivalently, allow for dye doping at a much higher concentration without loss of spatial resolutions. The intense fluorescence emission, aided by a suitable positive-feedback, for example a PhC-based microcavity, would enable polymer devices like lasers or amplifiers. In fabrications, we could routinely dope (or co-dope with a dye for device functions) with a trace dye for 3D micro-diagnosis.

4.6 Multi-Beam Interference

As we will discuss in Sect. 5, one of the most important applications of two-photon photopolymerization is the fabrication of PhCs. By direct laser writing, PhCs with arbitrarily designed lattices could be created, which would facilitate the deep understanding of PhC physics and stimulate its broad applications in photonics and optoelectronics. For several particular types of lattices, PhCs may be created more conveniently by means of multi-beam interference.

4.6.1 Photonic Crystal Hologram

It is already known that when two or more coherent laser beams interfere with each other, a stable spatial pattern with periodically distributed light intensity maxima and minima is formed. If the light intensity pattern is created inside a photopolymerizable material, and if the light intensities of each beam are properly chosen, it is possible that material at the maxima sites will be solidified, following the same interference pattern, while at volumes corresponding to light minima, the unpolymerized or less polymerized material can be removed during post exposure developing. As result, solidified interference patterns consisting of polymer skeleton and air voids function as 3D PhCs. This method has the following merits: (i) rapid production and relatively high accuracy due to the elimination of a scanning process; (ii) ease of batch production; (iii) suitability for large-volume samples up to a size of several millimeters; and (iv) system simplicity and low cost.

Both PhCs and holograms have periodical 3D dielectric functions and perform the complete reconstruction of the electromagnetic fields. It is reasonable to consider a PhC as a particular class of holograms [166, 167]. As a consequence of the periodicity of PhCs, the Fourier transform of the refractive index distribution function was well approximated by a small number of Dirac functions, implying that PhC holograms can be recorded by a small set of plane waves. The Fourier transformation manifests the relation of PhC lattices and the laser wavevectors that were involved in the interference: the wavevectors are the vectors of the reciprocal lattices. Therefore, PhC lattices can be designed by choosing a suitable number of beams, their geometrical arrangement, and their polarizations [166–169]. An SC lattice can be obtained by the interference of six beams, collinear or perpendicular to each other; a BCC lattice is realizable by four beams that are aligned along the axes of a standard tetrahedron; an FCC lattice, an ideal candidate for a large PBG effect, can also be realized by elaborately arranging four beams [39], about which we will give a detailed introduction shortly.

4.6.2 Layered Planar Hexagonal and Simple Square Lattices

The idea of constructing 3D PhC using multi-beam interference was proposed by Mei et al. in 1995 [36]. Berger et al. utilized this technology for 2D photoresist patterning [166–167]. 3D interference patterning in photopolymers was first reported independently by Shoji et al. [38] and Campbell et al. [37] in 2000.

Figure 38 shows the experimental scheme [40]. In the first step of the fabrication, Shoji et al. arranged three beams (442 nm, He-Cd laser) symmetrical to the sample surface normal, producing a structure with rods of uniform shapes arranged in a 2D triangular lattice. In the top-view SEM image (Fig. 39a), the bright spots correspond to the tip end of the micro-rods. The planar lattice constant is 1.0 μm, which can be adjusted by the incident angles of the beams. The rods were grown from the bottom to the top of the sample cell, and were 150 μm in length (Fig. 39b), distributed in an area of 500 μm diameter.

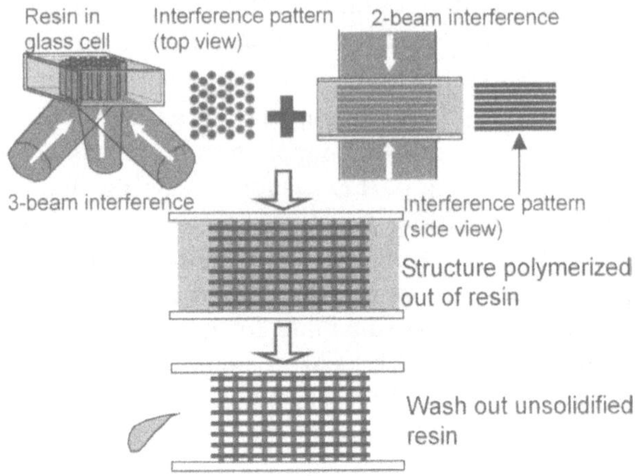

Fig. 38 The procedure of fabricating photonic crystals using two-step multi-beam interference technology. In the first step, a 2D planar triangular rod array was created by 3 symmetrical beam interference, and then the array was intersected by a series of polymerized planes that were produced by the second step two-beam interference

The above structure had translation symmetry along the rod axis, and there is no periodicity in this direction. For fabricating 3D PhC, two coherent beams were introduced from the cell top and bottom in the second step of the fabrication, where the rods were vertically intersected by 150 cross-sectional layers (Fig. 39c). So, by a combination of sequential three-beam and two-beam interferences, a 3D periodic structure was produced.

The two-step interference method, although technically a little complicated, has the freedom to design arbitrary lattice constants of different lattices. This is its significant merit compared with the single-step four beam interference by Campbell et al., which will be introduced in 4.6.3.

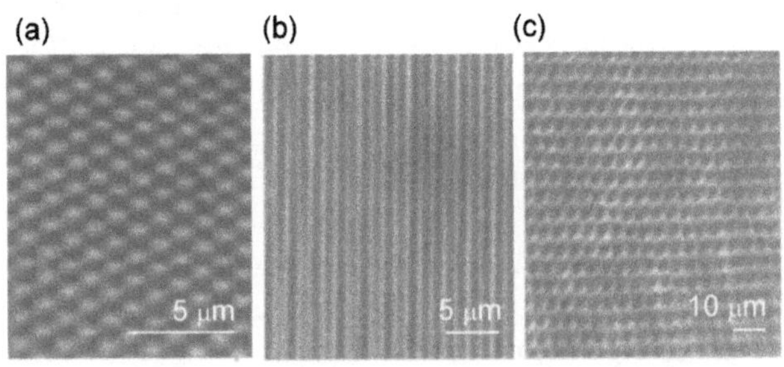

Fig. 39 SEM images of two-step multi-beam interference-produced PhCs. **a** top-view of the rod ends; **b** side view of the rods; and **c** side view after the intersecting layers were introduced

A similar rod array was produced by interference of four beams that were arranged with 90 °substrate-projection angles (120 ° for three beam as discussed above). In this case, rods were arranged with four-fold symmetry, so the cross-section has a simple square lattice with 700 nm periodicity [39]. With the additional fifth central beam, it is possible to induce cross-sectional layers. However, light distribution at the surface normal direction is not steep enough to give a sharp polymerized plane unless very tight focusing is employed. It is worth mentioning that, in this four-beam interference work, Kondo et al. [39] utilized a diffractive beam splitter for generating the four noncoplanar coherent beams. These beams are split and pass an identical light path, and were finally focused with an objective lens, undergoing no phase shift between each other. Therefore there is no need to induce any delay line to overlap the pulses in the time domain, and also the issue of pulse overlapping in the spatial domain was much simplified [168].

4.6.3 FCC Structure Realized with Four-Beam Interference

In principle, by increasing the beam number, even complicated 3D lattices can be produced. However, this not only brings about difficulties in arranging and aligning optical components, but also challenges the material properties. In order to completely remove the less polymerized resin, it is desired that the light intensity contrast in the interfering field should be as large as possible. The sub maxima caused by many beams would be problematic in developing. Campbell et al. [37, 169] design FCC lattices by elaborately arranging four beams. Figure 40 shows the calculated FCC lattice pattern. In Fig. 40(b) the four equal intensity beams propagate along (i) [2,0,1], (ii) [$\bar{2}$, 0,1], (iii) [0, 2, $\bar{1}$,] and (iv) [0, $\bar{2}$, $\bar{1}$] directions respectively. An intuitive description is that two beams are incident from the bottom of the substrate in the same incident plane, each of which has an incident angle of 63.4°; the other two beams are from the substrate top with their incident plane perpendicular to that of the bottom beams and each has the same incident angle. Adoption of a low-index reciprocal lattice vector allows close packing of the primitive units, giving rise to a lattice constant of $\sqrt{5}\lambda/2$.

If a suitable set of high-index wavevectors are used, relative large lattices are possible. For example, an FCC with a lattice constant of $3\sqrt{3}\lambda/2$ is generated (Fig. 40a) by introducing four beams from (i) [1,1,1], (ii) [5,1,1], (iii) [1,5,1], and [1,1,5] directions, or in another word, four beams are introduced from and convergent to a single side of substrate, three of them with interangle of 65 ° and three-fold symmetrically surrounding the central beam. The plane containing two of the three side beams forms a 13.3° dihedral angle with the substrate. As expected from the theoretical calculation, lattices corresponding to the above two beam configurations were success-

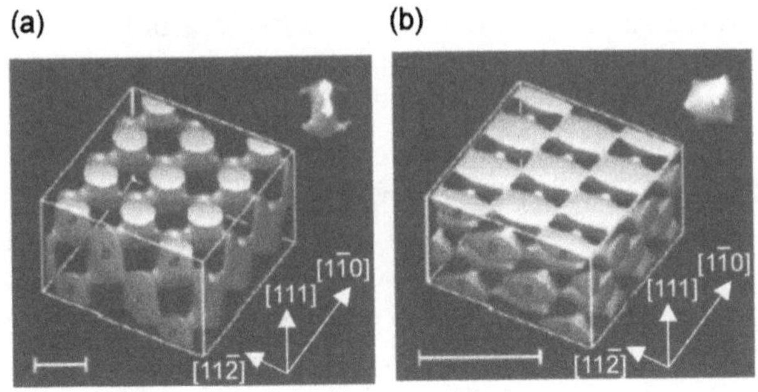

Fig. 40 Calculated constant-intensity surface in 4-beam laser interference patterns. The primitive units (contents of Wigner-Seitz unit cell) is shown inset in each case. **a** Scheme-1, high-index beam vectors interference, producing pattern of 922-nm lattice constant. **b** Scheme-2, low-index beam vectors interference, producing FCC pattern with 397-nm lattice constant. In both case, the use of 355-nm YAG laser was assumed. Scale bars: 500 nm

fully photopolymerized into a SU-8 photoresist using the third harmonic of a YAG laser, 355 nm.

Laser polarization is essential for designing interference patterns [169]. The lattice in Fig. 41a is created by interfering four beams with identical polarization so that all of them interact with each other. However, if the polarizations of the beams are (i)⊥(ii) and (iii)⊥(iv), and (i)∥(iii) or (i)∥(iv), light of perpendicular polarizations don't interact with each other, so that the interference pattern is a simple addition of two sets of two-beam interference planes (Fig. 41b). In both cases, the Scheme 1 beam geometry in Fig. 40 was utilized.

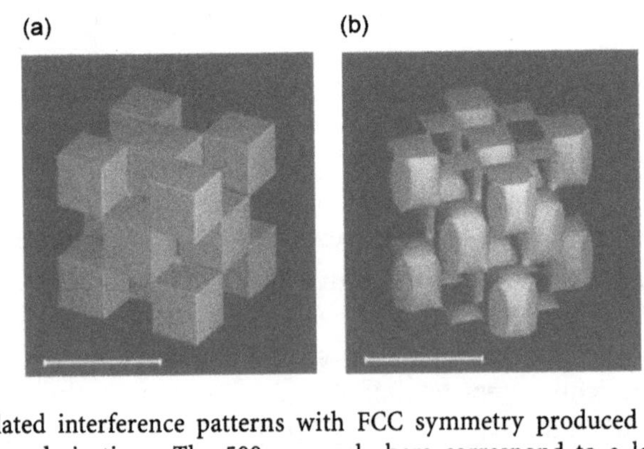

Fig. 41 Calculated interference patterns with FCC symmetry produced by four beams with different polarizations. The 500 nm scale bars correspond to a laser wavelength 355 nm; the lattice constant is 397 nm. **a** all 4 beams have identical linear polarization, and **b** beam arrangement corresponds to those in Scheme 1, but with polarizations (i)⊥(ii) and (iii)⊥(iv), and (i)//(iii) or (i)//(iv)

4.6.4
Application to Two-Photon Photopolymerization

All of the fabrication examples [39–41, 170–172] given above are from single photon photopolymerization. The same process is applicable to two-photon fabrication, which is, however, hindered by two issues intrinsic to TPA. First, TPA generally needs ultrashort laser pulses, typically femtosecond lasers. The short pulse width is always accompanied by a broad emission spectrum due to the frequency-time uncertainty. The wider spectral width sharply reduces the coherence length (l_c). A transform-limited 100 fs pulse possesses only around 40 μm l_c, as estimated by $l_c=0.624\lambda^2/(\Delta\lambda_{FWHM})$. This sharply contrasts with a nanosecond laser, which has a coherence length of centimeters, and a CW laser where it is even longer. In single photon multi-beam interference, the limiting factor of the achievable thickness is the power attenuation due to the linear absorption of resins. TPA can record deeper due to its penetration capability, but this merit is counteracted by the short coherence length. A solution is the use of picosecond lasers, which should have coherence lengths of the order of millimeters, and their transient high power may be still sufficient to launch TPA.

The second issue is that TPA uses a wavelength double that needed for linear absorption. TPA is helpful in reducing holographic linewidth, but the doubled wavelength becomes a detrimental factor in reducing the structure periodicity. We noted in 4.6.3 that the PhC lattice constant is proportional to the interference laser wavelength, $\sqrt{5}\lambda/2$ and $3\sqrt{3}\lambda/2$ for Schemes 2 and 1 respectively. Recording with a 355 nm laser gives rise to lattice constants of 397 nm (Scheme 2) and 922 nm (Scheme 1), while the corresponding two-photon writing produces lattice constants of 794 nm (Scheme 2) and 1844 nm (Scheme 1).

The spatial distribution of the square of light intensity is steeper than that for the light intensity itself. Therefore two-photon polymerized structures are relatively easy to develop due to a sharper contrast between the maximum and minimum of polymerization. To use this advantage, a picosecond pulse at visible wavelength range may be a good choice for two-photon lithographic recording.

4.7
Protein and Biomaterials

In principle, two-photon laser nanofabrication technology is applicable to any photopolymerizable material for structure fabrication, as long as the material's mechanical properties permit and if the light intensity is sufficient to launch TPA. Proteins or other biomaterials are important possible choices [172], considering the promising prospect of biocompatible micromachine or MEMS applications used in, for example, drug delivery, local treatment of

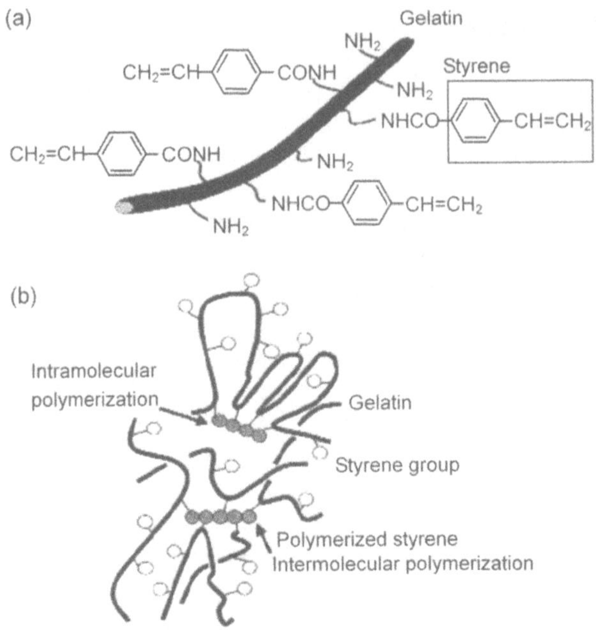

Fig. 42 Photocurable gelatin for application to femtosecond laser microfabrication. **a** Chemical structures of styrene-derived gelatin. **b** photogelation mechanism by formation of cross-linked gelatin networks via intermolecular crosslinking and intramolecular polymerization

tissue or cell, cell positioning, and biosensors. Therefore, in this section we introduce work related to this field.

Figure 42a is a multifunctional styrene-derivative gelatin, in which styrene groups are multiply attached via the amino groups of lysine residues of gelatin. It was developed [173] for photocurable tissue, and the photocuration occurs under visible light irradiation. We also succeeded in writing sub-micron-structures using two-photon-induced photogelation, for which water-soluble carboxyl camphorquinone was utilized as a photoinitiator. Upon two-photon excitation, photo produced radicals lead to the formation of a crosslinked gelatin network through intramolecular polymerization and intramolecule photocrosslinking of styrene groups in gelatin molecules (Fig. 42b).

Pitts et al. [175] explored photocuring of two different size proteins, bovin serum albumin (BSA, molecular weight MW=66 Kg/mol) and fibrinogen (MW= 340 Kg/mol). Rose Bengal (RB) was used as an initiator, whose fundamental absorption was at 550 nm. The photocrosslinking was attributed to one of the two possible photooxidation mechanisms: (i) the two-photon excited photoinitiator is excited to triplet state, and then energy transfer occurs to the ground state triplet molecular oxygen, generating singlet molecular oxygen; or (ii) abstraction of hydrogen directly from a protein molecule. In the former case, the oxygen reacts with an oxidizable amino acid residue of the protein. This irreversible mechanism generates an electron-deficient protein that may react with another protein's amino acid residue at or near the protein surface. In the latter case, direct hydrogen abstraction from the protein molecule by the photoexcited initiator allows for direct protein cross-linking. A further experiment shows that at low RB concentration, the fibrinogen grows 2~10 times faster than for BSA, which was interpreted as being because the fibrinogen has a relatively larger molecular weight, so fewer cross-linking reactions are needed to reach optically detectable dimensions. However, this effect doesn't hold true when the RB concentration is sufficiently high to negate the effects of photo activator diffusion or fewer protein reactive sites. The existence of a marked dependence on the RB concentration for both BSA and fibrinogen implies that no dye is produced during the photocuring. This supports the hydrogen abstraction mechanism because in the first mechanism, the dye is recoiled by subsequent optical pumping.

Both of the above two researches targeted the drug delivery devices or sustained release devices. It is believed that the ability to fabricate with both hydrogel materials and proteins at milli-, micro- and nanoscales would expand the utility of such devices that are currently millimeter-sized. Critical to creating useful release devices is the entrapment of target molecules. For two-photon curing, the task was simplified to include only the molecules of the fabrication solution. Pitts et al. find that alkaline phosphatase maintains bioactivity after entrapment in the protein structure. As sustained release devices, it is found the rates of diffusion of fluorescently labeled dextrans (10 and 40 kDa) from a two-photon fabricated BSA matrix vary with molecular weight and are linearly correlated with cross-link density. The release half-life of 10 kDa dextran-TMR from a BSA microstructure is equal to or less than 6 minutes while that for 40 kDa dextran-TMR is 25 minutes.

4.8 A High Efficiency Photoacid Generator and its Application to Positive-Tone Microfabrication

4.8.1 Large-δ and High Quantum Yield Photoacid Generators

We introduced the application of conventional UV photoacid generators (PAGs) such as diaryliodonium and triaylsulfonium cations to two-photon microfabrications in 3.3.7. However, the TPA cross-sections for these initiators are generally small, of order $\delta \sim 10 \times 10^{-50}$ s cm^4/photon. As for radical polymerization, a high-efficiency 3D fabrication would benefit from large δ

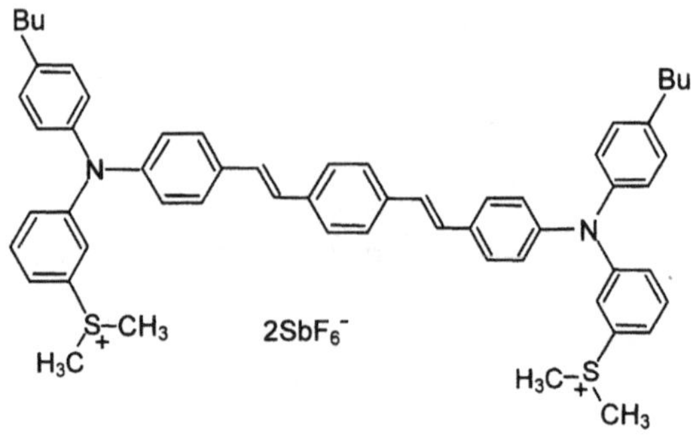

Fig. 43 Chemical structure of BSB-S_2

and large photochemical quantum yield, ϕ_{H+}[1–5, 104–106]. The problem in designing a large-δ molecule is that most of them have excited states of relatively low energy due to extended conjugation, which is insufficient to afford the energy for direct bond cleavage. Zhou et al. [176] use the strategy of photoinduced electron transfer from TPA dyes to covalently linked sulfonium groups. They realized [177] that (i) two-photon-excited bis(diphenylamino)-substituted bis(styryl)benzene dyes (δ~800×10^{-50} s cm^4/photon) should have ample reducing power to transfer an electron to the S-C σ^* orbital of a dimethylaryl sulfoium cation, which would cleave the S-C bond and generate acid; and that (ii) triacrylamine dialkylsulfonium salts are photosensitive in the near-UV region and generate protons with a photochemical quantum yield ~0.5. A compound, BSB-S_2, was synthesized and characterized (Fig. 43) [177].

A solution of BSB-S_2 in acetonitrile become acidic after irradiation into the lowest energy absorption band, and the quantum efficiency of proton generation was determined to be ϕ_{H+}~0.5±0.05. Measured from the two-photon excitation spectrum, the TPA cross-section of BSB-S_2 was significantly larger than conventional PAGs, and was determined at δ>100± 10^{-50} s cm^4/photon for 705–850 nm, with a peak at 710 nm, where δ~690± 10^{-50} s cm^4/photon. BSB-S_2 was proved to be a high efficiency initiator for photopolymerization of epoxide monomers. In addition to its use in negative type 3D fabrication similar to radical photopolymerization, another merit of photoacid generation is that it permits positive tone microfabrication based on polymer degradation.

4.8.2 Positive Tone Microfabrication

In mechanical manufacturing, components can be produced either by means of additive or subtractive processing. Microfabrication follows the

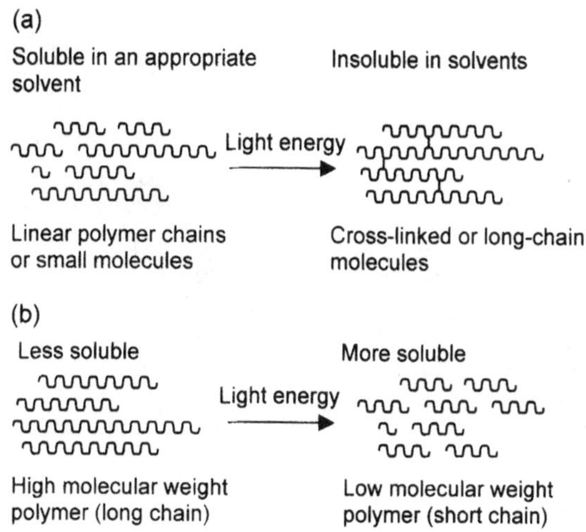

Fig. 44 Microfabrication using negative and postive photopolymers. **a** photopolymerization, and **b** photo-depolymerization

same logic, and negative and positive photopolymers are useful for each technical route. Photopolymerization is a typical additive-type fabrication method, where exposed volume becomes insoluble in post-exposure development. In a positive type fabrication, the precursory photopolymer is insoluble to a developing solvent but the exposure volume become soluble and removable from the matrix. Figure 44 shows the two schemes.

Zhou et al. [177] use a random copolymer of tetrahydropyranyl methacrylate (THPMA) and methyl methacrylate (MMA) polymer doped with BSB-S_2 as the PAG for microfabrication. At the laser focal spot, the THPMA groups were converted to carboxylic acid groups due to photogenerated acid-induced ester cleavage reactions, and were therefore rendered soluble in aqueous base developer. Figure 45 shows the 3D microstructure produced by this method. By two-photon fluorescence imaging, it was found that the buried channels are open and a continuous connection was made between the two cavities.

Different from two-photon photopolymerization, fabrication based on positive polymers should be called two-photon depolymerization. It could be an important complementary technology in 3D microfabrication, particularly those with many buried structures like those in a micro fluid system.

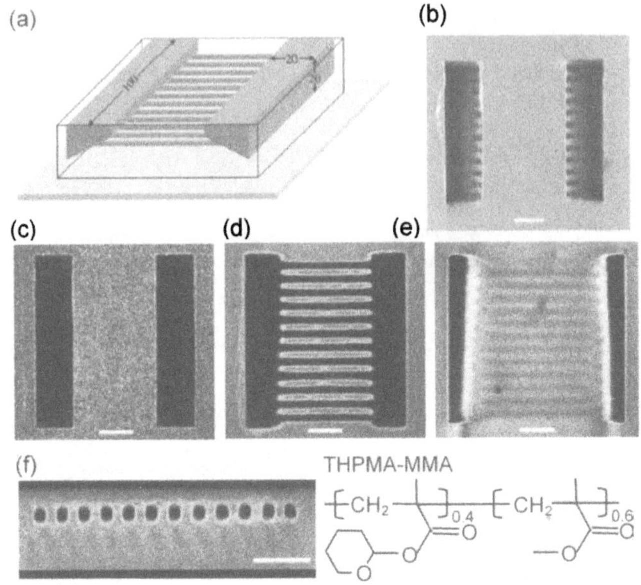

Fig. 45 A 3D microchannel structure produced by two-photon exposure of BSB-S_2 in THPMA-MMA. A 50-μm-thick film of 4 wt % BSB-S_2 in THPMA-MMA was exposed in the pattern of the target structure at 745 nm with tightly focused (1.4 N.A.) 82-MHz, 80-fs pulses from a Ti:sapphire laser at an average power of 40 mW and a linear scan speed of 50 μm/s. After irradiation, the film was baked for 1 min at 90°C. The target structure was then obtained by dissolving the exposed resin in aqueous 0.26 M tetramethylammonium hydroxide. **a** Target structure consisting of two rectangular cavities (width, 100 μm; length, 20 μm; depth, 20 μm) with sloped side walls that are connected by 12 channels (length, 50 μm; cross-section, 4 μm by 4 μm) lying 10 μm below the surface and spaced apart by 8 μm (center to center). **b** Scanning electron micrograph of the final structure, viewed normal to the substrate. (**c** to **e**) Two photon fluorescence images of the final structure (viewed normal to the substrate) **c** at the surface of the film, **d** 10 μm below the surface, and **e** 19 μm below the surface. **f** Two-photon fluorescence cross-sectional image of the buried channels. The scale bar in **b** through **f** corresponds to 20 μm. [177]

5 Applications

Two-photon photopolymerization, a micro-nanotechnology still in its infancy and a subject of active research, has potential use in fields such as photonics, optoelectronics, biology, micromachines and MEMS, and so forth. Among many, here we introduce applications that already have proof of concept.

5.1 Photonic Crystals and PhC-Based Optoelectronic Devices

Photonic crystals (PhCs) [92–94] are microstructures with a periodical distribution of refractive indexes. They are the optical analogue of semiconductors, where a bandgap is open due to the electron wave modulation by periodic Coulombic potential. In a PhC, the multiple interference among waves scattered from each primitive unit may lead to a frequency region, called a *photonic bandgap* (PBG), where light propagation in all directions is forbidden. Incident from outside, PhC is highly reflective at the bandgap wavelength; and inside the structure, light emission will be suppressed or forbidden. These features provide a novel scheme for high-efficiency optoelectronic devices such as waveguides with sharp bends and low-threshold lasers, for new nonlinear optical applications such as superprisms, and for the enhancement of nonlinear processes like harmonic generation.

The period of PhCs should match the wavelength of interest. For application in the visible and NIR communication wave range, the lattice constant is expected to be several hundreds of nanometers. This length scale is not large enough to use mechanical processing like hole drilling, and it is not small enough that well-developed semiconductor expitaxial growth technologies like MOCVD or MBE are applicable. Although excellent performances have been theoretically predicted [92–94], realization of well-defined 3D lattices is technically challenging. Up to now, a number of technologies have been explored. Self-organizing colloidal particles [96–98], and filling the interval to form a reverse opal [178] have been conducted using various materials; hole drilling has been carried out using lithography with an ensuing wet etching, as has electrochemical etching of porous silicon [179], and electron beam lithography plus dry etching [180]; log pile structures were implemented by sequentially stacking micromachined 2D grids of silicon or gallium arsenide [181], and by repeatedly deposition and etching of silicon [182]. None of the above technologies are ideal. The self-organization technology tends to give close-packing lattices, and therefore leaves little room to tailor the lattice type; almost no satisfactory 3D lattices have been created from hole drilling technologies due to the restraint of the small depth-to-width ratio; semiconductor processing methods are expensive and complicated, and haven't produced structures with more than a few periods. Compared with above technologies, two-photon photopolymerization has the intrinsic capability to tailor 3D structures of arbitrary lattices, which will open the door to PhC-based polymer optoelectronic devices.

5.1.1 Two-Photon Polymerized PhC Structures and Bandgap Effects

There are at least two advantages to fabricating PhCs using two-photon photopolymerization technology. First of all is the potential to produce PhCs of arbitrarily designed lattices. PhCs of varied lattice types, lattice constant, and filling factors are realizable just by scanning different CAD patterns. This simplicity in fabrication permits a systematic study of PhC physics and suits various requirements for a practical system. Secondly, there is the diversity of usable materials and functions. The progress of molecular material engineering has made it possible to synthesize polymers with performances similar to or better than their inorganic counterparts. By introducing functional groups to unsaturated monomer or oligomer units in a molecular structure, or just by doping the functional polymers into known photopolymerizable materials, optical, electronic, magnetic, and mechanical functions can be imparted to devices.

Sun et al. [15] were first to propose using two-photon photopolymerization technology for creating 3D PhCs, and they observed a pronounced PBG effect. They used a commercially available resin, Nopcocure 800 (San Nopco), consisting of a radical photoinitiator and acrylic acid ester. The linear absorption of the resin extends from the UV to around 370 nm. Due to a low TPA cross-section, 10^{-56}cm^{4}s photon^{-1}, the fabrication was conducted using regeneratively amplified laser pulses. Shown in Fig. 46a is a schematic illustration of the log-pile structure [183]. It consists of parallel rod layers with a stacking sequence that repeats every fourth layer with a period Λ. Within each layer, the rods are arranged in a simple 2D array with rod spacing a. The rods in the next layers are rotated by an angle α with respect to rods in the previous layer. The array in the second neighboring layer is shifted by half the spacing a relative to rods in the first plane in a direction perpendicular to the rods. Different variations of logpile structures were realized by parameter assignment of a, c (period along the packing direction) and α (60~90°). The general logpile structure has the symmetry of a face centered tetragonal (FCT). When $\frac{c}{a} = \sqrt{2}$, the lattice can be derived from an *FCC* unit cell with a basis of two rods.

Since the generally used resins are just optical adhesives, organic solvents like acetone or methanol work quite well for developing. However, compared with the micromechnical structures as described previously, photonic crystals consist of holes of high aspect ratio. Therefore a much longer soaking duration is needed for complete removal of unsolidified components. Figure 46b is the top view optical microscope image of a fabricated logpile structure. The same structure, after being cleaved and coated with a thin-layer of Au film, was observed under SEM (Fig. 46c). It is shown that a well-defined 3D spatial structure had been achieved, and the rods were arranged

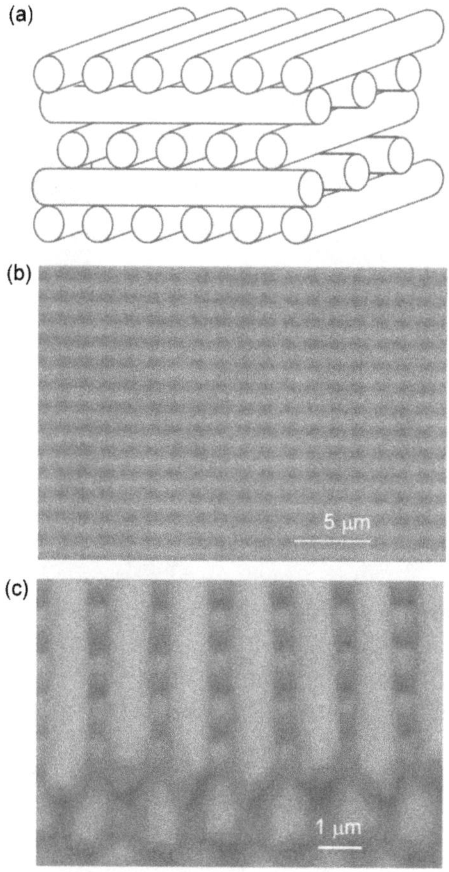

Fig. 46 Logpile PhC structure. **a** an illustration, and **b** a structure fabricated using two-photon photopolymerization of resin, top-view optical microscopic image, and **c** side-view SEM image

regularly in the same plane with half a period offset to the nearest layers with the same orientation.

To reveal signatures of PBG, transmission spectra of the PhC samples were measured using a Fourier-transform infrared (FTIR) spectrometer combined with a microscope imaging system. Plotted in Fig. 47 are the transmission characteristics of 20-layer logpile structures with different in-plane rod spacing: 1.2 μm, 1.3 μm, and 1.4 μm. Each sample has a dimension of 40 μm×40 μm, all fabricated under the same laser pulse energy, 90 nJ/pulse, giving rise to a lateral rod diameter of 1.0 μm. All spectra were normalized to the transmission of the uniformly solidified bulk resin.

The transmittance dips under normal incidence occur at wave numbers of 2550, 2510, and 2450 cm^{-1}, respectively. The increase of the wavelength of

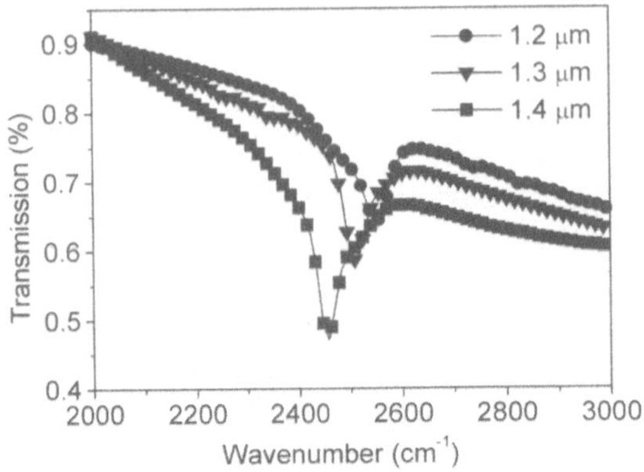

Fig. 47 Transmission spectra of two-photon polymerized logpile PhC structures with different in-plane lattice constants

transmission minima versus lattice constants is an expected feature since the frequency scales as $1/n$ in a medium of refractive index n. The mean dielectric constant increases with the filling ratio of resin and the latter is roughly estimated according to $f=\pi r/4d$, where r and d are the rod radii and the inter-rod spacing. An average attenuation of 1.3 dB per unit cell was not satisfactory. Several factors should be responsible for this: fluctuation of rod diameter arising from the shrinkage of polymerized resin and from ambient

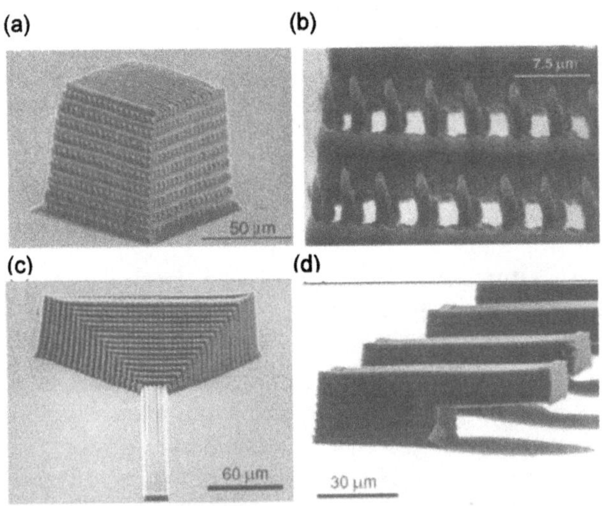

Fig. 48 PhC structures produced by photopolymerizing a resin with high-efficiency radical initiators. **a** SEM image of the entire structure, **b** magnified top view, **c** tapered PhC waveguide, and **d** array of cantilevers, as example micromechanical devices

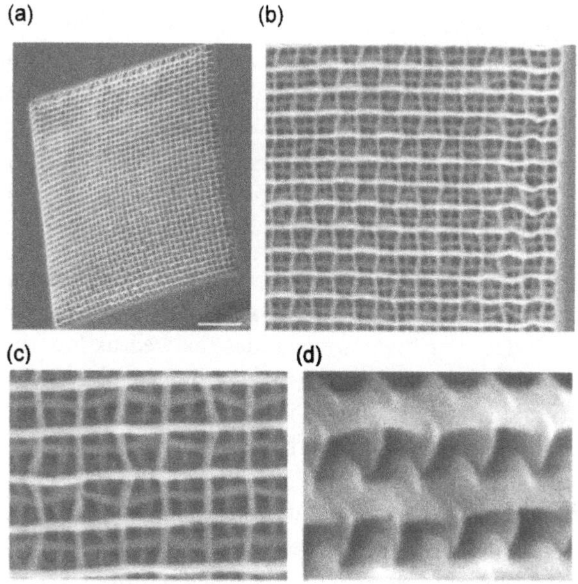

Fig. 49 SEM image of logpile structure with 500-nm layer spacing and 1.5 μm in-plane rod spacing. **a** Entire view of the 60 μm×60 μm structure, scale bar 10 μm. **b**, **c**, and **d** magnified top views

vibrations, and most importantly, the incomplete removal of unsolidified or low-degree polymerized resin at the bottom layers.

A similar logpile structure (Fig. 48a) was reported by Cumpston et al. [22] soon after the first demonstration of working photopolymerized PhCs. Although there is no observed bandgap effect, this work is known for use of highly efficient two-photon photoinitiators as introduced in 3.3. The resin they adopted consisted of a polymer binder, a crosslinkable acrylate monomer, and D-π-D chromophore **8** or **9** (Fig. 6) as the photoinitiator. They tuned laser pulses of 150 fs duration at a repetition frequency of 76 MHz to wavelengths near the TPA peak of the initiator. Figure 48a and b are views of the logpile structure they fabricated. Figure 48c shows a tapered optical waveguide, of which the cross-section varies along the length from a 100 μm×100 μm square aperture to a 2 μm×10 μm rectangular aperture. Tapered optical waveguides have the potential to reduce optical loss in the coupling of waveguide components with disparate cross-sections.

In the work of Sun et al. [15, 17], the PBG appears at around 4 μm, which is mainly determined by the periodicity along the light propagation direction. By reducing the lattice constant, it is possible to move the bandgap center to shorter wavelengths. Straub et al. [184] observed a bandgap effect at 1.5 μm ~ 2.3 μm from logpile structures of layer spacing ranging from 350 nm to 500 nm. Figure 49 shows SEM images of a 40-layer 60 μm×60 μm structure scanned with 60 μm/s of 540 nm, 140 fs, 76 MHz and 1.0 mW laser pulses. Subsequent layers were stacked at intervals of 500 nm and with an in-plane rod spacing of 1.5 μm. Because the rods were produced by single scanning, instead of multi-lateral scanning [15, 17], their cross-section was measured at 860 nm×280 nm, reflecting the shape of two-photon PSF.

5.1.2 Defects for PhC Functions

Up to now most successful unadulterated and defect-contained PhCs have been fabricated in semiconductors [185, 186]. To mention just a few examples, line, air-bridge, and bent waveguides, as well as low-threshold lasing have been reported. Recent progress in functional molecular synthesis stimulates the development of organic photonics. Organic structures employing various geometries, such as microdisk, microring, sphere, vertical-cavity surface-emitting, distributed Bragg reflection (DBR), distributed feedback (DFB), and fiber grating, have been used in lasers and amplifiers [187]. One problem that was encountered in designing organic LD and light emitting diode (LED) structures is the inherent absence of mirrors that provide positive feedback for lasing. In semiconductor structures cleaved facets play this role. Although this problem can be partly solved using DFB and DBR structures, the emission efficiency is still unsatisfactory. Therefore it is interesting to investigate whether the use of PhCs provides a novel mechanism to circumvent this issue.

Like the role of impurities in semiconductors, defects could be essential for tailoring PhC properties. For instance, PhC-based waveguides can guide light through sharp turns with negligible power loss, which would enable photonic integrated circuits. Embedding an optically active medium into a PhC structure inhibits light emission into undesirable modes, and allows the realization of highly efficient optical emitters, such LEDs and thresholdless LDs. The essential requirement to reach this end is the possibility of forming waveguide channels and microcavities by introducing defects into periodic structures. In the two-photon photopolymerization method, 3D structures are written on the voxel basis. This gives rise to the feasibility of individual addressing, that is, that each voxel could be produced in the desired way, implying that not only complete periodic structures, but also defects – either missing rods or extra portions – could be created.

PBG effects have been observed in two-photon polymerized PhCs [15]. The research was naturally extended towards the formation of planar microcavities by introducing defects into the logpile PhC (Fig. 50a). The fabricated structure [17] consisted of 20 layers of rods. The planar defect was introduced by skipping the exposure of every other rod in the tenth layer, located in the middle of the PhC structure. This was simply accomplished by closing the laser beam while drawing the particular lines.

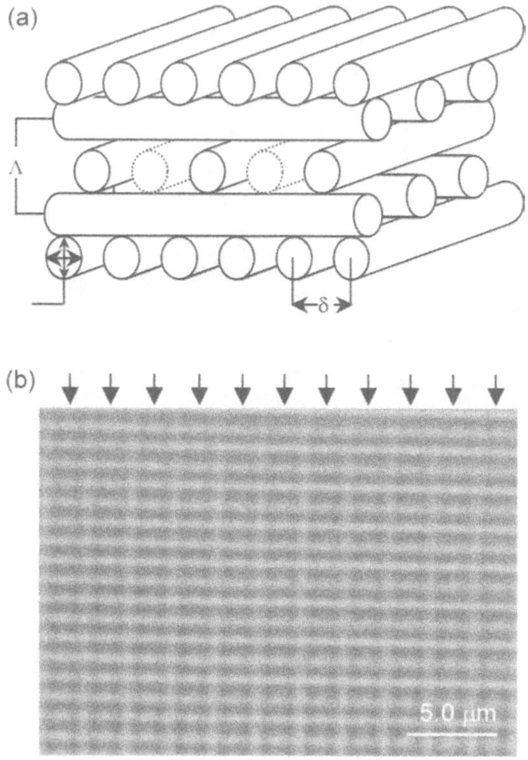

Fig. 50 Logpile PhC structure containing planar defects. **a** an illustration of the concept, and **b** optical microscopic image of the fabricated structure. The arrows indicate where no rods exist due to exposure skipping

Figure 50b shows an optical micrograph of the defected layer. The positions of the missing rods are indicated by the arrows. The structure has a rod diameter of $2r$~0.8 μm (assuming a cylindrical rod shape), in-plane period δ~1.3 μm, and rod length L~40 μm. For the PBG midgap wavelength $\lambda_M \approx 4.0$ μm, so the geometric thickness of the defected layer $2r$~0.8 μm is somewhat smaller than $\lambda_M/2$, indicating that localized modes penetrated significantly into the PhC mirrors.

Figure 51 shows the transmission spectra of the investigated sample (solid line), and the reference sample without defect (dashed line) for unpolarized light along the stacking direction <001>, as indicated schematically in the top inset. In the reference sample, a pronounced transmission dip at 3.98 μm with a spectral width of approximately 91 nm (FWHM) is seen. The dip magnitude also implies the existence of a photonic pseudogap rather than full PBG in this structure, as can be expected in the case of a relatively low refractive index contrast. This assumption is also confirmed by our numerical simulations (see description below). In spite of incomplete PBG, sig-

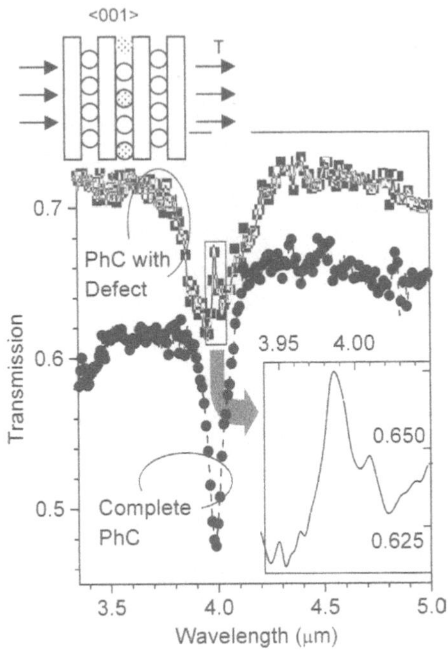

Fig. 51 Transmission spectra of the defect-free and defect-contained PhC logpile structures. The appearance of a sharp peak was assigned due to PhC microcavity resonance

natures of evolving defect modes can be nevertheless seen in the defected sample. The transmission spectrum of this sample reveals a similar, but slightly broader PBG transmission dip at the same spectral position, and a pronounced transmission peak within it.

The physical origin of the peak can be sketched in terms of multiple reflections of light between two PBG mirrors surrounding the defected layer, forming a planar microcavity. Hence, the peak marks the formation of the microcavity resonance. Light trapping by the cavity is usually characterized by the quality factor $Q=\omega_0 E/P$, where E is the energy stored in the cavity, ω_0 is the resonant frequency, and $P=dE/dt$ is the dissipated power. The transmission data allows us to estimate the quality factor using the expression $Q=\lambda_R/\Delta\lambda\approx130$, where λ_R and $\Delta\lambda$ are the peak center wavelength and width, respectively, assuming the measured values of $\Delta\lambda$=29.8 nm (FWHM) and λ_R=4.01 μm. It is helpful to note here that semiconductor microcavities formed from high refractive index materials may have quality factors as high as 1600. In our case, relatively low Q value means that the planar defect exerts a localizing perturbation on the light modes at the midgap, but their spatial spreading and escape from the defected region is also significant. As a result, the effective mode volume exceeds the geometric volume of the de-

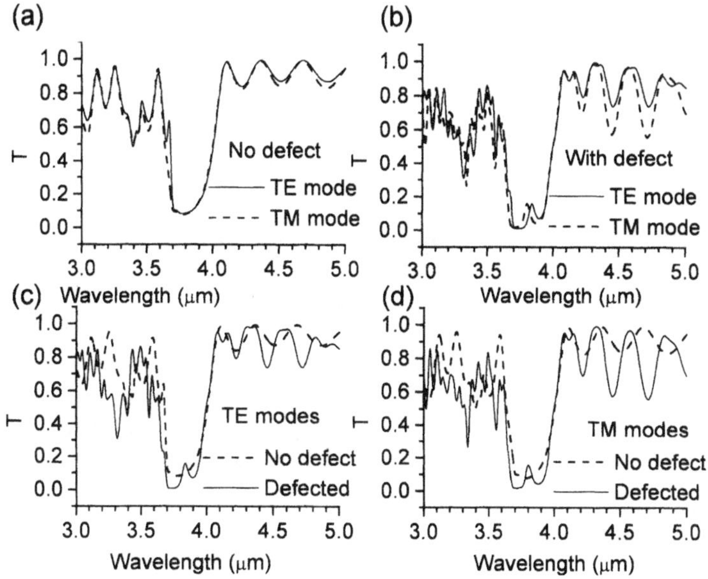

Fig. 52 Transmission spectra of the defect-free and defect-contained logpile PhC structures for different detecting laser polarizations. The existence of defect modes were theoretically proved

fect and may even become comparable to the entire illuminated volume of the sample.

Formation of the PBG and microcavity resonance was also confirmed by numerical modeling. Figure 52 shows the transmission spectra of reference (**a**) and defected (**b**) samples, calculated using the transfer-matrix technique. Transmission for TE (broken line) and TM (solid line) linearly polarized modes was considered separately in anticipation that the absence of some oriented rods in the defected sample would result in different conditions for the propagation of TE and TM modes. The calculated transmission spectrum of the reference sample is shown in Fig. 52a.

Spectral positions of the calculated transmission dips are close to those observed experimentally, but the calculated dips are somewhat deeper and broader (gap to midgap ratio is about 9%). A similar result is obtained for the defected sample (Fig. 52b), but in this case a distinct peak within the dip, marking the microcavity resonance, is seen for each polarization. The peaks are centered at 3.801 μm (TM) and 3.838 μm (TE), have Lorentzian line shapes, and almost identical amplitudes (about 16%) and Q-factors (about 85). As expected, there is a slight displacement between the peaks of different polarizations due to the anisotropic nature of the defect. Transmission at the maximum of the resonance peak is about 16%. Altogether, the numeric simulations qualitatively reproduce the experimental data, routinely yielding PBG at slightly shorter wavelengths than the measured ones. This is likely to be related to our assumption about cylindrical shape of the rods, whereas in reality they are oval-shaped, with elongation along the layer stacking direction, resulting in higher lattice constant, and scaling up of the PBG center wavelength.

5.1.3 Photopolymerization Created Waveguide Channels in PhC Templates

Defects, essential to PhC functions as discussed in 5.1.2, may be produced as an intrinsic part of PhCs in the structural design and fabrication processes if PhCs are fabricated by direct laser writing. However, self-organization of colloidal particles and multi-beam interference, the two technologies which are promising for industrial manufacturing, have no individual addressing capability. Hence, how to induce defects into perfect periodic lattices is an important issue to solve. A promising method is writing defect structures in ready PhC templates using two-photon photopolymerization. In this method, liquid resin is infiltrated into the interval of a PhC structure, and then a particular pattern could be written, provided that the PhC materials are transparent to the fabricating wavelength. After washing out unpolymerized resin, the polymerized structure would be connected and retained inside the original lattice (Fig. 53a) [188].

Lee et al. [189] realized the above idea on PhCs produced by self-assembling colloidal particles. The colloidal crystals were achieved by gravity sedimentation of 1.58 μm spherical silica particles from a solution of dimethylformamide (DMF) and water. They infiltrate a precursor solution of 0.1 mM of the initiator, 9-fluorenone-2-carboxilic acid, and 0.1 M of the coinitiator, triethanolamine, in the monomer, triethylolpropane triacrylate. Polymerization was conducted using 780 nm, 80 fs, 82 MHz laser pulses. Figure 53a illustrates the procedure of their experiments. To reveal information about the polymerized structure inside the colloidal template, a Rhodamine solution in DMF was filled into the colloidal assembly after the removal of unsolidified liquid.

Figure 53b shows the single-photon fluorescence image of the entire polymerized structure; a portion of it was magnified and is shown in Fig. 53c. The lack of fluorescence in the dark region indicates that the particle interval there was sealed by resin and prevents the dye penetration. The existence of polymerized features was further confirmed by the SEM image of Fig. 53d, taken after the silica particles were removed by chemical etching in HF acid.

The defined-defect-contained colloidal crystals could be utilized as a mold for producing inverse PhC structures with ideal lattices and with sufficiently high refractive index contrast. The artificially induced defects could be removed by further processing like selective chemical etching or sinter-

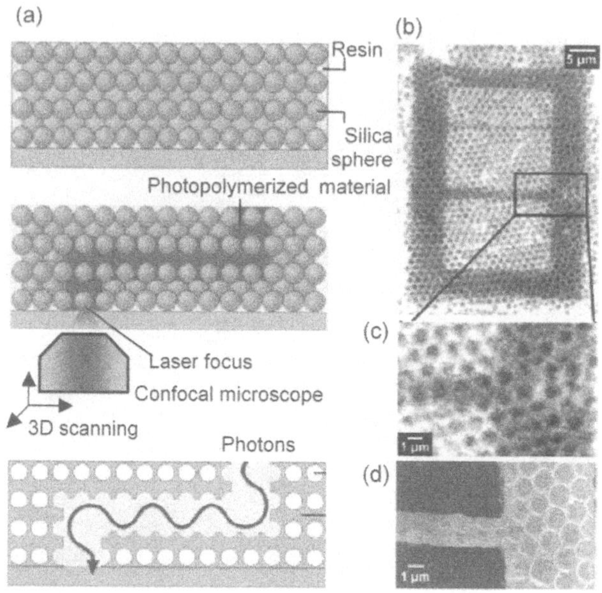

Fig. 53 Defect induced inside a colloidal PhC template by photopolymerizing the immersed resin. **a** Schematic of the experimental procedure and its future use, **b** fluorescence image of the PhC structures that contain the polymerized defect, **c** magnified view of **b**, and **d** SEM image

ing. This technology may broaden the use of colloidal PhCs [189] or multi-beam interference polymerized PhCs due to the new possibility of artificially designing defects.

5.2 Functional Micromachines and Microelectromechanical Systems and their Optical Actuating

A number of micro components have been fabricated, shown in Figs. 11, 15, 45, 48d, as well as in Fig. 54, a micro-gearwheel pair (Fig. 54a), a micro-gearwheel affixed to a shaft (Fig. 54b), a microchain (Fig. 54c), and a human skeleton (Fig. 54d).

These structures are good proof of the fabrication capability of two-photon photopolymerization technology. Even more complicated devices can be produced, which should be composed of two types of components classified according to their functions, static parts for support, connection or confinement, and movable parts. Both need precise shaping, positioning and jointing during photofabrication. For movable components, an essential issue to address prior to fabrication is finding a suitable actuating mechanism. Appropriate electric, optical, thermal, magnetic, and chemical effects need to

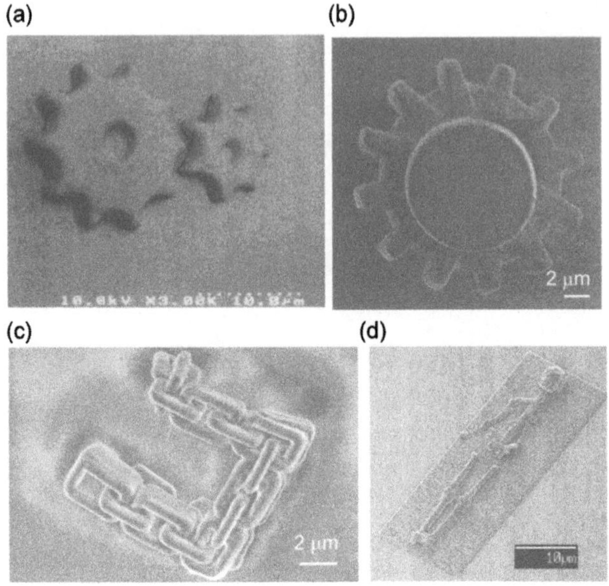

Fig. 54 Micromechanical structures produced by two-photon photopolymerization **a** a microgearwheel pair, **b** a microgearwheel affixed to a shaft, **c** a microchain, and **d** a micro human skeleton

be found to achieve this end. Electrically controlling micro systems, the requirement of MEMS, is most desirable, however, there is a long way to go before introducing conductive polymer into structures and integrating polymer devices on an IC-contained semiconductor chip or developing polymer ICs. Optical force provides a simple solution, which is the currently most practical mechanism for actuating micro-nanodevices.

5.2.1 Optical Driving of Micromechanical Devices

5.2.1.1 Optical Trapping Force

The technique of optical manipulation has been employed as a unique means of controlling microdynamics of small objects without physical contact since the pioneering works by Ashkin [190–192]. For a better understanding of the actuating mechanism, we will briefly introduce how the laser trapping force originates [190, 191]. Assume that a transparent sphere with refractive index higher than its surrounding medium is situated slightly off a laser beam axis (Fig. 55). Consider beam *a* in a typical beam ray pair *a* and *b*. The light beams undergo Fresnel reflection and refraction at the medium/sphere and sphere/medium interfaces. Forces exerted on the sphere

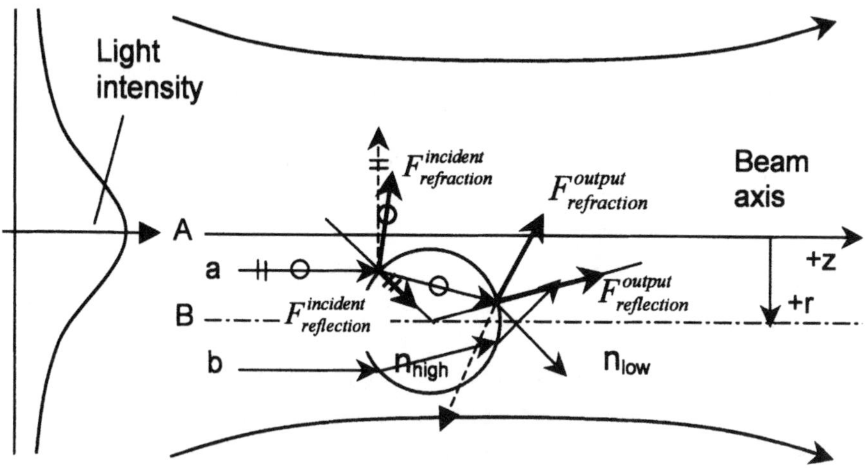

Fig. 55 The origin of optical trapping force, which provides various mechanisms for micro-nanodevice optical actuating

can be deduced from the momentum changes of photons in the beam due to elastic scattering. Four forces appear, $F^{incident}{}_{reflection}$, $F^{incident}{}_{refraction}$, $F^{output}{}_{reflection}$, $F^{output}{}_{refraction}$, as shown in the figure, and all give the bead acceleration in the beam propagation direction, $+z$. The two reflection forces at each interface are much smaller than those from refraction, and moreover $F^{incident}{}_{reflection}$ and $F^{output}{}_{reflection}$ cancel radially. The two refraction forces add radially to the $-r$ direction. Therefore the net radial force exerted by ray a is inward towards the beam axis direction ($-r$). The same analysis conducted for ray b leads to a net radial force outward. Consider the fact that a radial laser beam intensity distribution is always of Gaussian shape, so ray a is stronger than b. Summing all beams (a's and b's) leads to net forces interacting on the sphere in two directions. First, the bead is attracted inward towards the high intensity region. This force arises in nature from the gradient intensity distribution, therefore also called gradient force. Actually it is already known that in a nonuniform medium, for example in a waveguide, light will be guided to propagate along a high-index path. Here we see that if the media have sufficiently small mass, they can be reversibly attracted by light. The second force is simple, pushing the sphere to move along the beam propagation direction.

It is easily understood that if the light is tightly focused so that intensity gradients are nearly symmetrically constructed at two sides of the focal spot along the optical axis, the sphere is trapped at the high intensity focal spot. For an absorptive media, absorption of a large amount of directional photons (those in a laser beam) causes a net increase of momentum of the medium along the light propagation direction. In this case, the force in the laser beam direction is also called scattering force. The laser irradiation pressure (gradient and scattering forces) as discussed above forms the basis of light actuation of micromechanical devices, which falls into three categories: windmill rotation, rotating by photon angular momentum transfer, and push-pull due to 3D trapping. In the following sections, we will describe how these optical powering approaches have been or will be applied for driving micromachines.

5.2.1.2
Windmill Rotation

Windmills rotate when facing wind, a phenomenon known since ancient times. It is not surprising that similar rotational phenomena have been frequently observed in laser-trapped particles. The rotational torque arises from the axial irradiation force as discussed above and from the asymmetrical or rotation-symmetrical shape of the particles. Actually it is believed that lifetimes of interplanetary dust grains are governed by bursting caused by solar irradiation-induced rotation! The rotation rate is proportional to the trapping laser's power, and is related to the shape of the objects and the viscous drag from the ambient medium. In a micromachine, it is important to design a device structure of helical shape and of proper rotation symmetry so that the structure could be fixed (trapped) at a suitable position and with the desired orientation, for high stability and for minimizing the friction between the rotating parts and its axle (if there is one). The translation momentum from the "photon wind" needs to be efficiently converted to the spinning momentum of the object. A number of microcomponents satisfying the above requirements have been produced by various microfabrication technologies [193–195]. Shown in Fig. 56 is a two-photon photopolymerized eight-fold rotary symmetrical rotor [195].

The resin used is a commercially-available UV optical adhesive, Norland NOA 63. A 514 nm laser was utilized as the irradiation source. When applying a 994 nm trapping laser from a LD, the rotator was first 3D trapped with its axial line coinciding with the beam axis. Two opposite orientations, whether leaf-side or shaft-side facing the laser beam, produced stable equilibrium. With the rotator position stabilized, a rotation started. Due to the large viscous drag, the rotation rate was relatively low, only several tens of Hertz. It is found that the average rotation doesn't depend on the polarization of light, consistent with the actuating mechanism: the trapping is due to the gradient force and rotation arises from momentum transfer from the photon flux, that is, the windmill effect. Scattering force is proportional to the light intensity and acts in the light propagation direction. Gradient force is proportional to the gradient of the spatial light intensity and acts in the direction of the intensity gradient. In a further experiment, the authors demonstrated that the rotor drove cogwheels. Two neighboring and slightly

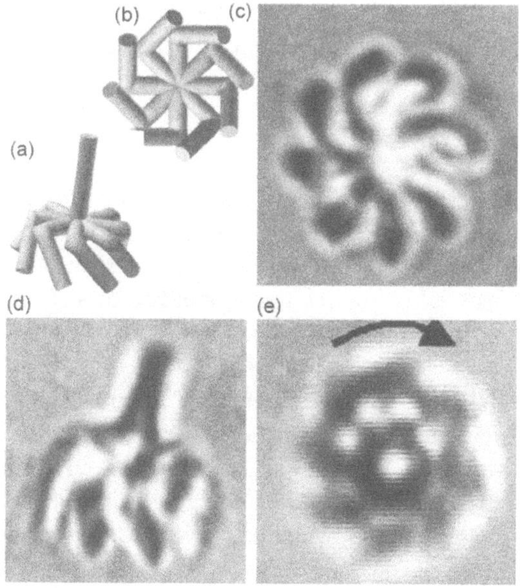

Fig. 56 Design and two-photon fabricated rotator driven by a windmill mechanism. **a** Design and **c** photograph of the rotor in an arbitrary position when it is tumbling freely, viewed from identical directions. **b** Design and **d** equivalent photograph of the rotor when it is trapped in focus but held against the cover glass thereby preventing rotation and yielding a sharp image. **e** Photograph of the spinning rotor trapped in focus and rotated by the light

touching gearwheels were affixed to separate axes that were polymerized on the substrate. When the rotating rotor was moved to one of the two engaged cogwheels, all of the three gearwheels turned.

5.2.1.3 Photon Angular Momentum Transfer

Light itself can carry angular momentum. Another machine rotating mechanism is therefore to couple the photon angular momentum to the object to be rotated. Microparticle rotating experiments have been carried out (i) using elliptically polarized light [196], (ii) by rotating the asymmetric laser beam, and (iii) using a laser beam with a helical phase structure interacting with absorptive particles [197]. When absorbing CuO particles (1–5 μm) were trapped in a focused "donut" laser beam, they rotated due to the helical phase structure of the beam. Changing the polarization of the light from plane to circular caused the rotation frequency to increase or decrease, depending on the sense of the polarization with respect to the helicity of the beam.

Although no suitable examples have been found, it is believed that these mechanisms are in principle applicable to micromachine driving.

5.2.1.4
Push-Pull Random Structures

A microparticle can be three-dimensionally trapped at the laser focus, in which case it follows the movement of the focal spot if the beam is moved. This phenomenon isn't specific to isolated particles but also applicable to a portion of an object. This implies that the entire object may be pushed or pulled in random directions if only a part of a structure is captured by a laser focus.

Compared to the last two technologies, (i) the push-pull method doesn't need designing machine to create special shapes, which is needed by the windmill driving mechanism, and (ii) the actuating isn't limited to rotating, but to any random 3D movement. As an example, a micro-oscillator system was driven by this mechanism, which will be discussed in the following section.

5.2.2
Mechanics of Two-Photon Polymerized Nanodevices

Like numerous electronic devices, such as personal computers and cellular phones, that have gained cost advantage from integrating most of their functions onto a single chip, mechanical micro-nanodevices and their integrated systems are expected to spur the next revolution in the manufacturing industry in the post-IC era [128, 129]. Therefore, developing micro-nanodevice fabrication technology and studying their performance becomes an immediate and urgent task for the research community. It is already a well-known principle in the aircraft manufacturing industry that real-size systems that are proportionally scaled up from designed models don't work. This is because the surface area and mass (or volume) of an object do not proportionally increase with dimensions; they follow different laws (square and cubic laws respectively). The same principle applies when the size of devices are scaled down to micro nanometer sizes. For example, if the feature size of a device is reduced from millimeters to nanometers, the surface-to-mass ratio increases by 10^6 times. Therefore, in the nano realm, mass and inertia are no longer important, while physical, mechanical and electric characteristics such as stress and tension, thermal transfer, phase transition, fluid phenomena, and achievable field strength abide by rules much different from current experience and would dominate [128, 129]. Without related knowledge, it is impossible to properly design, fabricate and operate nanodevices and their integrated systems.

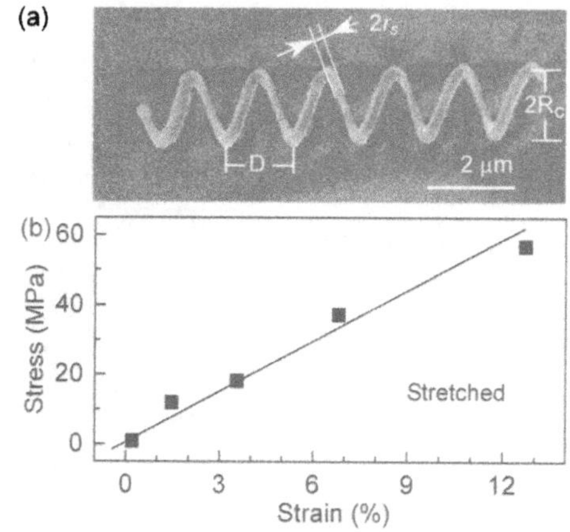

Fig. 57 **a** SEM image of a nanospring, and **b** the stress-strain plot of a resin fiber produced by UV photopolymerization

Two-photon photopolymerization has been recognized as an important method for producing micromechanical and MEMS devices. It provides a good opportunity to explore nanodevice mechanics. Sun et al. [13, 19] fabricate a nano-spring using this technology as shown in Fig. 57a.

The spring has a spiral radius of r_s=150 nm, a coil radius of R_c=1 μm, and a pitch of D=2 μm. According to these parameters, the spring constant, k, can be calculated according to the relation:

$$k = \frac{G_s r_s^4}{16 N R_c^3} \tag{23}$$

where G_s is the shear modulus of the solidified polymer, and N=4 is the number of active coils. G_s was determined from the elastic elongation of a fiber-like sample that was uniformly polymerized under UV exposure. From the stress-strain plot (Fig. 57b), the Young's modulus of the polymer was deduced to be E=0.46 GPa. Based on the relation $E=2G_s(1+\upsilon)$, G_s=0.15 GPa is obtained, where υ=0.49 is the Poisson's ratio of the material. This immediately gives rise to a spring constant of k=4.8 mN/m according to Eq. 23.

In order to know characteristics of nanodevice operation, it is critical to find an appropriate mechanism to run real-size devices. To achieve this end, the push-pull method as introduced in 5.2.1 was adopted. It is conjectured that if a spring end is trapped, moving the laser focus would prolong, compress and bend the spring itself (Fig. 58a). However, to enhance the trapping force, a microball was polymerized at the open end of the spring, and the other end was affixed to a photopolymerized anchor. Figure 58b shows the micro-oscillator system, where the spring was positioned 10 μm above the glass substrate, and had the same spiral and core diameters as those shown

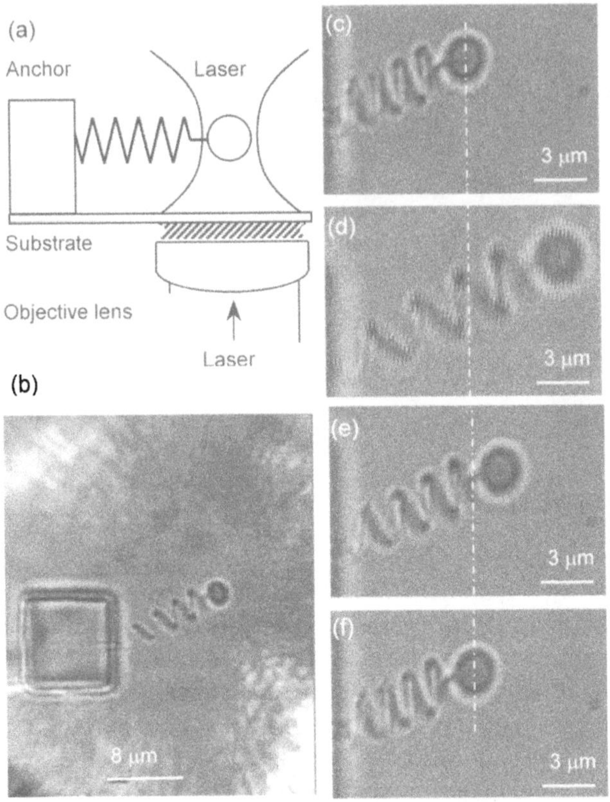

Fig. 58 A functional micro-oscillator system. **a** A designed scheme for mechanically activating the oscillator using a laser trapping force. **b** A photograph of a fabricated micro-oscillator, where the end bead was being trapped by the laser. The micro-spring is **c** in its natural state, **d** pulled by a length, **e** released, and **f** recovered to its original state 20 s after release

in Fig. 57a. The bead's diameter was $2r$=3 μm. In operation, the oscillator was kept in ethanol in order that the buoyancy would balance the gravity, eliminating the bead-substrate friction. The same laser system as was utilized for TPA photopolymerization, but with a wavelength tuned to 820 nm, was used for the laser trapping.

When the laser focus was carefully adjusted (Fig. 58b), the bead was found to be three-dimensionally trapped and able to be freely manipulated. The spring was pulled by moving the trapped bead, and then it was released by blocking the laser, initiating an oscillation. The spring was observed to be prolonged (Fig. 58d) from its original length (Fig. 58c), and restored (Fig. 58e) to its original state after the laser was turned off (Fig. 58f). Elongations of up to 7 μm over many cycles didn't cause any elasticity failure, as evidenced by the fact that the spring always returned to its original length.

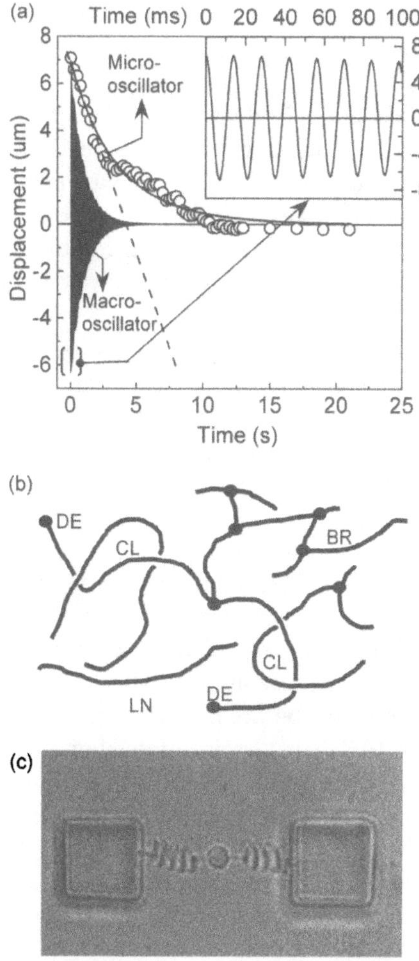

Fig. 59 **a** The bead displacement versus time for the spring recovery from the elongated state, for the micro-oscillator (circles) and a linearly 1000-times-scaled up macro-oscillator. **b** Crosslinking status of polymerized resins. **c** An oscillator consisting of a double spring

From the bead movement, it is possible to roughly estimate the spring constant. Notice that the movement isn't a reciprocating oscillation, but an over-damped one (solid line in Fig. 59a). The viscous friction should play an important role. Since the bead velocity is slow (v=1.5 μm/s at maximum, the dashed line in Fig. 59a), it is reasonable to assume the viscous resistance, a non-conservative force that is always opposed to its direction of motion, is proportional to the speed, in other words $f_{vis}=cv$, where c is a coefficient. In this case the spring oscillation can be described by the following equation:

$$\frac{\partial^2 x}{\partial t^2}+2\mu\frac{\partial x}{\partial t}+\omega^2 x=0 \tag{24}$$

where $\mu=\frac{c}{2m}$, and $\omega=\sqrt{\frac{k}{m}}$ is the circular frequency of the natural vibration of the spring; m (1.57×10^{-14} kg) is the mass of the bead, and k is the spring constant to be determined. The viscous resistance exerted to the bead was assumed to be $f_{vis}=6\pi\eta rv$ (Stokes Law), where $\eta=1.084\times10^{-3}$ Pa·s (25°C) is the liquid viscosity. Then by fitting the general solution with the experimental data (circles in Fig. 59a), the spring constant was derived, surprisingly, to be $k=1\times10^{-8}\text{Nm}^{-1}$

Here we see a five-order difference in the measured (10^{-8} N/m) and calculated (10^{-3} N/m) spring constants. To interpret the origin of the huge discrepancy, we noticed that the bead was rested before release at a stretched state of Δx=7 μm, almost the maximum of the prolongation. Calculations and dragging experiments showed that the trapping force imparted to the bead was not more than F=10 pN, which was quite consistent with reported values. Then the spring constant was directly obtained by Hook's law, $F=k\Delta x$, and was $k=10^{-6}$ N/m. This value is more reliable, since in the static state the influence from the viscosity was excluded.

For describing the motion of the micro-oscillator, we used Eq. 24, where the mass of the spring and the viscous resistance between the spring and the liquid were ignored. This may be the major source of the negative deviation of two orders, ($10^{-6}\rightarrow10^{-8}$ N/m). It is interesting to compare the performances of devices of different dimensions. When experimented in identical circumstances, an oscillator linearly scaled up by a factor of 1000 shows a typical damping oscillation (Fig. 59a), and ignoring the dimension of the spring causes a discrepancy of not more than 20%. It is the scaling effect that caused utterly different movement behaviors of nanodevices.

Materials that are assembled into micro-nanomachines are condensed in a way different from bulk matter or nanoparticles. At the mesoscale of near or sub-100 nm, the machine feature size, interactions from any single molecular bond are already less pronounced, but the collective effects of molecular arrays, orientation and molecule aggregation still play a critical role in the material's mechanical property. Like ambient factors, the material attributes themselves also influence the behaviors of nanodevices. In the calculation that used Eq. 23, physical characteristics of the spring such as modulus of elasticity and density of TPA-polymerized resin, which are determined by the degree of crosslinking, degree of crystallinity, and the value of T_g and T_m [1], were assumed identical to that polymerized under UV exposure. In photopolymers, the main contribution to the elastic strength is from the crosslinking of polymer chains. Young's modulus is related to the degree of crosslinking via $E=3nRT$, where n is the crosslink density, defined as the number of network chain segments (Fig. 59b) per unit volume, R is the gas constant, and T the temperature. First, different from the UV exposure used for polymerizing the sample fiber, where the sample had been sufficiently illuminated and therefore fully crosslinked, in TPA fabrication the exposure duration for each voxel was short (~1 ms), which was comparable with the reaction decay time for monomer and oligomer polymerization. As a result, the polymer is more likely to be present in forms of linear (LN, in Fig. 59b) or branched (BR) chains, and a relatively small proportion is crosslinked (CL). Second, since the spiral radius is small (150 nm), the shear modulus was reduced due to a large percentage of surface volume. The surface layer is abundant in dangling ends (DE, in Fig. 59b) that are not attached to any other chain sections. G_s or E is also decreased by some weak points or segments of elasticity that were induced by the lateral fluctuation in cross-linking degree. Third, removal of remnant monomers, oligomers, and less polymerized segments by the developer, due to its relatively deep infiltration, reduced the steric hindrance, facilitating the stretching of the spring. The increase of three orders of magnitude ($10^{-6} \rightarrow 10^{-3}$ N/m) of the spring constant were interpreted by the above three factors, showing that at the nanoscale, device characteristics are also significantly affected by material status.

The ambient factors that govern the nano-oscillator operation may be minimized by actuating the spring in air or vacuum conditions. Shown in Fig. 59c is a micro-oscillator consisting of two springs, by which the bead was suspended without making contact with the substrate. Laser driving such devices may reveal much more valuable information about the mechanic and material charateristics of nanodevices.

5.2.3
Towards Photoactive Structures

The polymeric structures discussed above use passive materials; materials that have no response to light (their electron excitation status is unchanged). If photosensitive materials are used, either as an intrinsically bonded or extrinsically doped component of polymers, light can be applied to motivate the microstructures through material response: shape or volume variation for mechanical functions; refractive index change for photonic devices; and conductivity and light emission for optoelectronic devices. Various photo-induced isomerization effects are good candidates for this purpose. For example, azobenzene and derivatives [69, 198] have two geometrical isomers, the *trans* and the *cis* forms. An isomerization reaction is a light- or thermally-induced interconversion of the two isomers. The *trans* isomer is thermally more stable. When irradiated with UV light a *trans*→*cis* conversion is promoted and the *cis*-form concentration increases, which is characterized by an enhancement in visible absorption. Visible light irradiation pushes the reaction in the opposite direction. Figure 60a shows azobenzene as an example. An important class of photochromatic materials, diarylethenes also have

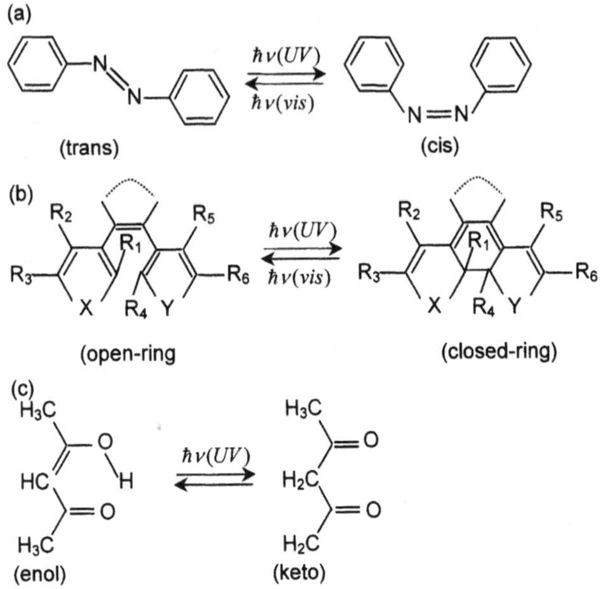

Fig. 60 Several promising photoactive materials that could be used for actuating nanodevices. **a** Photoisomerization due to *cis-trans* transition, **b** photoisomerization from ring close-open mechanism; and **c** optical tautomerization

two forms (Fig. 60b) [67, 198, 199]: open ring and closed ring, which can be interswitched by irradiation by UV and visible light, too. In both cases, the shape and size of molecules are different for the two forms, which may cause a volume change of material. However, these effects are too weak to use since the photosensitive molecules are distributed into a host polymer as a small percentage, <10% say, to avoid aggregation. It is easier to render intramolecular interval modification than impart appreciable volume change. Special measures should be taken to increase the concentration of photosensitive molecules in the matrix [199].

β-Diketone compounds, like acetylacetone, are known to exist in forms of enol and diketo tautomers (Fig. 60c). [200] A conversion from enol to diketo occurs with UV irradiation. Watanabe et al. [201] prepared a hydrogel of a cantilever shape by two-photon photopolymerizing a comonomer solution containing caryloylacetone, acrylamide, and *N,N'*-methylene bisacrylamide. They illuminated UV (244 nm) light from one side of the cantilever, which was then deflected due to the photoconversion-induced volume expansion. Compared with the previous two mechanisms, here the tautomerization reaction isn't reversible and needs an aqueous ambient for operation, therefore, the UV irradiation of the cantilever is more like a final step of the prototyping.

Functionalizing microdevices using photoactive materials is a promising direction for MEMS; however, a lot more research is needed to find suitable materials and actuating mechanisms.

6 Future Prognosis

Nanophotonics addresses detecting, controlling and fabricating features in a spatial volume less than that defined by the optical diffraction limit. There are basically two routes to access the SDL regime. First is the near-field, which works based on sharp-tip induced photon tunneling of an evanescent field, built either on a scanning tip end (apetured tip) or on a sample surface (apertureless tip). The other technology is multi-photon technology, which possesses the SDL spatial resolution due to nonlinear laser-matter interactions. Nanofabrication – production of nano-sized devices or devices with nanofeatures – is an important task of nanophotonics. Near-field fabrication, restrained by the nature of exponential decay of evanescent fields, gives rise only to 2D structures, which can't satisfy the requirement of three dimensional micro-nanodevices and systems.

Two-photon photopolymerization, as an important method of multiphoton laser micro-nanofabrication, is expected to play an essential role in producing polymer-based optoelectronic and MEMS devices. A lot of work has been done along this line as we have reviewed in the previous sections of this chapter. Future research in materials, optics and fabrication of functional devices are needed to further its use in diverse scientific research fields and industrial applications.

From a materials point of view, synthesis of two-photon chromophores with even higher two-photon cross-sections is required, so that photopolymerization may be induced by less expensive picosecond, nanosecond and even CW lasers for commercial applications. Also the high δ materials would enable polymerization simultaneously from multibeam split out of one output of a laser for batch production. Also desired is the use of copolymerization of functional molecules and resins, or doping of other functional components like nanocrystals into resins, so that particular functions could be imparted to polymerized devices. Uniform dispersion of the alien components into the matrix polymer is an important problem to solve. Finally, we need to synthesize functional polymers; in other words, instead of doping, we directly induce functional chromophores as a bonded component of monomers or oligomers, so that functions such as light-emission, polarization control, refractive index tuning and so forth become more pronounced.

From the optical standpoint, tailoring the point spread function to produce a spherical voxel shape is advantageous for precise 3D prototyping. Another important line of work is finding suitable mechanisms to construct a parallel production system. This may be accomplished by a diffraction beam splitter or by a microlens array. The latter is preferable because (i) foci number increases simply by expanding the area of a lens array since a single micro lens occupies a fixed size, for example, 250 μm diameter (ii) no further focusing is needed since each beamlet is naturally focused after passing through the array. Measures should be taken to ensure a uniform power distribution for sensitive control of polymerization from each beam. Without the capability of batch production, the cost of single-beam writing would severely hinder the versatile industrial use of the technology.

References

1. Odian G (1991) Principles of polymerization, 3rd edn. Wiley, New York
2. Fouassier JP, Rabek JF (1990) Lasers in polymer science and technology. CRC Press, Boca Taton, FL
3. Reiser A (1989) Photoreactive polymers: The science and technology of resists. Wiley, New York
4. Fouassier JP, Rabek JF (1993) (eds) Radiation curing in polymer science and technology. Elsevier, London
5. Fouassier JP (1995) Photoinitiation, photopolymerization and photocuring: Fundamentals and applications. Hanser, Munich Vienna New York
6. Nakagawa T, Marutani Y (1996) (eds) Layered manufacturing systems: the latest development of three-dimensional copying technnologies (in Japanese). Kogyo Chosakai, Japan
7. Zhang X, Jiang XN, Sun C (1999) Sensor Actuat 77:149
8. Jiang XN, Sun C, Zhang X, Xu B, Ye YH (2000) Sensor Actuat 87:72
9. Bertsch A, Lorenz H, Renaud (1999) Sensor Actuat 73:14
10. Maruo S, Nakamura O, Kawata S (1997) Opt Lett 22:132
11. Parthenopoulos DA, Rentzepis PM (1989) Science 245:843
12. Denk W, Strickler JH, Webb WW (1990) Science 248:73
13. Kawata S, Sun HB, Tanaka T, Takada K (2001) Nature 412:697
14. Tanaka T, Sun HB, Kawata S (2002) Appl Phys Lett 80:312
15. Sun HB, Matsuo S, Misawa H (1999) Appl Phys Lett 74:786
16. Sun HB, Kawakami T, Xu Y, Ye JY, Matsuo S, Misawa H, Miwa M, Kaneko R (2000) Opt Lett 25:1110
17. Sun HB, Mizeikis V, Xu Y, Juodkazis S, Ye JY, Matsuo S, Misawa H (2001) Appl Phys Lett 79:1
18. Sun HB, Tanaka T, Takada K Kawata S (2001) Appl Phys Lett 79:1411
19. Sun HB, Takada K, Kawata S (2001) Appl Phys Lett 79:3173
20. Sun HB, Tanaka T, Kawata S (2002) Appl Phys Lett 80:3673
21. Albota M, Beljonne D, Bredas JL, Ehrlich JE, Fu JY, Heikal AA, Hess SE, Kogej T, Levin MD, Marder SR, McCord-Maughon D, Perry JW, Rockel H, Rumi M, Subramaniam C, Webb WW, Wu XL, Xu C (1998) Science 281:1653
22. Cumpston BH, Ananthavel SP, Barlow S, Dyer DL, Ehrlich JE, Erskine LL, Heikal AA, Kuebler SM, Lee IYS, McCord-Maughon D, Qin JQ, Rockel H, Rumi M, Wu XL, Marder SR, Perry JW (1999) Nature 398:51

23. Zhou WH, Kuebler SM, Braun KL, Yu TY, Cammack JK, Ober CK, Perry JW, Marder SR (2002) Science 296:1106
24. Belfield KD, Schafer KJ, Liu YU, Liu J, Ren XB, Van Stryland EW (2000) J Phys Org Chem 13:837
25. Bhawalkar JD, He GS, Prasad PN (1996) Rep Prog Phys 59:1041
26. Lehmann O, Stuke M (1995) Science 270:1644
27. Wanke MC, Lehmann O, Muller K, Wen QZ, Stuke M (1997) Science 275:1284
28. Daneshvar K, Raissi M, Bobbio SM (2000) J Appl Phys 88:2205
29. Shiomi M, Yoshidome A, Abe F, Osakada K (1999) Int J Mach Tool Manu 39:237
30. Zhang YZ, Shi LK, Zhang PZ, Xu J (2000) Rare Metal Mat Eng 29:361
31. Sun HB, Xu Y, Juodkazis S, Sun K, Watanabe M, Matsuo S, Misawa H, Nishii J (2001) Opt Lett 26:325
32. Sun HB, Xu Y, Matsuo S, Misawa H (1999) Opt Rev 6:396
33. Menzel R (2001) Photonics: Linear and nonlinear interactions of laser light and matter, Springer, Berlin Heidelberg New York
34. Saleh BEA, Teich MC (1991) Fundamentals of photonics. Wiley, New York
35. Milonni PW, Eberly JH (1988) Lasers. Wiley, New York
36. Mei DB, Cheng BY, Hu W, Li ZL, Zhan DH (1995) Opt Lett 20:429
37. Campbell M, Sharp DN, Harrison MT, Denning RG, Turberfield AJ (2000) Nature 404:53
38. Shoji S, Kawata S (2000) Appl Phys Lett 76:2668
39. Kondo T, Matsuo S, Juodkazis S, Misawa H (2001) Appl Phys Lett 79:725
40. Allmen M, Blatter A (1995) Laser-beam interactions with materials. 2nd edn. Springer Berlin Heidelberg New York
41. Grigoryants AG (1994) Basics of laser material processing. CRC, New York
42. Shen YR (1984) The principles of nonlinear optics. Wiley, New York
43. Boyd RW (1992) Nonlinear optics. Academic, San Diego
44. Kieffer JC, Matte JP, Belair S, Chaker M, Audebert P, Pepin H, Maine P, Strickland D, Bado P, Mourou G (1989) IEEE J Quantum Elect 25:2640
45. Bado P (2000) Laser Focus World 36:73
46. Diels JC, Rudolph W (1996) Ultrashort laser pulse phenomena: fundamentals, techniques, and applications on a femtosecond time scale (Optics and Photonics). Academic, New York
47. Saeta P, Wang JK, Siegal Y, Bloembergen N, Mazur E (1991) Phys Rev Lett 67:1023
48. Glezer EN, Milosavljevic M, Huang L, Finlay RJ, Her TH, Callan JP, Mazur E (1996) Opt Lett 21:2023
49. Glezer EN, Mazur E (1997) Appl Phys Lett 71:882
50. Goeppert-Mayer M (1931) Ann Phys 9:273
51. Kaiser W, Garrett CGB (1961) Phys Rev Lett 7:229
52. Eberly JH, Lambropoulos P (1978) (eds) Multiphoton processes : proceedings of an international conference at the University of Rochester, Rochester, N.Y., June 6–9, 1977. Wiley, New York
53. Kano H, Kawata S (1996) Opt Lett 21:1848
54. Higdon PD, Torok P, Wilson T (1999) J Microsc-Oxford 193:127
55. Hell S, Stelzer EHK (1992) Opt Commun 93:277
56. Bhawalkar JD, He GS, Park CK, Zhao CF, Ruland G, Prasad PN (1996) Opt Commun 124:33
57. He GS, Xu GC, Prasad PN, Reinhardt BA, Bhatt JC, Dillard AG (1995) Opt Lett 20:435
58. Fisher WG, Partridge WP, Dees C, Wachter EA (1997) Photochem Photobiol 66:141
59. He GS, Bhawalkar JD, Zhao CF, Park CK, Prasad PN (1995) Opt Lett 20:2393

60. Smith NI, Fujita K, Nakamura O, Kawata S (2001) Appl Phys Lett 78:999
61. Schaffer CB, Nishimura N, Glezer EN, Kim AMT, Mazur E (2002) Opt Express 10:196
62. Strickler JH, Webb WW (1991) Opt Lett 16:1780
63. Kawata S, Kawata Y (2000) Chem Rev 100:1777
64. Liphardt M, Goonesekera A, Jones BE, Ducharme S, Takacs JM, Zhang L (1994) Science 263:367
65. Moerner WE (1987) (ed) Persistent spectral hole burning: Science and applications. Springer, Berlin Heidelberg New York
66. Kim MK, Kachru R (1989) Opt Lett 14:423
67. Irie M (2000) Chem Rev 100:1685
68. Toriumi A, Herrmann JM, Kawata S (1997) Opt Lett 22:555
69. Sekkat Z, Knoll WJ (1995) J Opt Soc Am B 12:1855
70. Toriumi A, Kawata S, Gu M (1998) Opt Lett 23:1924
71. Kawata Y, Ishitobi H, Kawata S (1998) Opt Lett 23:756
72. Meerholz K, Volodin BL, Sandalphon, Kippelen B, Peyghambarian, N (1994) Nature 371:497
73. Day D, Gu M, Smallridge A (2001) Adv Mater 13:1005
74. Day D, Gu M (1999) Opt Lett 24:288
75. Tanaka T, Yamaguchi K, Yamamoto S (2002) Opt Commun 212:45
76. Qiu JR, Kojima K, Miura K, Mitsuyu T, Hirao K (1999) Opt Lett 24:786
77. Miura K, Qiu JR, Fujiwara S, Sakaguchi S, Hirao K (2002) Appl Phys Lett 80:2263
78. Sun HB, Juodkazis S, Watanabe M, Matsuo S, Misawa H, Nishii J (2000) J Phys Chem B 104:3450
79. Yamasaki K, Juodkazis S, Watanabe M, Sun HB, Matsuo S, Misawa H (2000) Appl Phys Lett 76:1000
80. Watanabe M, Sun HB, Juodkazis S, Takahashi T, Matsuo S, Suzuki Y, Nishii J, Misawa H (1999) Jpn J Appl Phys 37: L1527
81. Watanabe M, Juodkazis S, Sun HB, Matsuo S, Misawa H, Miwa M, Kaneko R (1999) Appl Phys Lett 74:3957
82. Watanabe M, Juodkazis S, Sun HB, Matsuo S, Misawa H (2000) Appl Phys Lett 77:13
83. Watanabe M, Juodkazis S, Sun HB, Matsuo S, Misawa H (1999) Phys Rev B 60:9959
84. Miura K, Qiu JR, Inouye H, Mitsuyu T, Hirao K (1997) Appl Phys Lett 71:3329
85. Minoshima K, Kowalevicz AM, Hartl I, Ippen EP, Fujimoto JG (2001) Opt Lett 26:1516
86. Homoelle D, Wielandy S, Gaeta AL, Borrelli NF, Smith C (1999) Opt Lett 24:1311
87. Minoshima K, Kowalevicz AM, Ippen EP, Fujimoto JG (2002) Opt Express 10:645
88. Li Y, Watanabe W, Yamada K, Shinagawa T, Itoh K, Nishii J, Jiang YY (2002) Appl Phys Lett 80:1508
89. Kawamura K, Sarukura N, Hirano M, Hosono H (2001) Appl Phys Lett 78:1038
90. Kawamura K, Sarukura N, Hirano M, Ito N, Hosono H (2001) Appl Phys Lett 79:1228
91. Watanabe W, Kuroda D, Itoh K, Nishii J (2002) Opt Express 10:978
92. Yablonovitch E (1987) Phys Rev Lett 58:2059
93. John S (1987) Phys Rev Lett 58:2486
94. Joannopoulos JD, Meade RD, Winn JN (1995) Photonic crystals: Modeling the flow of light. Princeton Univ Press, Singapore
95. Ho KM, Chan CT, Soukoulis (1990) Phys Rev Lett 65:3152
96. Fukuda K, Sun H, Matsuo S, Misawa H (1998) Jpn J Appl Phys 37: L508
97. Sun HB, Song JF, Xu Y, Matsuo S, Misawa H, Du GT, Liu SY (2000) J Opt Soc Am B 17:476

98. Sun HB, Xu Y, Ye JY, Matsuo S, Misawa H, Song JF, Du GT, Liu SY (2000) Jpn J Appl Phys 39: L591
99. Wu PW, Cheng W, Martini IB, Dunn B, Schwartz BJ, Yablonovitch E (2000) Adv Mater 12:1438
100. Stellacci F, Bauer CA, Meyer-Friedrichsen T, Wenseleers W, Alain V, Kuebler SM, Pond SJK, Zhang YD, Marder SR, Perry JW (2002) Adv Mater 14:194
101. Horiyama M, Sun HB, Miwa M, Matsuo S, Misawa H (1999) Jpn J Appl Phys Lett 38: L212
102. Maruo S, Ikuta K (2000) Appl Phys Lett 76:2656
103. Maruo S, Ikuta K (2002) Sensor Actuat A-Phys 100:70
104. Pappas SP (1985) Radiation Phys Chem25:633
105. Pappas SP (1992) (ed) Radiation curing science and technology. Plenum, New York
106. Fouassier JP, Rabek JF (1993) (eds) Radiation curing in polymer science and technology. Elsevier, London
107. Scully MO, Zubairy MS (1997) Cambridge Univ Press
108. Hopfield JJ, Worlock JM, Park KJ (1963) Phys Rev Lett 11:414
109. Frohlich D, Staginnus B (1967) Phys Rev Lett 19:476
110. Pao YH and Rentzepis PM (1965) Appl Phys Lett 6:93
111. Chin SL, Bedard G (1971) Phys Lett 36A: 271
112. Papouskova Z, Pola J, Bastl Z, Tiaskal J (1990) J Mecromol Sci Chem A27:1015
113. Morita H, Sadakiyo T (1995) J Photochem Photobiol A 87:163
114. Lee KS, Lee JH, Choi HY, Cha M, Chung MA, Kim YJ, Jung SD (2001) Mol Cryst Liq Cryst 370:155
115. Chung MA, Lee KS, Jung SD (2002) ETRI J 24:221
116. Li CD, Luo L, Wang SF, Huang WT, Gong QH, Yang YY, Feng SJ (2001) Chem Phys Lett 340:444
117. Belfield KD, Schafer KJ, Mourad WJ (2000) J Org Chem 65:4475
118. Jortner J, Ratner M (1997) (eds) Molecular electronics. Blackwell Science, London
119. Belfield KD, Hagan DJ, Van Stryland EW, Schafer KJ, Negres RA (1999) Org Lett 1:1575
120. Belfield KD, Schafer KJ, Alexander MD Jr (2000) Chem Mater 12:1184
121. Adronov A, Frechet JMJ, He GS, Kim KS, Chung SJ, Swiatkiewicz J, Prasad PN (2000) Chem Mater 12:2838
122. Hu Y (2001) MRS Bull 26:595
123. Adronov A, Frechet JMJ, He GS, Kim KS, Chung SJ, Swiatkiewicz J, Prasad PN (2000) Chem Mater 12:2838
124. Rentzepis PM (1989) US Patent 07 342 978
125. Bradbury S, Bracegirdle B (1998) Introduction to light microscopes. Springer, Berlin Heildelberg New York
126. Born M, Wolf E (1999) Principle of Optics, 7th edn. Cambridge Univ Press, Cambridge
127. Witzgall G, Vrijen R, Yablonovitch E, Doan V, Schwartz BJ (1998) Opt Lett 23:1745
128. Madou MJ (2002) Fundamentals of microfabrication: The science of miniaturization, 2nd edn. CRC Press, New York
129. Lyshevski SE (2002) MEMS and NEMS: Systems, devices, and structures. CRC Press, New York
130. (Anon) (2002) Laser Focused World 38 11
131. Cartlidge E (2002) Phys World 15:10
132. Quake SR, Scherer A (2000) Science 290:1536
133. Xia YN, Whitesides GM (1998) Ann Rev Mater Sci 28:153

134. Kumar A, Whitesides GM (1993) Appl Phys Lett 63:2002
135. Chou SY, Krauss PR, Renstrom PJ (1995) Appl Phys Lett 67:3114
136. Kim E, Xia Y, Whitesides GM (1995) Nature 376:581
137. Terris BD, Mamin HJ, Best ME, Logan JA, Rugar D (1996) Appl Phys Lett 69:4262
138. Masuda H, Fukuda K (1995) Science 268:1446
139. Xia Y, Kim E, Zhao X-M, Rogers JA, Prentiss M, Whitesides GM (1996) Science 273:347
140. Wu ES, Strickler JH, Harrell WR, Webb WW (1990) SPIE Proc 1674:776
141. Maruo S, Kawata S (1998) J IEEE MEMS, 7:411
142. Wu PW, Dunn B, Yablonovitch E, Doan V, Schwartz BJ (1999) J Opt Soc Am B 16:605
143. Brodeur A, Chin SL (1999) J Opt Soc Am B 16:637
144. Stuart BC, Feit MD, Herman S, Rubenchik AM, Shore BW, Perry MD (1996) Phys Rev B 53:1749
145. vonderLinde D, Schuler H (1996) J Opt Soc Am B 13:216
146. Boiko Y, Costa JM, Wang M, Esener S (2001) Opt Express 8:571
147. Joshi MP, Pudavar HE, Swiatkiewicz J, Prasad PN, Reianhardt BA (1999) Appl Phys Lett 74:170
148. Kirkpatrick SM, Baur JW, Clark CM, Denny LR, Tomlin DW, Reinhardt BR, Kannan R, Stone MO (1999) Appl Phys A 69:461
149. Matsumoto K (1997) P IEEE 85:612
150. Sauer BB, McLean RS, Thomas RR (1998) Langmuir 14:3045
151. Tarun A, Daza MRH, Hayazawa N, Inouye Y, Kawata S (2002) Appl Phys Lett 80:3400
152. Flory PJ (1952) Principles of polymer chemistry. Cornell University Press, New York
153. Schulz GV (1997) Chem Ber 80:232
154. Bolon DA, Webb KK (1978) J Appl Polymer Sci 22:2543
155. Decker C, Faure J, Fizet M, Rychla L (1979) Photog Sci Eng 23:137
156. Hageman HJ (1985) Prog Organic Coatings 13:123
157. Booth MJ, Wilson T (2001) J Biomed Opt 6:266
158. Booth MJ, Wilson T (2001) J Microsc-Oxford 201:416
159. Booth MJ, Neil MAA, Juskaitis R, Wilson T (2002) P Natl Acad Sci USA 99:5788
160. Fujimoto M, Aoshima S, Hosoda M, Tsuchiya Y (1999) Opt Lett 24:850
161. DeVoe RJ, Kalveit H, Leatherdale CA, Williams TR (2002) Proc SPIE
162. Stamners J (1986) Waves in focal region. Adam Hilgar, Bristol
163. Gu M (1999) Advanced optical imaging Theory. Springer, Berlin Heidelberg New York
164. Bahlmann K, Hell SW (2000) Appl Phys Lett 77:612
165. T. Wilson (1990) (ed) Confocal microscopy. Academic, London
166. Berger V, GauthierLafaye O, Costard E (1997) J Appl Phys 82:60
167. Berger V, GauthierLafaye O, Costard E (1997) Electron Lett 33:425
168. Nakata Y, Okada T, Maeda M (2002) 81: 4239
169. Sharp DN, Campbell M, Dedman ER, Harrison MT, Denning RG, Turberfield A (2002) J Opt Quant Electron 34:3
170. Yang S, Megens M, Aizenberg J, Wiltzius P, Chaikin PM, Russel WB (2002) Chem Mater 14:2831
171. Segawa H, Yoshida K, Kondo T, Matsuo S, Misawa H (2003) J Sol-Gel Sci Techn 26:1023
172. Campagnola PJ, Delguidice DM, Epling GA, Hoffacker KD, Howell AR, Pitts JD, Goodman SL (2000) Macromolecules 33:1511
173. Nakayama Y, Matsuda T (1999) J Biomed Mater Res 48:511
174. Okino H, Nakayama Y, Tanaka M, Matsuda T (2002) J Biomed Mater Res 59:233

175. Pitts JD, Campagnola PJ, Epling GA, Goodman SL (2000) Macromolecules 33:1514
176. Zhou WH, Kuebler SM, Carrig D, Perry JW, Marder SR (2002) J Am Chem Soc 124:1897
177. Zhou WH, Kuebler SM, Braun KL, Yu TY, Cammack JK, Ober CK, Perry JW, Marder SR (2002) Science 296:1106
178. Blanco A, Chomski E, Grabtchak S, Ibisate M, John S, Leonard SW, Lopez C, Meseguer F, Miguez H, Mondia JP, Ozin GA, Toader O, van Driel HM (2000) Nature 405:437
179. Gruning U, Lehmann V, Engelhardt CM (1995) Appl Phys Lett 66:3254
180. Xu Y, Sun HB, Ye JY, Matsuo S, Misawa H (2001) J Opt Soc Am B 18: 1084
181. Noda S, Tomoda K, Yamamoto N, Chutinan A (2000) Science 289:604
182. Kawakami S (1997) Electron Lett 33:1260
183. Biswas R, Chan CT, Sigalas M, Soukoulis CM, Ho KM (1995) In: Soukoulis CM (ed) Photonic band gap materials. Kluwer Academic Press, London, p 23
184. Straub M, Gu M (2002) Opt Lett 27:1824
185. Noda S, Chutinan A, Imada M (2000) Nature 407:608
186. Noda S, Yokoyama M, Imada M, Chutinan A, Mochizuki M (2001) Science 293:1123
187. Salaneck WR, Seki K, Kahn A, Pireaux JJ (2001) (eds) Conjugated polymer and molecular interfaces: science and technology for photonic and optoelectronic applications. Marcel Dekker, New York
188. Taton TA, Norris DJ (2002) Nature 416:685
189. Lee WM, Pruzinsky SA, Braun PV (2002) Adv Mater 14:271
190. Ashkin A (1970) Phys Rev Lett 24:156
191. Ashkin A, Schutze K, Dziedzic JM, Eutenuer U, Schliwa M (1990) Nature 348:346
192. Ashkin A (1992) Biophys J 61:569
193. Higurashi E, Sawada R, Ito T (1998) Appl Phys Lett 71:2951
194. Gauthier RC (1995) Appl Phys Lett 67:2269
195. Galajda P, Ormos P (2001) Appl Phys Lett 78:249
196. Friese MEJ, Enger J, Rubinsztein-Dunlop H, Heckenberg NR (1996) Phys Rev A 54:1593
197. He H, Friese MEJ, Heckenberg NR, Rubinsztein-Dunlop H (1995) Phys Rev Lett 75:826
198. Sekkat Z, Knoll W (2002) (eds) Photorefractive organic thin films. Academic, New York
199. Kaneuchi Y, Kawai T, Hamaguchi M, Yoshino K, Irie M (1997) Jpn J Appl Lett, 36:3736
200. Masuda S, Sertowa N, Petkov I (1997) J Polymer Sci A 35:3683
201. Watanabe T, Akiyama M, Totani K, Kuebler SM, Stellacci F, Wenseleers W, Braun K, Marder SR, Perry JW (2002) Adv Funct Mater 12:611

Editor: Kwang-Sup Lee Received: July 2003

Author Index Volumes 101–170

Author Index Volumes 1-100 see Volume 100

Subject Index

Printing: Saladruck, Berlin
Binding: Stein+Lehmann, Berlin

Kolibri 360 GmbH
Ferdinand-Jühlke-Straße 7
99095 Erfurt, Deutschland
produktsicherheit@kolibri360.de